T0297846

CAMBRIDGE LIBRARY COLLECTION

Books of enduring scholarly value

Physical Sciences

From ancient times, humans have tried to understand the workings of the world around them. The roots of modern physical science go back to the very earliest mechanical devices such as levers and rollers, the mixing of paints and dyes, and the importance of the heavenly bodies in early religious observance and navigation. The physical sciences as we know them today began to emerge as independent academic subjects during the early modern period, in the work of Newton and other 'natural philosophers', and numerous sub-disciplines developed during the centuries that followed. This part of the Cambridge Library Collection is devoted to landmark publications in this area which will be of interest to historians of science concerned with individual scientists, particular discoveries, and advances in scientific method, or with the establishment and development of scientific institutions around the world.

A Handbook of Descriptive and Practical Astronomy

This handbook by George Frederick Chambers (1841–1915), a young and enthusiastic amateur astronomer, became a best-seller soon after its publication in 1861 and made Chambers' reputation as a popular astronomy writer. The work is divided into ten parts covering the following topics: the planets of our solar system; eclipses; gravity and tides; phenomena including aberration and refraction; comets; chronological astronomy; stars; astronomical instruments; the history of astronomy; and meteoric astronomy. It is richly illustrated with photographs and woodcuts depicting a wide variety of astronomical phenomena. Chambers went on to become one of the leading amateur astronomers of the nineteenth century. The usefulness and accessibility of his practical advice ensured that his published works became indispensable for thousands of less famous amateurs. The *Handbook*, reissued in many editions, remains his most popular work and is a key text in the history of Victorian popular astronomical writing.

Cambridge University Press has long been a pioneer in the reissuing of out-of-print titles from its own backlist, producing digital reprints of books that are still sought after by scholars and students but could not be reprinted economically using traditional technology. The Cambridge Library Collection extends this activity to a wider range of books which are still of importance to researchers and professionals, either for the source material they contain, or as landmarks in the history of their academic discipline.

Drawing from the world-renowned collections in the Cambridge University Library, and guided by the advice of experts in each subject area, Cambridge University Press is using state-of-the-art scanning machines in its own Printing House to capture the content of each book selected for inclusion. The files are processed to give a consistently clear, crisp image, and the books finished to the high quality standard for which the Press is recognised around the world. The latest print-on-demand technology ensures that the books will remain available indefinitely, and that orders for single or multiple copies can quickly be supplied.

The Cambridge Library Collection will bring back to life books of enduring scholarly value (including out-of-copyright works originally issued by other publishers) across a wide range of disciplines in the humanities and social sciences and in science and technology.

A Handbook of
Descriptive and
Practical Astronomy

GEORGE FREDERICK CHAMBERS

CAMBRIDGE UNIVERSITY PRESS

Cambridge, New York, Melbourne, Madrid, Cape Town, Singapore,
São Paolo, Delhi, Dubai, Tokyo

Published in the United States of America by Cambridge University Press, New York

www.cambridge.org
Information on this title: www.cambridge.org/9781108014755

This edition first published 1861
This digitally printed version 2010

ISBN 978-1-108-01475-5 Paperback

A HANDBOOK OF ASTRONOMY.

Fig. 44.   *Plate I.*

**APPEARANCE OF A PORTION OF THE SUN'S
SURFACE ACCORDING TO NASMYTH.**

JULY 29, 1860.

A HANDBOOK

OF

DESCRIPTIVE AND PRACTICAL

ASTRONOMY.

By GEORGE F. CHAMBERS, F.R.G.S.

"The Heavens declare the Glory of GOD : and the
Firmament sheweth his Handywork."—*Psalm* xix. 1.

LONDON:

JOHN MURRAY, ALBEMARLE STREET.

1861.

TO

GEORGE BIDDELL AIRY, Esq., M.A.

ASTRONOMER-ROYAL FOR ENGLAND,

K.C.M. Pruss., D.C.L., F.R.S., F.R.A.S.

&c. &c. &c.

This Volume is Respectfully Dedicated,

IN RECOGNITION OF HIS EMINENT PUBLIC SERVICES IN THE CAUSE OF

ASTRONOMY AND ITS KINDRED SCIENCES,—MORE PARTICULARLY

DURING THE TWENTY AND SIX YEARS HE HAS DIRECTED

THE OPERATIONS OF ONE OF THE MOST DISTINGUISHED

OBSERVATORIES OF MODERN TIMES,—

BY

THE AUTHOR.

PREFACE.

ENGLISH literature, abundant though it be in other respects, is undoubtedly very deficient in works on Astronomy. Our choice is limited either to purely elementary books, few in number, on the one hand; or to advanced treatises, of which there is a similar paucity, on the other. The present work is designed to occupy a middle position between these two classes; to be attractive to the general reader, useful to the amateur, and "handy" also, as an occasional book of reference, to the professional astronomer.

In pursuance of the plan laid down from the first, theoretical matter is, as a rule, excluded; but in many cases, it has been thought desirable not to abide with perfect strictness to the limitation. All *speculations* however, bearing on the origin of the created universe have been carefully avoided, sufficient mischief having already been done by the artful sophistries of those who delight in what they are pleased to term "Free Philosophical Enquiry."

The most recent discoveries in all branches of the science, will be found incorporated with information of older date. The Chapters on Comets may be instanced as an example; and the catalogues belonging to them, will, it is anticipated, be found serviceable to the professional computer.

The Engravings have been chosen, as far as possible, as exemplifying new results and discoveries; and the author desires to tender his best thanks to Mr. Warren De La Rue of Cranford; to Mr. Nasmyth of Penshurst; and to Dr. C. A. F. Peters of Altona, for their kindness in supplying him with the originals of many of the more important plates. With reference to the illustrations of the clusters and nebulæ, it may be well to mention, that in almost all cases, the *brilliancy* of the objects has been *exaggerated* by the engraver. This was an accident which it was found impossible to guard against.

The Author is also under obligations to the Astronomer Royal; to Mr. J. R. Hind; and to Mr. E. J. Stone of the Royal Observatory, Greenwich, for their kindness in looking over different portions of the work in its progress through the press, and in offering many useful suggestions.

In a work so abounding in figures, it can hardly be doubted that there are still some errors remaining; probably they will be but few, and unimportant, as the greatest possible care has been taken to secure accuracy in printing; and a Supplement has been added for the purpose of including the latest discoveries.

Finally, it is hoped, that this book may be the means of inducing some at least to interest themselves in the study of that noble science, which in so conclusive a manner shows forth the wonderful Wisdom, Power, and Beneficence of the Great Creator and Omnipotent Ruler of the Universe.

G. F. C.

East-Bourne, Sussex :
August 28, 1861.

SUPPLEMENT.

Page 6. We have omitted to mention that Wolf, who has given much attention to solar spots, has determined 11·11 years to be the period of the cycle.

Page 54. Exception having been taken to the name Titania, hitherto applied to planet, No. ⑥₂, on the ground that one of the satellites of Uranus already has that designation, Lieut. Gillis, with the consent of Mr. Ferguson, has renamed it *Echo*.

Page 54.—A new minor planet, No. ⑦₁, named Niobe, was discovered on August 13, 1861, by Luther, at Bilk.

The following are its elements, approximately, as determined by M. Tietjen :—

$$
\begin{aligned}
\text{Epoch} &= 1861, \text{ Aug. 28 B.M.T.} \\
\lambda &= 96^\circ \ 3' \\
\pi &= 221 \quad 9 \\
\delta &= 316 \ 11 \\
\iota &= 23 \quad 8 \\
\epsilon &= 0\text{·}16451 \\
\mu &= 780\text{·}57'' \\
a &= 2\text{·}7441
\end{aligned}
$$

Page 197.—Encke's comet was re-discovered by Förster at the Royal Observatory, Berlin, on Sept. 28, 1861 ; it was however very faint. Encke mentions that Struve detected it in 1828, 115 days before the perihelion, but that on the present occasion its discovery was effected 125 days before that epoch.

The following theoretical elements, for the present apparition, have been calculated by M. Powalky.

PP. = 1862 Feb. 6·2 B.M.T.

$$\begin{aligned}
& \quad\quad\quad\quad ° \quad\; ' \quad\; '' \\
\pi &= 158 \quad\; 0 \quad 50 \\
\Omega &= 334 \quad 30 \quad 50 \\
\iota &= 13 \quad\; 5 \quad\; 0 \\
q &= 0·3399 \\
\epsilon &= 0·84670 \\
\mu &= +
\end{aligned}$$

In consequence of the PP. occurring in next year, the comet will be designated as No. (i.) of 1862, notwithstanding that it was discovered long before the close of the present year.

Page 198, line 3. Add—This comet was not seen either in August, 1855, or in January, 1861, at both of which periods it was due.

Page 198, line 9. Add—A subsequent investigation by Seeling assigns, as the period of this comet, 2031·153 days, equal to 5·561 years, a value which somewhat exceeds that given in the Table on page 195.[1]

Page 199, line 16. Add—On July 22, 1861, M. Yvon Villarceau communicated to the Academy of Sciences an interesting memoir on this comet, of which the following is a concise analysis : —

The perturbations experienced by this comet are owing chiefly to the action of Jupiter, to which it is so near, that during the month of April of the present year its distance was only 0·36, or little more than *one third* of the Earth's distance from the Sun. Before and after this epoch Jupiter and the comet have continued, and will continue, so little distant from one another, as to produce the great perturbations to which the comet is at present subject.

From a table of the elements of the perturbations produced by Jupiter, Saturn, and Mars, in the interval between the appearance of the comet in 1857–8, and its return to its perihelion in 1864, M. Villarceau obtained the following results : — 1. The longitude of the perihelion will have diminished 4° 35′ to August 1863, and will remain sensibly stationary for about a year from that epoch. 2. The longitude of the node will have continually diminished to the amount of 2° 8′. 3. The inclination will have increased 1° 49′ to the middle of 1862, and will diminish 6′ during a year, continuing stationary during the year following. 4. The eccentricity, after having increased to the middle of 1860, will diminish rather quickly, and will remain stationary from 1863–5 to 1864–6.

---

[1] *Ast. Nach.*, No. 1319, Aug. 2, 1861.

"But of all these perturbations," says M. Villarceau, "the most considerable are those of the mean motion and the mean anomaly. After having increased from 5″ to July, 1860, the mean motion diminishes 9″ in one year, and nearly 12″ in the year following, remaining stationary in the last year, and with a value 15″, 5 less than at its origin. The perturbations of the mean anomaly, after having gradually increased till 1860, will increase rapidly till 1861, when they will amount to 10° 28′; and setting out from this, they will increase 9′, and in 1863 and 1864 they will have resumed the same value which they had in 1861."

The effect of the first of these perturbations will be to increase the time of the comet's revolution about 69 days, and of the second, to hasten by 49 days the return of the comet to its perihelion in 1864. It will pass its perihelion on February 26, whereas without the influence of these perturbations it would have passed it on April 15.

M. Villarceau has shown that it will be very difficult to see it on its return. From October 25, 1863, to April 22, 1864, its distance from the Sun will be less than 16° or 18°, so that it cannot then be seen, its lustre being only 0·037 at the first of these dates, and 0·089 at the second. On August 22, 1864, it will have increased to 0·035, the difference of longitude between it and the Sun being then 69° When Sir T. Maclear observed the comet at the Cape in the beginning of 1858, its lustre was 0·190, when it was described as very feeble.

Page 207. Since the main portion of this work was placed in the hands of the printer, another grand comet has come and gone; which, though surpassed by preceding ones in some respects, yet is unequalled as regards the immense *angular* extent of its tail.

It was first discovered by Mr. Tebbutt in New South Wales, on May 13, (prior to its perihelion passage on June 11,) but it was not detected in this country till June 29. On the following evening, June 30, and for several weeks subsequently, it was pretty generally seen: from the published observations we make some selections.

Sir J. Herschel observed it in Kent. He says:—

"The Comet, which was first noticed here on *Saturday* night, June 29, by a resident in the village of Hawkhurst (who informs me that his attention was drawn to it by its being taken by some of his family for the Moon rising), became conspicuously visible on the 30th, when I first observed it. It then far exceeded in brightness any comet I have before observed, those of 1811 and the recent splendid one of 1858 not excepted. Its total light certainly far surpassed that of any fixed star or planet, except perhaps Venus at its maximum. The tail extended from its then position about 8 or 10° above the horizon to within 10 or 12° of the Pole-star, and was therefore about 30° in length. Its greatest breadth, which diminished rapidly in receding from the head, might be about 5°. Viewed through a good achromatic, by Peter Dollond, of $2\frac{3}{4}$ inches aperture and 4 feet focal length, it exhibited a very condensed central light, which might fairly be called a nucleus; but, in its then low situation, no other physical peculiarities could be observed. On the 1st instant it was seen early in the evening, but before I could bring a telescope to bear on it clouds intervened, and continued till morning twilight. On the 2nd (Tuesday), being now much better situated for observation, and the night being clear, its appearance at midnight was truly magnificent. The tail, considerably diminished in breadth, had shot out to an extravagant length, extending from the place of the head above *o* of the Great Bear at least to π and ρ Herculis; that is to say, about 72°, and perhaps somewhat further. It exhibited no bifurcation or lateral offsets, and no curvature like that of the comet of 1858, but appeared rather as a narrow prolongation of the northern side of the broader portion near the comet than as a thinning off of the latter along a central axis, thus imparting an unsymmetrical aspect to the whole phenomenon.

"Viewed through a 7-feet Newtonian reflector of 6 inches aperture the nucleus was uncommonly vivid, and was concentrated in a dense pellet of not more than 4″ or 5″ in diameter (about 315 miles). It was round and so very little *woolly* that it might *almost* have been taken for a small planet seen through a dense fog; still so far from *sharp* definition as to preclude any idea of its being a solid body. No sparkling or star-light point could, however, be discerned in *its* centre with the power used (96), nor any separation by a darker interval between the nucleus and the cometic envelope. The gradation of light, though rapid, was continuous. Neither on this occasion was there any *unequivocal* appearance of that sort of fan or sector of light, which has been noticed on so many former ones.

"The appearance of the 3rd was nearly similar; but on the 4th the fan, though feebly, was yet certainly perceived; and on the 5th was very distinctly visible. It consisted, however, not in any vividly radiating jet of light from the nucleus of any well-defined form, but in a crescent-shaped cap formed by a very delicately graduated condensation of the light on the side towards the Sun, connected with the nucleus, and what may be termed the *coma* (or spherical haze immediately surrounding it), by an equally delicate graduation of light, very evidently superior in intensity to that on the opposite side. Having no micrometer attached, I could only estimate the distance of the brightest portion of this crescent from the nucleus at about 7′ or 8′, corresponding at the then distance of the comet to about 35,000 miles.

On the 4th (Thursday) the tail (preserving all the characters already described on the 2nd) passed through α Draconis, and τ Herculis, nearly over η and ϵ Herculis, and was traceable, though with difficulty, almost up to α Ophiuchi, giving a total length of 80°. The northern edge of the tail, from α Draconis onwards, was perfectly straight,—not in the least curved,—which, of course, must be understood with reference to a great circle of the heavens.

"Viewed, on the 5th, through a doubly refracting prism well achromatised, no certain indication of polarisation in the light of the nucleus and head of the comet could be perceived. The two images were distinctly separated, and revolved round each other with the rotation of the prism without at least any marked alternating difference of brightness. Calculating on Mr. Hind's data, the angle between the Sun and Earth and the comet must then have been 104°, giving an angle of incidence equal to 52°, and obliquity 38°, for a ray supposed to reach the eye *after a single* reflection from the cometic matter. This is not an angle unfavourable to polarisation, but the reverse. At 66° of elongation from the Sun (which was that of the comet on the occasion in question), the blue light of the sky is very considerably polarised. The constitution of the comet, therefore, is analogous to that of a cloud; the light reflected from which, as is well known, at that (or any other) angle of elongation from the Sun, exhibits no signs of polarity."

The Rev. T. W. Webb, of Hardwick, Herefordshire, has published a very complete account of his observations, which we transcribe with a few verbal alterations and omissions.

"The following observations, though interrupted by unfavourable weather, and less complete than might be wished in several respects, are offered to the public as being made with the advantage of an achromatic telescope, the object-glass of which, $5\frac{1}{2}$ inches in clear aperture, is the workmanship of the celebrated American optician, Alvan Clark. Astronomers will judge of its competency when they are informed that, with a power of about 460, it shows a distinct division between the components of η Coronæ Borealis. The values of its magnifying powers will be understood as merely approximate.

"June 30. The comet was first perceived about 10h. 15m., a few degrees above the N.N.W. horizon, with a nucleus as vivid as that of 1858, and a tail far more extensive as the night advanced, when it could be traced beyond the zenith, as far as the space marked out by α Lyræ, β and γ Draconis, and δ Cygni,—consequently about 90° from the head, though for great part of the distance as an extremely faint stream. Its outline was convex towards the left or west side as far as Polaris, which it involved; but near that star its inflexion ceased, and the rest of its course was straight, not as a tangent to the upper part of

June 30.

July 2.

July 2.

July 3.

July 4.

THE GREAT COMET OF 1861.

the curve, but in a fresh and independent direction, pointing to β and γ Lyræ: a similar change in the curvature of the tail was repeatedly noticed in 1858. It grew gradually broader in receding from the head, to a width of 3° or 4°, but very ill-defined, and without that contrast between the density and distinctness of its sides, which was so remarkable in the Donati. The right-hand side was thought to strike off from the coma for a few degrees at a smaller angle with the horizon, but could not be traced as a separate branch. The first aspect of the head in the 7-feet achromatic with the comet-finder (power about 27) was magnificent; a minute brilliant point was situated a little way from the vertex of a bright parabolic arc, and a great part of the field was filled with irregular curved flakes and trains of misty light, the whole presenting a picture like that of a miniature full moon, shining in a sky diversified with light cirrus clouds. A power of 110 brought out the details finely, especially a bright jet issuing from the nucleus towards the vertex of the parabola, and forming the most luminous part of the arc; this bore magnifying very well even up to 460. With the higher powers the nucleus was a softly defined golden disc, having, with 460, a very confused edge, with no increase of brightness in the centre, and an intensity of light supposed to be between that of Venus and Jupiter. Notwithstanding its small elevation above the horizon, it was distinctly circular, without any perceptible tendency to a phasis, though in a position highly favourable for such an appearance. Its diameter was very roughly guessed at about 2″. At 11h. 15m., a sketch taken with the comet-finder, of which Fig. 152 is a copy, will give some idea of the appearances, but those who have tried to draw such details will allow for great imperfection; and it must be borne in mind that no attempt has been made to represent the extent of the head, or its outer boundary, if such a term can be applied to a haze melting imperceptibly away.

"The figure is telescopic, or inverted, with respect to the natural object, but placed vertically as regards the comet's passage through the field of view. A is the nucleus; B, the jet of light; C c, the parabolic arc (as I then understood it); D, a fainter brush; E, a moderately bright cloud, the most conspicuous part of an irregular parabolic envelope, F f; G, a dark impression or indentation between two luminous clouds, H and I, the former of which was by far the brightest and largest; they both formed part of another envelope, which became confused at its sides with the interior one, F f; K L were two narrow faint envelopes, which became united at their vertex. The diffused coma had a much greater extent. The darkest part of the head was the vacancy between the letters C and F.   *   *   *

"About 11h. 45m. my wife pointed out to me a faint ray about 3° or 3½° broad, stretching beneath the square of Ursa Major, having ψ Ursæ in its lower edge, and Cor Caroli about 1° above its upper, and traceable about half-way from the latter star to Arcturus; it pointed to the comet, but in the twilight no connection could be made out. About 20m. later it had risen higher, so as to stand midway between ψ and γ Ursæ; its further end had become much more distinct, and its termination was plainly visible near ε Boötis: some time afterwards I could not see it, and supposed it might have been a cirrus cloud. Subsequently, however, I have been favoured by G. Williams, Esq., of Liverpool, with a communication and sketch, from which it appears that he saw the same ray about 12h. 30m. passing between α and β Ursæ Majoris, and another, somewhat brighter, extending to Cassiopea; which may have been hid from me by trees. Though supposing they might be cirrus clouds, he was induced, like myself, to record them, from the peculiarity of their direction towards the comet, and thus fortunately we seem to have obtained ocular proof that at that moment the tail was rising, as it were, off the earth, or at least from its immediate vicinity. Had I then known of Mr. Hind's computation as to the position of the tail, I might have perceived that the rising of the beam in the sky, which at the moment I thought was a motion the wrong way for the comet's course (and so was confirmed in my idea that it must be a cloud), was exactly the perspective result of the retreat of the tail from the Earth.

"Several falling stars were seen during this night's observation.

"July 1. A densely clouded night.

"July 2. Clouds rendered this evening's work less satisfactory.

"The tail is now slightly bent the other way, concave to the left; it can be traced nearly to Vega, about 80°; its edges are still alike undefined; in the comet-finder, at some distance behind the nucleus, its central regions appeared slightly less luminous than its sides; an indication of its hollow structure. 460 showed the nucleus much smaller than two nights previously, with a sector proceeding from it in a different direction from that of the ray, but as there was little time it was not drawn. Fig. 153 is the appearance in the comet-finder. c having now become equally faint with D, the misconception of the previous observation was rectified; it was evident that c and D are respectively the sides of a very narrow envelope, and that the continuation of C is really to be found in f. E has disappeared, with the preeminence of H, the two envelopes which brightened up in these spots, and the space included between them, are nearly equalised as to light. The dark area immediately exterior to the slightly indented boundary, H I, rendered the latter one of the most conspicuous features of the coma; this dusky space was lost towards the right hand in confused haze; on the opposite side it curved round, and was at first thought to join the darkness between C and F, but it was uncertain whether it was not interrupted by a prolongation of the light from I

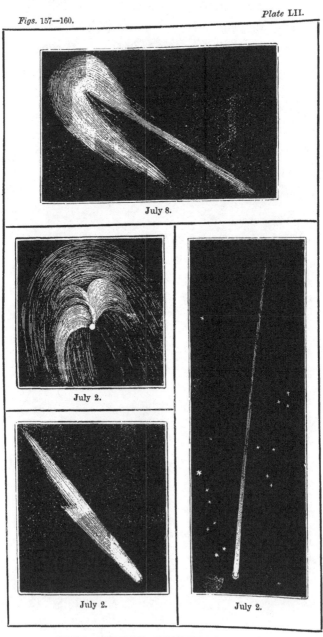

July 8.

July 2.

July 2.

July 2.

THE GREAT COMET OF 1861.

and H, forming a continuous envelope; and such I have little doubt was the case. K and L could no longer be separately distinguished.

" The axis of such symmetry as the coma possessed seemed to coincide with D, but the direction of the tail was better represented by c; this might have been owing to the curvature of the latter, but I observed a somewhat analogous oblique distribution of light in the comet of 1858. 110 diluted most of these features too much, but brought out beautifully a vivid sector, hardly discernible from the irradiation of the nucleus with the lower power. It was brightest along the edge directed towards the Sun, which indeed could hardly be distinguished from the nucleus at its origin; its "angle of position" (i. e. with the meridian) was guessed about 190°; the other edge shorter, less distinct, and probably a little curved,—about 315°; a portion of the sector to the left of the nucleus was fainter than the rest. (Fig. 154.) This bright area seemed to radiate outwards like a fan; its extent might be one-seventh of the distance, A G, in Fig. 152.

" July 3. Strong twilight and much vapour. 460 shows the nucleus small, dull, and ill-defined; with 110 it is sharp and vivid, and the luminous fan has a greater angular extent than last night, its edges standing at about 170° and 330°; but the south edge has lost its brilliancy, and the strongest portion, which but little surpasses the rest, is a ray at 225°. Edges sometimes appear a little concave. (See Fig. 155.)

" July 4. Clouds, haze, and wind troublesome, with occasional clear spaces. Comet much diminished, though on a far darker background than at its first appearance, but still a grand object to the naked eye. The tail is more dense for some distance, but its edges very undefined, and equally so. It is now unexpectedly nearly straight, to another observer's eye as well as my own; though, on taking in its whole length, I fancy—but perhaps only fancy—something like a double curvature; first left, then right. Beyond the first 12° or 15° it is very dim, but I think it reaches ρ Herculis, about 60° Nucleus as last night with 460, probably not exceeding 0·5″. 110 shows it still very vivid with a fan strikingly like an electric brush, reminding me strongly of a similar appearance in Halley's comet at its last return. This is now well seen in the comet-finder, and must have gained in extent, as it probably reaches one-third of the distance A G. It is also very conspicuous in the micrometer, power about 55, with which the sketch, Fig. 156, was made. The south radius of the fan, a b, is very ill marked; a beam, a c, is very evident, with a thicker and denser end at c; between this and d is a darker space running in towards the nucleus, and the circular outline, d e, seems a little indented towards e, a e appearing somewhat concave.

" These details, however, except a c, are rather precarious. Measures of position give a b 207°, a c 251°, a e 340°. The radius, c a, is about

15s. in passing behind the edge of the micrometer, giving nearly $1\frac{1}{2}'$ of arc. All round the sector, and rather beyond its north end, lies a darker space, about one-third its breadth; and outside of this the general coma, with some traces of streakiness, as though it were composed of narrow envelopes not separately distinguishable. The diameter of the whole head, from comparison with the field of the comet-finder, must equal about 20'. The nucleus seems to stand at the vertex of a large ill-marked parabola, of which $a\,e$ forms a portion, and within which a less degree of light indicates a hollow structure. The left side (inverted) of the coma in passing into the tail is rather the denser, but there is little difference, and nothing that could be called two streams of light, as in the Donati; nor has there been throughout a trace of the very curious ray directed towards the Sun, which has been called an "anomalous tail," and which was noticed in 1680, and very conspicuous in the spring of 1824 [the comet of December, 1823].

"July 5. Clouds and haze, but space enough open to show there has been little change. The fan is more symmetrical, from opening out further towards the S., and the ray, $a\,c$, is broader and less distinct. The dark surrounding space is nearly obliterated; the streaky aspect of the coma, and that of the tail, are unaltered.

"July 6. Cloudy; but an opening shows the tail turned again a little to the left, in which another observer concurred.

"July 8. Generally clear sky; nucleus more diffused than ever with 460; merely a condensed patch at the end of a dull yellow sector; even the micrometer does not bring it to a point. Sector, with this power, much as in the last observation, but the circular outline is almost lost, and it fades down into the coma, without any surrounding vacancy as before. A mean of 3 measures of position gives the bright beam, $a\,c$, in Fig. 156, an angle of 271°, implying, from its increase, *if it can be sufficiently trusted*, a swinging motion, similar to that observed by Bessel in Halley's comet. In the comet-finder the preceding side of the tail, according to the comet's orbital motion, appears as a faint narrow ray, estimated at 5' in breadth, and extending for several degrees; the other side of the tail is equally wide, but fainter, and perhaps 30' long, fading gradually away; it comes off from the coma with a kind of bend, as if the longer ray had occupied the axis of the tail, and a corresponding minor branch had been missing on the opposite side of the coma; in fact, my impression was that the tail had a bright central streak, especially as I thought I perceived it in the small finder also; yet this seems invalidated by the next observation. It is represented in Fig. 157. The length, to the naked eye, was 14° or 15°.

"July 10. A transparent night. The comet is a beautiful object to the naked eye, distinctly white, as it has been in fact for some time; in the comet-finder tinged with bluish-green, a singular change from its first aspect. With this eye-piece, 110 and 460, the respective appearances are much as before, with decreasing intensity; the 6 envelopes of June 30 have all disappeared, the sector has no circular boundary, and the ray, *a c*, is very obscure; it seems inclined much as before. I now distinctly see that the comparatively vacant space is in the centre of the tail; the longer side is rather broader than on July 8, and less defined, the shorter and fainter stream can be traced for nearly 2° The bend in coming off from the coma is very visible in the finder, and perceptible even with the naked eye. I estimated its length at 12° or 15°; my wife traced it further. The whole coma, in the comet-finder, subtends an angle of 18' or 20', and gives me the general impression of being more condensed and luminous than it was on June 30.

"July 15. Rather hazy. Powers 27, 55, and 110, all give the head a faint bluish-green tinge; the nucleus seems to have a greenish-yellow hue; brownish-yellow, from want of light, with 460, which shows it extremely nebulous, but still exhibits a dim sector. 55 and 110 show no darker space in the tail, but it can be distinctly made out with the comet-finder, though not near the nucleus, and may be traced about 1° from the head; the streams on each side are now nearly alike in density and definition — possibly the orbitally preceding may be the more distinct. The bend near the head still seems to exist. The diameter of the coma is about 13'.

"July 16. Fine clear night, but a Moon a little past quadrature; tail much diminished, and hollow structure difficult to be made out. With 460 my impression is confirmed that the coma has become more dense, as it is very conspicuous around and beyond the sector, which has hardly a separate existence, and does not now seem, as last night, with this much clearer air, diminished in angular extent. The nucleus passed to-night near a small star, which must have been only 2' or 3' from it; I did not watch the nearest appulse.

"July 23. Clear night: comet very much diminished, and colourless apparent: nucleus I thought yellowish; still pointed with 55 and 110, a nebulosity with 460; through which a denser centre peeped by averted vision. Hollowness of tail scarcely perceptible."

Figures 158, 159, and 160 represent the comet as it appeared to us on the evening of July 2. The first shows the conformation of the coma, and coincides in the main with Mr. Webb's. The second shows a peculiar fan-like appendage attached to the main tail. This secondary portion was about 20° long, and the breadth of the tail and fan, at the extremity of the latter, was about 7°. The third gives an idea of the

naked eye appearance of the tail, which was determined to be
more than 105° long.[1]

In a letter published at the time in one of the London
morning papers, Mr. Hind stated that he thought it not only
possible, but even probable, that in the course of Sunday,
June 30, the Earth passed through the tail of the comet at a
distance of perhaps two-thirds of its length from the nucleus.

The head of the comet was in the ecliptic at 6 P.M. on June
28, distant from the Earth's orbit 13,600,000 miles on the
inside, its longitude, as seen from the Sun, being 279° 1′.
The Earth at that moment was 2° 4′ behind that point, but
would arrive there soon after 10 P.M. on Sunday, June 30. The
tail of a comet is seldom an exact prolongation of the radius
vector, or line joining the nucleus with the Sun; towards the
extremity it is almost invariably curved; or, in other words,
the matter composing it lags behind what would be its situation if it travelled with the same velocity as the nucleus.
Judging from the amount of curvature on the 30th, and the
direction of the comet's motion as indicated by his orbit
already published, Mr. Hind thinks the Earth would very
probably have encountered the tail in the early part of that
day, or, at any rate, it was certainly in a region which had
been swept over by the cometary matter shortly before.

In connection with this subject, he adds that on Sunday
evening, while the comet was so conspicuous in the northern
heavens, there was a peculiar phosphorescence or illumination
of the sky, which he attributed at the time to an auroral
glare; it was remarked by other persons as something unusual,
and, considering how near we must have been on that evening
to the tail of the comet, it may perhaps be a point worthy of
investigation whether such an effect can be attributed to our
proximity thereto. If a similar illumination of the heavens
has been remarked generally on the Earth's surface, it will be
a significant fact.

Mr. Lowe, of Highfield House, also confirmed Mr. Hind's

---

[1] All the illustrations of this comet have been kindly supplied to us
by the Editor of the *London Review*, who devotes much of his space to
the consideration of scientific subjects.

statement of the peculiar appearance of the heavens on June 30. The sky, he says, had a yellow auroral, glare like look, and the Sun, though shining, gave but feeble light. The comet was plainly visible at a quarter to 8 o'clock (during sunshine), while on subsequent evenings it was not seen till an hour later. In confirmation of this, he adds that in the parish church the vicar had the pulpit candles lighted at 7 o'clock, a proof that a sensation of darkness was felt even with the Sun shining. Without being aware that the comet's tail was surrounding our globe, yet being struck by the singularity of the appearance, he recorded in his day-book the following remark : — " A singular yellow phosphorescent glare, very like diffused Aurora Borealis, yet being daylight such Aurora would scarcely be noticeable." The comet itself, he states, had a much more hazy appearance than at any time since that evening.

Mr. Warren De La Rue attempted to photograph the comet. After 3 minutes' exposure in the focus of his 13-inch reflector the comet had left no impression upon a sensitized collodion plate, although a neighbouring star, π Ursæ Majoris — close to which the comet passed on the night of the 2nd (Tuesday) — left its impression twice over, from a slight disturbance of the instrument.

Mr. De La Rue also, at that time, fastened a portrait camera upon the tube of his telescope, and, with the clock motion in action, exposed a collodion plate for 15 minutes to the open view of the comet without any other effect than the general blackening of the surface by the skylight, accompanied with impressions of other fixed stars in the neighbourhood.

Of the polarization of the light of the comet, M. Secchi says : —

" The most interesting fact I observed is this : the polarization of the light of the comet's tail and of the rays near the nucleus, was very strong, and one could even distinguish it with the band polariscope ; but the nucleus presented no trace of polarization, not even with Arago's polariscope with double coloured image. On the contrary, on the evenings of July 3, and following days, the nucleus presented decided indications, in spite of its extreme smallness, which, on the evening of July 7, was found to be hardly one second.

"I think this a fact of great importance, for it seems that the nucleus on the former days shone by its own light, perhaps by reason of the incandescence to which it had been brought by its close proximity to the Sun.

"During the following days the tail has been constantly diminishing, but it is remarkable that it has always passed near to α Herculis, and that it reached to the Milky Way up to July 6. It would seem that the two tails were nearly independent, and that on July 5 the length and straightness had gone off from the large one, and that this bent itself to the southern side. Last night (July 7) the long train was hardly perceptible. The light was polarized in the plane of the tail."

Observations on the polarization of the light of the comet were also made by M. Poey, at Passy. This gentleman observed the polarization in Donati's comet at the Havanna in 1858, in which case the light was polarized in a plane passing through the Sun, the comet, and the observer; but, in the present comet, "the plane of polarization seemed to pass sensibly perpendicular to the axis of the tail," which, he thinks, may have been owing to atmospherical refraction.

The general elements of the comet will be found under their proper heads in Catalogue I.; but we add a few additional particulars. The comet was at its least distance from the Earth on the same day that it was first generally seen — June 30 — when it was only 13,000,000 miles from us, but after that evening it gradually receded both from the Earth and the Sun, and on August 1 it was 88,000,000 miles from the former, which distance was increased by September 1 to 156,750,000 miles, and by October 1 to 209,250,000 miles.

On July 2, Hind determined the following : —

				Miles.
Diameter of the Nucleus	.	.	.	400
Length of the Tail 70° or	.	.	.	16,000,000

And on July 11 : —

				Miles
Diameter of the Nucleus	.	.	.	850
Breadth of the Coma	.	.	.	150,000
Length of the Tail .	.	.	.	13,000,000

The following set of elements are by M. Auwers, and are to be preferred to those given in the Catalogue : —

$$PP = 1861 \quad \text{June } 11\cdot55 \text{ G.M.T.}$$

$$
\begin{aligned}
&\quad\quad\quad\ \text{o} \quad\ \ \prime\\
\pi &= 249 \quad 7\\
\Omega &= 278 \quad 58\\
\iota &= \ \ 85 \quad 28\\
q &= \quad 0\cdot8223\\
\epsilon &= \quad 0\cdot98845\\
\mu &= \ +\\
p &= \ 601 \text{ years.}
\end{aligned}
$$

Page 490.—The following variable stars have been announced since the catalogue was completed : —

α Argûs,   R.A. 6h. 20m. 53s. Decl.— 52° 36′.[1]

* in Cygnus,   R.A. 19h. 41m. 28s. Decl.+ 26° 56′, (D.) observed by Anthelm in 1670.[2]

40196 Lalande, R.A. 20h. 39m. 23s. Decl.— 5° 52′, (D.) discovered by Goldschmidt, 1861.

Page 501.—Sir W. K. Murray, 7th Baronet of Ochtertyre, died on Oct. 15, 1861. He is succeeded by his eldest son Patrick, but we have not heard whether the present baronet will maintain the observatory.

Page 502.—M. Valz has resigned the directorship of the Marseilles Observatory, and intends erecting a private one, at his own residence, in the vicinity of that city.

---

[1] *Ast. Nach.* 1311, May 29, 1861.
[2] *Month. Not.* R.A.S. vol. xxi. p. 231.

CONTENTS.

BOOK I.

A SKETCH OF THE SOLAR SYSTEM.

CHAPTER I.

THE SUN.

CHAPTER II.

THE PLANETS.

CHAPTER III.

VULCAN.

CHAPTER IX.

THE MINOR PLANETS.

CHAPTER X.

JUPITER.

CHAPTER XI.

SATURN.

CHAPTER XII.

URANUS.

CHAPTER XIII.

NEPTUNE.

BOOK II.

ECLIPSES AND THEIR ASSOCIATED PHENOMENA.

CHAPTER I.

GENERAL OUTLINES.

CHAPTER II.

ECLIPSES OF THE SUN.

CHAPTER III.

THE TOTAL ECLIPSE OF THE SUN OF JULY 28, 1851.

CHAPTER IV.

THE ANNULAR ECLIPSE OF THE SUN OF MARCH 14–15, 1858.

CHAPTER V.

THE TOTAL ECLIPSE OF THE SUN OF JULY 18, 1860.

CHAPTER VI.

HISTORICAL NOTICES.

CHAPTER VII.

ECLIPSES OF THE MOON.

CHAPTER VIII.

CHAPTER IX.

TRANSITS OF THE INFERIOR PLANETS.

CHAPTER X.

OCCULTATIONS.

BOOK III.

THE TIDES.

———

CHAPTER I.

CHAPTER II.

BOOK IV.

MISCELLANEOUS ASTRONOMICAL PHENOMENA.

———

CHAPTER I.

CHAPTER II.

BOOK V.

COMETS.

CHAPTER I.

GENERAL REMARKS.

CHAPTER II.

PERIODIC COMETS.

CHAPTER III.

REMARKABLE COMETS.

CHAPTER IV.

COMETARY STATISTICS.

CHAPTER V.

HISTORICAL NOTICES.

BOOK VI.

CHRONOLOGICAL ASTRONOMY.

CHAPTER I.

CHAPTER II.

CHAPTER III.

CHAPTER IV.

CHAPTER V.

BOOK VII.

THE STARRY HEAVENS.

CHAPTER I.

CHAPTER II.

DOUBLE STARS, ETC.

CHAPTER III.

CHAPTER IV.

CLUSTERS AND NEBULÆ.

BOOK IX.

A SKETCH OF THE HISTORY OF ASTRONOMY. 366

BOOK X.

METEORIC ASTRONOMY.

CHAPTER I.

CHAPTER II.

CHAPTER III.

APPENDICES.

LIST OF ILLUSTRATIONS.

WOODCUTS IN TEXT.

PLATES.

PRINCIPAL AUTHORITIES.

AIRY, G. B., *Ipswich Lectures.* 8vo. London, 1849.
ARAGO, D. J. F., *Astronomie Populaire.* 4 vols. 8vo. Paris, 1854—8.
  Popular Astronomy. Trans. W. H. Smyth and R. Grant. 2 vols. 8vo. London, 1855—8.
  Leçons d'Astronomie. 18mo. Bruxelles, 1837.
ARATUS, *Diosemeia.*
ARISTOTELES, *Opera.* Ed. Acad. Reg. Boruss. 4 vols. 4to. Berolini, 1831—6.
Astronomische Nachrichten. 4to. Altona, v. y.

BAILLY, J. S., *Histoire de l'Astronomie Ancienne.* 4to. Paris, 1781.
  Histoire de l'Astronomie Moderne. 3 vols. 4to. Paris, 1785.
  Traité de l'Astronomie Indienne et Orientale. 4to. Paris, 1787.
BAILY, F., *Astronomical Tables and Formulæ.* 8vo. London, 1827—9.
BAYER, *Uranometria: omnium Asterismorum, continens Schemata nova methodo delineata.* Fol. Ulmæ, 1604.
BESSEL, F. W., *Tabulæ Regiomontanæ, reductionum observationum Astronomicarum, 1750—1850 computatæ.* 8vo. Regiomonti, 1830.
  Fundamenta Astronomiæ pro anno 1755 deducta ex observationibus viri incomparabilis James Bradley in speculâ Astronomicâ Grenovicensi, &c. Fol. Regiomonti, 1818.
BIOT, J. B., *Traité Elementaire de l'Astronomie Physique.* 3rd Ed. 4 vols. Paris, 1844—7.
BODE, J. E., *Uranographia, sive Astrorum descriptio.* Fol. Berolini, 1801.
BONNYCASTLE, J., *Introduction to Astronomy.* 8vo. London, 1803.
BRADY, J., *Clavis Calendaria.* 2nd Ed. 2 vols. 8vo. London, 1812.
BRAHE, TYCHO, *Astronomiæ Instauratæ Progymnasmata.* 2 vols. 4to. Francofurti, 1610.
BREWSTER, SIR D., *More Worlds than One.* 16mo. Edinburgh, 1854.
  Treatise on Optics. 12mo. London, 1853.
British Almanac and Companion. 12mo. London, v. y.
BRITISH ASSOCIATION *Reports.* 8vo. London, v. y.
BRÜNNOW, *Spherical Astronomy*, Trans. Main. 8vo. Cambridge, 1860.

Connaissance des Temps. 8vo. Paris, v. y.
Comptes Rendus de l'Académie des Sciences. 4to. Paris, v. y.
COOPER, E. J., M.P., *Cometic Orbits.* 8vo. Dublin, 1852.
Cosmos, Revue des Sciences. 8vo. Paris, v. y.
COSTARD, REV. G., *History of Astronomy.* 4to. London, 1767.

DELAMBRE, J. B., *Histoire d'Astronomie — Ancienne — du Moyen Age — et Moderne* 5 vols. 4to. Paris, 1817—21.
  Astronomie, théoretique et practique. 3 vols. 4to. Paris, 1814.
DREW, J., *Manual of Astronomy.* 2nd Ed. 16mo. London, 1853.

EMERSON, W., *System of Astronomy, containing the Investigation and Demonstration of its Elements.* 8vo. London, 1769.
Encyclopædia Britannica. 21 vols. 4to. Edinburgh, 1853—60.
ENGLEFIELD, SIR H., *On the Determination of the Orbits of Comets, according to the Methods of Boscovitch and La Place.* 4to. London, 1793.
English Cyclopædia, Arts and Sciences division. 8 vols. 4to. London, 1859—61.

FERGUSON, J., *Astronomy.* 2nd Ed. 4to. London, 1757.
FLAMSTEED, Rev. J., *Account of his Life,* by F. Baily. 4to. London, 1837.

GALBRAITH, REV. J., and HAUGHTON, REV. S., *Manual of Astronomy.* 18mo. London, 1857.
Manual of Optics. 18mo. London, 1857.
GALILEO, G., *Opere.* 15 vols. 8vo. Firenze, 1842—56.
GASSENDI, P., *Omnia Opera.* 6 vols. Fol. Florentiæ, 1727.
GRANT, R., *History of Physical Astronomy.** 8vo. London, 1852.
GRAVIER COULVIER-, *Recherches sur les Etoiles Filantes.* 8vo. Paris, 1847, *et seq.*
Greenwich Observations. 4to. London, v. y.
GREGORY, J., *Optica promota, seu abdita Radiorum reflexorum et refractorum mysteria, geometricè enucleata.* Reprint, 4to. London, 1663.

HALLEY, E., *Astronomical Tables, with Precepts both in English and Latin.* 4to. London, 1752.
HERODOTUS, *Helicarnasseus,* Ed. Rev. G. Rawlinson. 4 vols. 8vo. London, 1859—60.
HERSCHEL, SIR J. F. W., *Outlines of Astronomy.* 6th Ed. 8vo. London, 1859.
Results of Astronomical Observations made during the Years 1834—8 at the Cape of Good Hope. 4to. London, 1847.
HEVELIUS, J., *Cometographia, totam naturam cometarum exhibens.* Fol. Gedani, 1668.
Mercurius in Sole visus. Fol. Gedani, 1662.
Selenographia, sive Lunæ descriptio. Fol. Gedani, 1647.
HIND, J. R., *Astronomical Vocabulary.* 16mo. London, 1852.
Illustrated London Astronomy. 8vo. London, 1853.
Solar System. 12mo. London, 1851.
The Comets. 12mo. London, 1852.
The Comet of 1556. 12mo. London, 1857.
HOOK, R., *Attempt to prove the Motion of the Earth from Observation.* 4to. London, 1674.
HUGENIUS, C., *Cosmotheoros, sive de terris Cœlestibus, eorumque ornatu conjecturæ.* 4to. Hagæ Comitum, 1698.
Systema Saturnium. Hagæ Comitum, 1659.
Opera Varia. 2 vols. 4to. Lugduni Batavorum, 1724.
HUMBOLDT, A. VON, *Cosmos,* Trans. E. C. Ottè. 5 vols. 8vo. London, 1849—58.
HUTTON, C., *Mathematical and Philosophical Dictionary.* 2nd Ed. 2 vols. 4to. 1845.

JOHNSTON, A. K., *Atlas of Astronomy,* Ed. Hind. 8vo. Edinburgh, 1855.
Physical Atlas. Fol. Edinburgh, 1849.

KEITH, T., *Treatise on the Globes,* Ed. Rowbotham. 8vo. London, 1844.
KEPLER, J., *Ad Vitellionem Paralipomena, quibus Astronomiæ pars optica traditur.* 4to. Francofurti, 1604.
De Stellâ novâ in Pede Serpentarii. 4to. Pragæ, 1606.

---

* This volume *contains materials* for a very complete history of every branch of Astronomy.

KEPLER, J., *De Motibus Stellæ Martis, ex observationibus Tychonis Brahe.* Fol. Pragæ, 1609.
    Epitome Astronomiæ Copernicanæ Libri tres priores de doctrinâ sphericâ. 8vo. Lentiis ad Danubium, 1618.
    Opera Omnia, Ed. Ch. Fusch. Frankfurt, 1859, *et seq.*

LALANDE, J. De, *Astronomie.* 4 vols. 4to. Paris, 1792.
    Bibliographie Astronomique, avec l'Histoire de l'Astronomie depuis 1781 *jusqu'à* 1802. 4to. Paris, 1803.
    L'Art de vérifier les Dates des Faites Historiques, des Chartes, des Chroniques, et d'autres anciens Monuments, depuis la Naissance de Notre Seigneur, par le moyen d'une Table Chronologique. 3 vols. fol. Paris, 1783.
LAPLACE, P. S. DE, *Exposition du Système du Monde.* 4to. Paris, 1798.
    Traité de Mécanique Céleste. 5 vols. 4to. Paris, 1798—1827.
LARDNER, D., *Handbook of Astronomy.* 2 vols. 12mo. London, 1853 ; also 2nd Ed. Ed. E. J. Dunkin. 1 vol. 12mo. London, 1860.
    Museum of Science and Art. 12 vols. 12mo. London, 1854—6.
L'Institut, Journal Universel des Sciences. Fol. Paris, v. y.
LONG, R., *Astronomy.* 2 vols 4to. Cambridge, 1742.
LOOMIS, E., *History of Astronomy, especially in the United States.* 12mo. New York, 1856.
    Practical Astronomy. 8vo. New York, 1855.
LUBIENITZ, S. DE, *Theatrum Cometicum.* 2 vols. fol. Amsterdam, 1668.
LUCRETIUS, *De Rerum Naturâ,* Trans. Watson, vol. 26, Bohn's *Class. Library.* 8vo. London, 1851.

MAILLA, J. A. M. DE, *Histoire Générale de la Chine.* 4to. Paris, 1776, *et seq.*
MAIN, Rev. R., *Rudimentary Treatise on Astronomy,* Weale's series, vol. 96. 12mo. London, 1852.
MANILIUS, *Astronomicon, in usum Delphinum.* 4to. Paris, 1676.
MILNER, Rev. T., *Gallery of Nature.* 2nd Ed. 8vo. London, 1859.
MONTUCLA, J. F., *Histoire des Mathématiques,* Ed. Lalande. 4 vols. 8vo. Paris, 1799—1802.
MORGAN, A. DE, *The Book of Almanacs.* Oblong 8vo. London, 1851.
MOSELEY, REV. H., *Lectures on Astronomy.* 12mo. London, 1850.

NARRIEN, J., *Origin and Progress of Astronomy.* 8vo. London, 1833.
Nautical Almanac. 8vo. London, v. y.
NICHOL, J. P., *Architecture of the Heavens.* 9th Ed. 8vo. London, 1851.
    Cyclopædia of the Physical Sciences. 2nd Ed. 8vo. London, 1860.

OLMSTED, D., *Mechanism of the Heavens.* 8vo. Edinburgh.

PEARSON, REV. W., *Introduction to Practical Astronomy.* 3 vols. 4to. London, 1824—9.
PETAVIUS, D., *Uranologion : Systema variorum auctorum de Sphærâ.* Folio. Lutet. 1630.
Philosophical Journal. 8vo. Edinburgh, v. y.
Philosophical Magazine. 8vo. London, v. y.
PINGRÉ, A., *Cométographie ; ou Traité historique et théoretique des Comètes.* 2 vols. 4to. Paris, 1783.
PLUTARCHUS, *Opera,* Ed. Reiske. 12 vols. Lipsiæ, 1778.
PTOLEMÆUS CLAUDIUS, *Almagestum.* Fol. Colon., 1515.

QUETELET, A., *Elements d'Astronomie.* 12mo. Paris, 1847.

REES, A., *Cyclopædia of Arts, Sciences, and Literature.* 45 vols. 4to. London, 1819.

Rios, J. De Mendoza, *A complete Collection of Tables for Navigation and Nautical Astronomy.* 4to. London, 1805.

Royal Astronomical Society, *Memoirs.* 4to. London, v. y.
  Monthly Notices. 8vo. London, v. y.
Royal Irish Academy, *Transactions.* 8vo. Dublin, v. y.
Royal Society, *Philosophical Transactions.* 4to. London, v. y.
Royal Society of Edinburgh, *Transactions.* 4to. Edinburgh, v. y.

Scheiner, C., *Rosa Ursina, sive Sol ex admirando Facularum et Macularum suarum phenomeno varius, &c.* Fol. Bracciani, 1630.
Schmidt, J. F. J., *Das Zodiacallicht.* 8vo. 1856.
  Resultate aus zehnjährigen Beobachtungen uber Sternschnupfen. 1852.
  Resultate aus eilfährigen Beobachtungen der Sonnenflecken. 1857.
Small, R., *Account of the Astronomical Discoveries of Kepler.* 8vo. London, 1804.
Smith, Robert, *A complete System of Opticks.* 2 vols. 4to. Cambridge, 1738.
Smyth, W. H., *Cycle of Celestial Objects.* 2 vols. 8vo. London, 1844.
  Speculum Hartwellianum. 4to. London, 1860.
Smyth, C. P., *Teneriffe; an Astronomer's Experiment.* 8vo. London, 1858.
Snooke, W. D., *Brief Astronomical Tables for the expeditious Calculation of Eclipses.* 8vo. London, 1852.
Struve, F. G. W., *Stellarum duplicium, &c. Mensuræ Micrometricæ per Magnum Fraunhoferi Tubum,* 1824—37, *speculâ Dorpatensi institutâ, &c.* Fol. Petropolis, 1837.
  Etudes d'Astronomie Stellaire. 8vo. St. Petersbourg, 1847.

Theophrastus, *Opera Omnia,* Ed. Heinsius. Fol. Lugduni Batavorum, 1613.
Thomson, D. P., *Introduction to Meteorology.* 8vo. Edinburgh, 1849.

Vince, Rev. S., *Complete System of Astronomy.* 3 vols. 4to. Cambridge, 1797—1808.

Webb, Rev. T. W., *Celestial Objects for common Telescopes.* 16mo. London, 1859.
Wing, V., *Astronomia Britannica.* Fol. London, 1669.

Xenophon, *Anabasis,* Trans. Watson, vol. 62, Bohn's *Class. Library.* 8vo. London, 1854.

Errata.

Page 22, in table, col. 7, line 5, *for* $\frac{1}{16}$ *read* $\frac{1}{16 \cdot 8}$.

  ,, 125, line 17, *for* ἐαπίνης *read* ἐξαπίνης.

  ,, 283, in table, col. 3, line 26, *for* κ Tauri *read* χ Tauri.

  ,,   ,,   ,,   ,, 4, ,, 20, ,, 44 Herculis *read* 95 Herculis.

  ,,   ,,   ,,   ,, 4, ,, 27, ,, κ Cygni   ,,   χ Cygni.

  ,, 299, line 32, *for* spots   *read* knots.

  ,, 389, ,, 34, ,, 75   ,, 877.

  ,, 497, ,, 4, ,, Calendar ,, Catalogue.

SUPPLEMENT, page x. in elements of Niobe : —

    For λ *read* M.

    ,, δ ,, Ω.

A HANDBOOK

OF

DESCRIPTIVE AND PRACTICAL ASTRONOMY.

BOOK I.

A SKETCH OF THE SOLAR SYSTEM.

———◆———

CHAPTER I.

THE SUN. ☉

———

"O ye Sun and Moon, bless ye the LORD: praise Him and
magnify Him for ever." — *Benedicite.*

———

*Distance of the Sun. — Its Diameter. — Its Volume. — Its Brightness. —
Solar Spots. — Confined within a certain Limit. — Sun's Axial Rota-
tion. — Spots visible to the naked Eye. — Scheiner. — Solar Spots and
Terrestrial Temperatures. — Physical Character of the Spots. — Faculæ.
— Luculi. — Observations of Nasmyth.*

THE Sun as the centre of the system will first occupy our
attention. The distance of the Earth from the Sun, which is
usually employed by astronomers as a unit of measurement,
has been ascertained with great accuracy, from the transit of
Venus over the disc of the latter in 1769, to be 95,298,260
miles, a distance which every successive transit will render
more and more exactly known. Having ascertained the true
mean distance of the Earth from the Sun, it is not difficult
to determine, by trigonometry, the true diameter of the latter
body, its apparent diameter being known from observation;
and as the most reliable results prove that the Sun, in the
above position, subtends an angle of about 32′, it follows that
its true diameter is about 887,000 miles; the volume of
this enormous globe, therefore, exceeds that of the Earth
1,400,000 times; in other words, it would take 1,400,000
Earths to make up a globe of the same size as the Sun. The

Sun's mass, or attractive power, exceeds that of the Earth 355,000 times, and is 476 times greater than the masses of all the planets put together.

We thus see that the Sun is eminently worthy of the important position it holds as the centre of our system. It is necessary to bear in mind that the Sun is a fixed body, so far as regards ourselves; therefore, when we say the "Sun rises," or the "Sun sets," or the Sun moves through the signs of the zodiac once a year, we are stating a conventional untruth; it is *we* that move and not the Sun, the apparent motion of the latter being merely an optical illusion. The Sun is of a spherical figure, is surrounded by an extensive but very rare atmosphere, and is self-luminous, emitting light and heat, which is transmitted even to the planet Neptune, a distance of 2,800 millions of miles! How much further we know not.

Every one knows the dazzling brilliancy of the solar orb, on which account it is necessary to use coloured glasses for observing it (a practice introduced soon after the invention of the telescope), but its light- and heat-giving powers diminish as the distance increases, so that at Neptune it is reduced to a minimum. It has been estimated that the direct light of the Sun is equal to that of 5563 wax candles of moderate size, supposed to be placed at a distance of one foot from the object. The light of the Moon being probably only equal to that of one candle at a distance of 12 feet, it follows that the light of the Sun exceeds that of the Moon 300,000 times.[1]

When telescopically examined, the surface of the Sun's disc is frequently found covered with dark spots or *maculæ*, each surrounded by a fringe of a lighter shade, called a *penumbra*; this, however, is not always the case, several spots being occasionally included within the limits of one penumbra. They are for the most part confined to a zone

---

[1] To show the great power of the calorific rays of the Sun, we may mention that in constructing the Plymouth breakwater, the men, working in diving bells, at a considerable distance below the surface, had their clothes burnt by coming under the focus of the convex lenses placed in the bell to let in the light.

Figs. 45, 46.

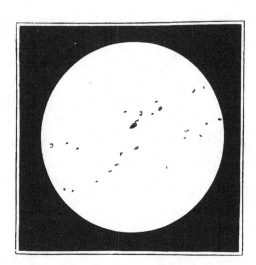

GENERAL TELESCOPIC APPEARANCE.

FACULÆ.

THE SUN.

extending 35° on each side of the solar equator, and are neither permanent in their form nor stationary in their position ; frequently appearing and disappearing with great suddenness. This, however, is by no means always the case, for some have been known to remain visible for several weeks or even months, as was the case with a fine group in the autumn of 1859. Solar spots seem to have been discovered by J. Fabricius and Galileo independently, early in the year 1611 ; the evidence on the subject is, however, very conflicting. By the latter observer the spots were seen as far as 29° from the solar equator. Scheiner noticed some at a distance of 30°, which he termed, in an elaborate work he published' in 1630[1], the "royal zone; " finally, Carrington, in 1858, saw a spot 44° 53', and C. H. Peters, in 1846, one, 50° 55 , from the solar equator. It is a singular circumstance that these spots are confined to two belts on either side of the Sun's equator, never being seen either under or very near that circle. It is on record that there have been periods of considerable length during which no spots have been visible ; such was the case, we are told, from 1650—1670, and during 1724. On the other hand, Scheiner states that with the exception of a few days in December 1624, the solar disc was never clear between the years 1611 and 1629.

We may here take occasion to advert to a very remarkable phenomenon seen on September 1, 1859, by two English observers, whilst engaged in scrutinising the Sun. A very fine group of spots was visible at the time, and suddenly at 11h. 18m. two patches of intensely bright white light were seen to break out in front of the spots. It was at first thought to be due to a fracture of the screen attached to the object-glass of the telescope, but such was not the case. The patches of light were evidently connected with the Sun itself; they remained visible for about 5 minutes, during which

---

[1] *Rosa Ursina, &c.* Alluding to this enormous book, Delambre says : "There are few books so diffuse and so void of facts. It contains 784 pages; there is not matter in it for 50 pages."— *Hist. Ast. Mod.*, vol. i. p. 690. Either printing must have been cheap, or authors very rich in those days.

time they traversed a space of about 35,000 miles. The brilliancy of the light was dazzling in the extreme, but the most noteworthy circumstance was the marked disturbance which (as was afterwards found), took place in the magnetic instruments at the Kew Observatory, simultaneously with the appearance in question, followed about 16 hours afterwards by a great magnetic storm.[1]

When a spot is observed for any length of time, it is seen at first on the eastern limb, disappearing in a little less than a fortnight on the western side; after an interval of nearly another fortnight, the spot, if still in existence, will reappear on the eastern side, and in like manner traverse the disc as before: some spots have been observed to pass seven or eight times over the Sun, as above described. This phenomenon can only be accounted for, on the supposition that the Sun rotates on its own axis in 25d. 7h. 48m. carrying the spots with it.[2] Long-continued observation shows that the number of spots visible in different years is subject to a sensibly periodic variation, and M. Schwabe, of Dessau, who has directed close attention to the subject for more than 30 years, thinks the cyclical period is about 10 years; a discovery both interesting and important, though at present inexplicable.

The following is a table of Schwabe's results[3] : —

Year.	Days of Observation.	Days of no Spots.	New Groups.	Mean diurnal Variation in Declination of the Magnetic Needle.
1826	277	22	118	
1827	273	2	161	
1828	282	0	225	
1829	244	0	199	
1830	217	1	190	
1831	239	3	149	
1839	270	49	84	
1833	247	139	33	

[1] Carrington and Hodgson, *Month. Not.* R.A.S., vol. xx. pp. 13–16.
[2] Kepler first suspected this.
[3] *Month. Not.* R.A.S., vol. xvi. p. 63.

Year.	Days of Observation.	Days of no Spots.	New Groups.	Mean diurnal Variation in Declination of the Magnetic Needle.
1834	273	120	51	
1835	244	18	173	9·57
1836	200	0	272	12·34
1837	168	0	333	12·27
1838	202	0	282	12·74
1839	205	0	162	11·03
1840	263	3	152	9·91
1841	283	15	102	7·82
1842	307	64	68	7·08
1843	312	149	34	7·15
1844	321	111	52	6·61
1845	332	29	114	8·13
1846	314	1	157	8·81
1847	276	0	257	9·55
1848	278	0	330	11·15
1849	285	0	238	10·64
1850	308	2	186	10·44
1851	308	0	151	
1852	337	2	125	
1853	299	3	91	
1854	334	65	67	
1855	313	146	79	
1856	321	193	34	
1857	324	52	98	
1858	335	0	188	
1859	343	0	205	
1860	332	0	211	

The numbers in the fifth column are given by Dr. Lamont
in a memoir on the subject. Those from 1835–40 have been
deduced from the *Göttingen Observations*, the remainder from
his own. He adds, that the observations of Colonel Beaufoy
from 1813–1820, and the still earlier observations of Gilpin
and Cassini, indicate a similar cycle.

Instances of solar spots visible to the naked eye are not
rare.   In June 1843, M. Schwabe observed one 2′ 47″, or

77,820 miles in diameter. More recently large spots were visible on March 15, 1858 [1], the day of the celebrated eclipse, on September 30, 1858 [2], and on January 26, 1859. [3]

The observation of solar maculæ was one of the first discoveries resulting from the invention of the telescope, though it is not improbable that they were seen before that time. Adelmus, a Benedictine monk, makes mention of a black spot being seen on the Sun on March 17, 807. It is also stated that a similar spot was seen by a Spanish Moor named Averroës, in the year 1161. [4] An instance of a solar spot is recorded by Hakluyt. He says, that in December 1590, the good ship *Richard of Arundell* was on a voyage to the coast of Guinea, and that her log states that " on the 7 at the going downe of the sunne, we saw a great blacke spot in the sunne, and the 8 day both at rising and setting we saw the like, which spot to our seeming was about the bignesse of a shilling, being in 5 degrees of latitude, and still there came a great billow out of the southerboard." [5] The spot was also seen on the 16th.

The natural purity of the Sun seems to have been an article of faith with the ancients, on no account to be called in question; so that we find that when Scheiner reported what he had seen to his superior (he was a Jesuit), the idea was treated as a delusion. " I have read," replied the superior, " Aristotle's writings from end to end many times, and I can assure you that I have nowhere found anything in them similar to what you mention. Go, my son, and tranquillize yourself; be assured that what you take for spots in the Sun, are the fault of the glasses or of your own eyes." Scheiner

---

[1] *Month. Not.* R.A.S., vol. xviii. p. 193, and elsewhere. Its breadth from W. to E. was 4′ or 113,000 miles.

[2] *Ast. Nach.*, 1172. Its breadth from W. to E. was 5′ 21″, or 150,000 miles.

[3] W. R. Dawes, in *The Times*, Jan. 28.

[4] Commentary on the *Almagest*, quoted by Copernicus, *De Revol. Orb. Cel.* lib. x.

[5] *The Principal Navigations, Voiages, Traffiques, and Discoueries of the English Nation, &c.*, vol. ii. p. 131. London, 1599.

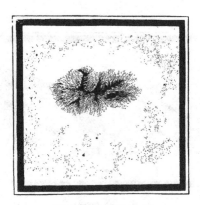

1826: July 2. (*Capocci.*)

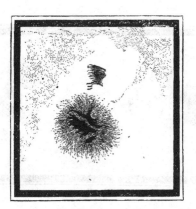

1826: September 29. (*Capocci.*)

SPOTS ON THE SUN.

May 23. (*Birt.*)

May 27.

A SPOT ON THE SUN: 1861.

in the end, though permitted to publish his opinions, was obliged to do so anonymously, so great was (and we regret to say, still is), the bigotry of the Church of Rome, both in things spiritual and temporal.

It has been thought that the prevalence of large masses of spots might give rise to a depression in the temperature for the time being, and thus affect the fertility of the soil. Modern observation, however, would lead us to infer that the contrary was rather the case, an elevation of temperature being contemporaneous with the prevalence of spots.

The outlines of some spots remain nearly constant for many weeks together, whilst others alter in form from hour to hour. Sir W. Herschel states, that on Feb. 19, 1800, he was watching a group, but that on looking away, even for a moment, they could not be found again.[1] Sir J. W. Lubbock has noticed the same kind of thing.

Respecting the physical nature of the spots little can be said, because nothing is known. It seems, however, probable that they are nothing more than some kind of excavation in the solar photosphere, and this opinion is confirmed by the changing appearance they present whilst traversing the Sun's disc, for after passing the centre, the penumbra is impaired on the side nearest to it, and gradually disappears ; then the nucleus is nipped on the same side till it vanishes, and then finally, the penumbra begins to contract, narrows to a line and is carried out of sight. Scheiner suggested they were solid bodies revolving round the Sun, but this is obviously impossible, and the idea was soon rejected. In addition to the spots, streaks of light may frequently be remarked upon the surface of the Sun, more luminous than the surrounding portion; these are termed *faculæ* (torches), and are generally found near the spots, or where the spots have previously existed, or afterwards appeared. Sir W. Herschel saw a facula on Dec. 27, 1799, 2 46″, or 77,000 miles in length.[2]

---

[1] *Phil. Trans.*, vol. xci. p. 293. 1801.
[2] *Phil. Trans.*, vol. xci. p. 284. 1801.

These phenomena were first described by Galileo, in his third letter to Welser.[1] With powerful optical assistance it is found that the surface of the Sun is also covered with irregular specks of light, giving the disc a resemblance to the skin of an orange, and termed *luculi* (little lights) : but the physical character of these and also of the faculæ is unknown, and conjecture is nearly valueless.

Nasmyth has recently put forth some remarkable observations on the structure of the Sun's surface, which he finds is formed all over of long narrow filaments resembling *willow leaves*. An engraving, reduced from a drawing which he has kindly forwarded for this work, forms the frontispiece.

---

[1] *Istoria et Dimostrazioni intorno alle Macchie Solari*, p. 131. Roma, 1613.

CHAPTER II.

THE PLANETS.

Motions of the Planets.— Features common to them all.— Kepler's Laws. — Explanation of the First Law.— Elements of a Planet's Orbit.— Explanation of the Second Law.— Explanation of the Third Law.— Popular Illustration of the Extent of the Planetary System.— Bode's Law.— Tables of the Major Planets.— Two Groups of Planets.— Singular Coincidences.— Planetary Conjunctions.— Conjunctions recorded in History.— Different Systems.

AROUND the Sun as a centre, certain bodies, called *planets*[1], revolve, at greater or less distances: they may be divided into two groups. (i.) The inferior planets, comprising Vulcan, Mercury, and Venus; and (ii.) the superior planets, including Mars, the Minor Planets[2], Jupiter, Saturn, Uranus and Neptune, the Earth being the boundary between the two. If viewed from the Sun, all the planets, inferior and superior, would appear to spectators placed on the Sun, to revolve round that luminary in the order of the zodiacal signs; such, however, is not the case when the motions of the planets are watched from one of their number itself in motion; it consequently happens that the *apparent* motions of the inferior planets on the one hand, and the superior on the other, differ considerably. The former are never seen in those parts of the heavens in opposition to the Sun, but are sometimes to the east and sometimes to the west of it. Twice in every revolution an inferior planet is in conjunction with the Sun (Fig. 1); in *inferior conjunction*, when it comes between the Earth and the Sun, and in *superior conjunction* when the Sun intervenes

---

[1] From πλάνητης, a wanderer.

[2] The term *asteroids*, formerly applied to these bodies, is now fallen into disuse.

between the Earth and the planet. When it attains its greatest distance from the Sun, east or west, it is said to be at its greatest elongation—east or west, as the case may be.

Fig. 1.

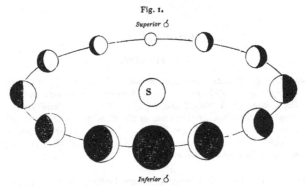

Phases of an Inferior Planet.

Although the planet's *true* path is always in the order of the signs, yet there are periods when it is *stationary;* and even when its motion appears *retrograde* or reversed : these peculiarities are owing, as we have before said, to the motion of the Earth itself in its orbit, and also obtain in the case of the superior planets. It sometimes (though very rarely) happens that an inferior planet, when in inferior conjunction, passes directly between the Earth and the Sun, and is consequently projected on the disc of the latter, which it crosses from west to east : this phenomenon is termed a *transit.*[1] In the case of any planet, the intervals between two successive conjunctions or oppositions as seen from the Earth, is called its *synodical revolution*, which more nearly approaches in length to the Earth's sidereal period the greater the distance of the planet from the Sun, though it must in all cases somewhat exceed the latter.

The planets possess certain characteristics common to them all, and thus enunciated by Hind [2] : —

---

[1] *Trans*, across, and *ire*, to go.
[2] *Illust. Lond. Ast.*, p. 40.

(1.) *They move in the same invariable direction round the Sun; their course, as viewed from the north side of the ecliptic, being contrary to the motion of the hands of a watch.*

(2.) *They describe oval or elliptical paths round the Sun, not however differing greatly from circles.*

(3.) *Their orbits are more or less inclined to the ecliptic, and intersect it in two points, which are the nodes; one half of the orbit lying north, and the other south of the Earth's path.*

(4.) *They are opaque bodies like the Earth; and shine by reflecting the light they receive from the Sun.*

(5.) *They revolve upon their axes in the same way as the Earth; this we know by telescopic observation to be the case with many planets, and, by analogy, the rule may be extended to all. Hence they will have the alternation of day and night, like the inhabitants of the Earth; but their days are of different lengths to our own.*

(6.) *Agreeably to the principles of gravitation, their velocity is greatest at those parts of their orbit which lie nearest the Sun, and least at the opposite parts which are most distant from it; in other words, they move quickest in perihelion*[1], *and slowest in aphelion.*[2]

From a long series of observations on the planets, Kepler found that certain definite laws might be deduced relative to their motions, which may be thus summed up:—

(1.) *The planets move in ellipses, having the Sun for one of the foci.*

(2.) *The planets describe equal areas of their orbits in equal times.*

(3.) *The squares of the periodic times of the planets are proportional to the cubes of their mean distances.*

These laws hold good for all the planets and all the satel-

---

[1] περί, near, and ἥλιον, the Sun.

[2] ἀπό, from, and ἥλιον. The fact here referred to is more strikingly manifest in the case of a comet, owing to the greater eccentricity of cometary orbits: thus it has been calculated that the velocity of comet vi., 1858 (Donati's), at perihelion is 127,000 miles per hour, but at aphelion, only 480 miles per hour.—(Hind, Letter in *The Times*, Oct. 25, 1858.)

lites. We have already referred, in general terms, to the fact stated in the first law; it may, however, be desirable to say, that the orbit of a planet, with reference to its form, magnitude,

Fig. 2.

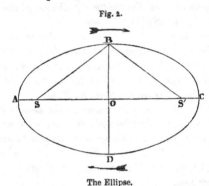

The Ellipse.

S and S′ are the *foci;* A C, are the *major axis:* B D, the *minor axis;* O, the *centre.*
Or, astronomically,—
O A, is the *semi-axis-major,* or *mean distance:* O B, the *semi-axis-minor;* the ratio of O S, to O A, is the *eccentricity:* the least distance, S A, is the *perihelion distance:* the greatest distance, S C, the *aphelion distance:* the time of describing the whole ellipse is the *periodic time.*

and position, is determined by the five following data or *elements :*—

(1.) *The longitude of the perihelion,* or the longitude of the planet, when it reaches this point, denoted by the symbol π.

(2.) *The longitude of the ascending node* of the planet's orbit, as seen from the Sun. — ☊ .

(3.) *The inclination of the orbit,* or the angle made by the plane of the orbit with the ecliptic. — ι.

(4.) *The eccentricity.* — ε. This is sometimes expressed by the angle ϕ of which ε is the sine.

(5.) *The semi-axis-major,* or mean distance. — a.

The second of Kepler's laws will be understood from the following diagram :—

Let A B C D be the elliptic path of a planet, and let it move from C to D in the same time that it takes to pass from A to B; then the area C S D will be equal to the area A S B.

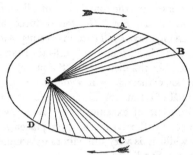

Fig. 3. Illustration of Kepler's Second Law.

In the third law we find a very curious coincidence, which may be thus expressed :—*If the squares of the periodic times of the planets be divided by the cubes of their mean distances from the Sun, the quotients thus obtained are the same for all the planets.* The following table exemplifies this: it should be remarked, however, that the want of *exact* uniformity in the fourth column is owing to error in the observations :—

Planet.	a	p	$\dfrac{p^2}{a^3}$
Vulcan	0·143	19·7	132716
Mercury . . .	0·38710	87·969	133421
Venus	0·72333	224·701	133413
Earth	1·00000	365·256	133408
Mars	1·52369	686·979	133410
Ceres	2·77692	1679·855	132210
Jupiter . . .	5·20277	4332·585	133294
Saturn	9·53878	10759·220	133401
Uranus . . .	19·18239	30686·821	133422
Neptune . . .	30·03680	60726·710	133405

This law also holds good for the satellites [1], as the reader may see, by making the calculation for himself.

---

[1] Newton has shown (*Principia*, prop. 59) that this is not rigorously true when the mass of the planet is considerable as compared with the

These three laws are the foundation of all astronomy, and it was from them that Sir I. Newton deduced his theory of gravitation.   Arago says: " These interesting laws, tested for every planet, have been found so perfectly exact, that we do not hesitate to infer the distances of the planets from the Sun from the duration of their sidereal revolutions; and it is obvious that this method of estimating distances possesses considerable advantages in point of exactness; for it is always easy to determine precisely the return of each planet to a point in the heavens, while it is very difficult to determine exactly its distance from the Sun."

Sir J. Herschel discusses the theoretical considerations connected with these laws, with great perspicuity, but, in a manner somewhat too advanced for these pages: we must therefore rest content with a bare notification of the fact, and a recommendation to the reader to consult his work.[1]

The following will assist the reader in obtaining a correct notion of the magnitude of the planetary system : choose a level field or common; on it place a globe 2 feet in diameter, for the Sun[2]; Vulcan will then be represented by a small pin's head, at a distance of about 27 feet; Mercury by a mustard seed, at a distance of 82 feet; Venus by a pea at a distance of 142 feet; the Earth also by a pea, at a distance of 215 feet; Mars by a large pin's head, at a distance of 327 feet; the minor planets by grains of sand, at distances varying from 500 to 600 feet: if space will permit, we may place a

---

Sun.   In this case the periodic time is shortened in the proportion of the square root of the number expressing the masses of the Sun and planet; and in general, whatever be the masses of any two bodies revolving round each other under the influence of gravitation, the square of their periodic time will be expressed by a fraction, whose numerator is the cube of the mean distance, *i. e.* the semi-axis-major of their elliptic orbit, and whose denominator is the sum of the masses. When one of the masses is incomparably greater than the other, Kepler's laws hold good; but when this is not the case, the above modification requires to be taken.

[1] *Outlines of Ast.,* pp. 322–7.

[2] The distances following are all reckoned from the *centre* of this ideal Sun.

moderate-sized orange nearly $\frac{1}{4}$ mile distant from the starting point, to represent Jupiter; a small orange $\frac{2}{5}$ of a mile, for

Fig. 4.

Comparative Sizes of the Planets.

Saturn; a full-sized cherry $\frac{3}{4}$ mile distance, for Uranus; and lastly a plum $1\frac{1}{4}$ miles off, for Neptune, the most distant planet yet known.

According to this scale, the daily motion of Vulcan in its orbit would be $4\frac{2}{3}$ feet; of Mercury, 3 feet; of Venus, 2 feet; of the Earth, $1\frac{7}{8}$ feet; of Mars, $1\frac{1}{2}$ feet; of Jupiter, $10\frac{1}{2}$ inches; of Saturn, $7\frac{1}{2}$ inches; of Uranus, 5 inches; of Neptune, 4 inches.

This shows that the orbital velocity of a planet decreases, the greater its distance from the Sun.

Extending the above scheme, we find that the aphelion distance of Encke's comet would be 880 feet; the aphelion distance of the 6th comet of 1858, 6 miles; and the distance of the nearest fixed star, 7,500 miles.

Fig. 5.

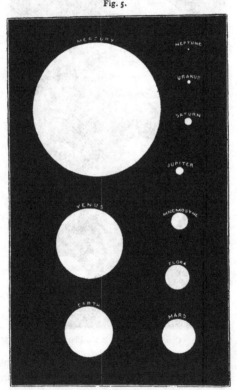

Relative apparent Size of the Sun, as viewed from the different Planets.

Connected with the distances of the planets, Prof. Bode, of Berlin, in 1778 published the following singular numerical relations existing between them; which although not dis-

covered by him, but by Prof. Titius of Wittemberg, usually
bears his name :

Take the numbers —

     0.   3.   6.   12.   24.   48.   96.   192.   384;

each of which (the second excepted) is double the preceding;
adding to each of these numbers, 4, we obtain

     4. 7. 10. 16. 28. 52. 100. 196. 388;

which numbers approximately represent the distances of the
planets from the Sun, as exhibited in the following table :—

Planets.	True Distance from ☉.	Distance by Bode's Law.
Mercury	3·87	4·00
Venus	7·23	7·00
Earth	10·00	10·00
Mars	15·23	16·00
Ceres	27·66	28·00
Jupiter	52·03	52·00
Saturn	95·39	100·00
Uranus	191·82	196·00
Neptune	300·37	388·00

Bode having examined these relations, and noticing the
void between 16 and 52 (Ceres and the other minor planets
not being then known), ventured to predict the discovery of
new planets; and it was this conjecture that guided the investigations of subsequent observers.[1] In the above table the
greatest deviation between the assumed and the true distance
is in the case of Neptune; it is possible, however, that when
more complete observations of this planet shall have been

---

[1] As far back as 450 B.C. Democritus of Abdera thought it probable
that eventually new planets would, perhaps, be discovered. (Seneca,
Quæst. Nat., lib. vii. cap. 3 and 13.) Kepler was of opinion that some
planets existed between the orbits of Mars and Jupiter, but too small to
be visible to the naked eye. The same philosopher conjectured that
there was another planet between Mercury and Venus.

THE MAJOR PLANETS.

Planet.	Symbol.	λ	π	Ω	ι	φ	ε	Sec. Var. π	Sec. Var. Ω	Sec. Var. ι
		° ′ ″	° ′ ″	° ′ ″	° ′ ″	° ′ ″		″	″	″
Mercury	☿	112 16 4	74 20 42	45 57 38	7 0 5	11 49 55	0·2054925	+ 643·56	− 782·27	+18·1828
Venus	♀	146 44 56	128 43 6	74 51 41	3 23 29	0 23 37	0·0068722	− 205·60	−1805·80	− 4·5522
Earth	⊕	100 53 30	99 30 29	0 0 0	0 0 0	0 57 43	0·0167917	+1177·81	...	...
Mars	♂	233 57 34	332 22 51	47 59 38	1 51 6	5 20 25	0·0931125	+1582·43	−2328·44	− 0·1523
Jupiter	♃	81 54 49	11 7 38	98 25 45	1 18 52	2 45 37	0·0481626	+ 663·86	−1577·57	−22·6087
Saturn	♄	123 6 29	89 8 20	111 56 7	2 29 36	3 13 6	0·0561501	+1943·07	−2266·46	−15·5131
Uranus	♅	173 30 37	167 30 24	72 59 21	0 46 28	2 40 23	0·0466686	+ 238·62	−3597·76	+ 3·1331
Neptune	♆	335 8 58	47 14 37	130 6 52	1 46 59	0 29 58	0·0087195	...	...	...
Sun	☉	...	...	...	...	...	...	0 ′ ″	0 ′ ″	...
Moon	☽	118 17 8	266 10 7	13 53 17	5 8 47	3 8 37	0·0548442	40 39 45	16 19 42	0

Symbol.	Sec. Var. ε	Semi-Axis Major. ⊕ = 1	Daily Hel. Motion.	Sidereal Period.		Equinoctial P.	Synodic P.	Distance from ☉		
				d.	y.	d.	d.	Max.	Min.	Mean.
								Miles.	Miles.	Miles.
☿	+0·00003867	0·3870985	4 5 32	87·969	0·240	87·968	115·87	44,474,532	29,304,996	36,889,765
♀	−0·00006271	0·7233317	1 36 7	224·700	0·615	224·695	583·92	69,405,183	68,459,186	68,932,185
⊕	+0·00004163	1·0000000	0 59 8	365·256	1·000	365·242	...	96,898,533	93,697,986	95,298,260
♂	+0·00009017	1·5236915	0 31 26	686·979	1·880	686·929	779·92	158,740,615	131,669,546	145,205,101
♃	+0·00015935	5·2027982	0 4 59	4332·584	11·862	4330·610	398·8	519,697,422	471,938,170	495,817,796
♄	−0·00031240	9·5388852	0 2 0	10759·219	29·458	10746·732	378·0	950,078,822	857,993,171	909,035,997
♅	−0·00025075	19·1827302	0 0 42	30686·820	84·018	30589·357	369·7	1,913,418,905	1,744,744,676	1,828,080,791
♆	...	30·0362807	0 0 21	60126·710	164·622	59744·710	367·5	2,887,363,962	2,837,446,479	2,862,405,220
☉	0 ′ ″	...	0 ′ ″	...	...	...	...	...	...	...
☽	0	—	13 10 35	27·32166	0·082	27·32158	29·53058	—	—	237,639

Symbol	Distance from ⊕; Sup. ☌ Max. Miles	Min. Miles	Mean Miles	Distance from ⊕; Inf. ☌ Max. Miles	Min. Miles	Mean Miles	Apparent Diameter From ⊕ Max.	Min.	Mean.	From ☉ Mean.	Real Diameter ⊕=1.	Miles.
☿	141,373,065	123,002,982	132,188,025	67,591,537	49,223,454	58,408,485	11·5	4·5	8·0	16	0·373	2,950
♀	166,303,716	162,157,172	164,230,445	24,839,347	24,292,803	26,366,075	62·0	9·5	35·7	30	0·984	7,800
⊕	...	...	...	...	...	...	...	...	...	17·2	1·000	7,925
♂	255,639,188	225,377,632	240,503,361	65,044,669	34,771,013	49,996,841	23·5	3·3	13·4	10	0·518	4,113
♃	616,595,955	565,636,156	591,116,056	435,999,336	375,049,617	400,519,536	46·0	30·0	38·0	37	11·255	89,203
♄	1,056,977,355	951,631,157	1,004,334,257	866,380,836	761,095,638	813,737,737	20·5	14·6	17·6	16	9·745	77,230
♅	2,010,317,438	1,816,440,662	1,923,379,051	1,819,720,919	1,645,844,143	1,732,782,501	4·3	3·5	3·9	4	5·353	34,500
♆	2,984,202,495	2,931,144,465	2,957,791,480	2,793,665,976	2,740,547,946	2,767,206,960	2·7	2·6	2·7	2	4·731	37,500
☉	—	—	—	—	—	—	32 35	31 31	32 2	· ·	111·933	887,076
☽	—	—	—	—	—	—	33 31	29 1	31 26	4·6	0·272	2,160

Symbol	Surface ⊕=1	Square Miles.	Volume ⊕=1	Cubic Miles.	☉=1	Mass ⊕=1	Tons.	Density ⊕=1	Water =1
☿	0·139	27,327,000	0·051	13,500,000,000	$\frac{1}{4865751}$	0·073	443,000,000,000,000,000,000,000	1·40	7·94
♀	0·971	191,098,440	0·952	247,329,000,000	$\frac{1}{401211}$	0·885	5,371,000,000,000,000,000,000,000	0·94	5·33
⊕	1·000	196,625,795	1·000	259,800,000,000	$\frac{1}{354936}$	1·000	6,069,000,000,000,000,000,000,000	1·00	5·67
♂	0·269	52,935,571	0·138	38,842,000,000	$\frac{1}{2880337}$	0·132	800,950,000,000,000,000,000,000	1·03	5·84
♃	125·900	24,998,261,636	1466·054	385,293,000,000,000	$\frac{1}{1047787}$	339·258	2,058,950,000,000,000,000,000,000,000	0·24	1·36
♄	85·735	16,737,988,662	787·094	204,487,000,000,000	$\frac{1}{35516}$	101·524	616,149,000,000,000,000,000,000,000	0·13	0·74
♅	19·013	3,738,575,250	82·821	21,517,000,000,000	$\frac{1}{24900}$	14·255	86,514,000,000,000,000,000,000,000	0·17	0·97
♆	22·464	4,417,031,250	105·624	27,441,000,000,000	$\frac{1}{18780}$	18·900	114,704,000,000,000,000,000,000,000	0·18	1·02
☉	12611·0	2,471,664,929,326	1402468·164	364,345,641,000,000,000	1	354936·000	2,154,106,584,000,000,000,000,000,000,000	0·19	1·07
☽	0·074	14,568,221	0·020112	5,200,000,000	$\frac{1}{31134986}$	0·0114	69,000,000,000,000,000,000,000	0·56	3·18

THE MAJOR PLANETS — *continued.*[1]

Symbol	Visible Diam. of ☉ — '	"	⊕=1	Light and Heat of ☉ ⊕=1	Axial Rotation h. m. s.	of Axis ° '	Polar Compression	Force of Gravity Fall: Feet in 1 sec.	Force of Gravity ⊕=1	Orbital Velocity Miles per hour	Orbital Velocity Feet per sec.	Orbital Velocity ⊕=1	Velocity of Rotation at Equator Miles per hour	Velocity of Rotation at Equator Feet per sec.	Time req'red for falling into the Sun, were influence of gravitation suspended. Days.
☿	82	27	2.580	6.67	24 5 30	63 ?	$\frac{1}{29}$	8.52	0.53	109,429	160,496	1.607	385	565	15.6
♀	44	7	1.380	1.91	23 21 23	73 32	very small	14.79	0.92	80,066	117,430	1.175	1,044	1,531	39.7
⊕	31	55	1.000	1.00	23 56 4	66 32	$\frac{1}{299}$	16.08	1.00	68,993	99,870	1.00	1,039	1,524	64.6
♂	20	56	0.656	0.43	24 37 23	59 42	$\frac{1}{62}$	8.36	0.52	55,163	80,906	0.810	521	764	121.5
♃	6	8	0.194	0.037	9 55 21	86 55	$\frac{1}{16\cdot}$	43.42	2.70	29,867	43,805	0.438	28,060	41,155	766.8
♄	3	21	0.105	0.011	10 29 17	58 41	$\frac{1}{9\cdot22}$	19.14	1.19	22,050	32,340	0.323	23,045	33,799	1900.0
♅	1	40	0.052	0.003	9 30 (?)	? ?	?	12.06	0.75	15,547	22,802	0.228	11,409	16,733	5382.0
♆	1	4	0.033	0.001	? ?	? ?	?	13.67	0.85	12,443	18,220	0.182	?	?	...
☉	...		...		(d. h. m.) 25 7 48	82 30	?	459.24	28.56	...	...	...	4,559	6,687	—
☽	...		...		27 7 43	88 30	?	2.48	0.15	2,273	3.334	0.033	10	15	...

[1] In page 20, λ, is "the mean longitude at epoch," which in this case is January 1, 1800.

made, the above difference may be somewhat reduced. We may sum up Bode's law as follows:—*That the interval between the orbits of any two planets is about twice as great as the inferior interval, and only half the superior one.*

Separating the major planets into two groups, taking Mercury, Venus, the Earth, and Mars as belonging to the interior; and Jupiter, Saturn, Uranus, and Neptune, to the exterior group, we shall find they differ in the following respects:—

(1.) The interior planets, with the exception of the Earth, are not, as far as we know, attended by any satellites, while the exterior planets *all* have satellites. We cannot but consider this as one of the many instances to be met with in the universe, of the beneficence of the Creator—that the satellites of these extreme planets are designed to compensate for the paucity of the light their primaries receive from the Sun, owing to their great distance from that luminary.

(2.) The mean density of the first group considerably exceeds that of the second, the approximate ratio being 5 : 1.

(3.) The mean duration of the axial rotations, or mean length of the day, of the interior planets, is much longer than that of the exterior ones; the average in the former case being 23h. 59m. 45s., but in the latter only 9h. 58m. 20s.

The following singular coincidences deserve to be mentioned:—

(1.) Multiply the Earth's diameter (7912 miles) into 110, and we get 870,320 = $\pm$ the $\odot$'s diameter in miles.

(2.) Multiply the Sun's diameter (870,320 miles) into 110, and we get 95,735,200 = $\pm$ the mean distance of the $\oplus$ from the $\odot$.

(3.) Multiply the Moon's diameter (2160 miles) into 110, and we get 237,600 = $\pm$ the mean distance of the $\mathfrak{C}$ from the $\oplus$.

A phenomenon of considerable interest, both of itself and on account of its rarity, is a conjunction of two or more planets, or their grouping together in a limited space. The earliest record we possess of such an occurrence is derived from the Chinese annals. It is stated that a conjunction of Mars, Jupiter, Saturn, and Mercury in the constellation *Shi*

was assumed as an epoch by the Emperor Chuen-hio; and it
has been found by MM. Desvignoles and Kirch that such a
conjunction actually did take place on February 28, 2446
B.C., between 11° and 18° Piscium.[1] Another calculator,
De Mailla, fixes upon Feb. 9, 2641 B.C., as the date of the
conjunction in question; and that the above four planets, with
the Moon, were comprised in an arc of 12°, from 15° to 27°
Piscium. He gives the following positions[2]:—

	R. A.		
	°	′	″
Mercury	344	56	16
Mars	356	45	11
The Moon	353	18	21
Jupiter	347	2	12
Saturn	354	39	47

The following are some other instances of this kind:—

In the years 1507, 1511, 1552, 1564, 1568, 1620, 1624,
1664, 1669, 1709, and 1765, the three most brilliant planets—
Venus, Mars, and Jupiter — were very near each other.

On Nov. 11, 1524, Venus, Jupiter, Mars, and Saturn were
very close to each other, and Mercury was only 16° distant.

On Nov. 11, 1544, Venus, Jupiter, Mercury, and Saturn
were enclosed in a space of 10°.

On March 17, 1725, Venus, Jupiter, Mars, and Mercury
appeared together in the same field of the telescope.

On Dec. 23, 1769, Venus, Jupiter, and Mars were very
close to each other.

On Jan. 29, 1857, Jupiter, the Moon, and Venus were in a
straight line with one another.

On July 21, 1859, Venus and Jupiter were very close to
each other: the actual conjunction took place at 3h. 44m. A.M.,
at which time the distance between the two planets was only
13″. They accordingly appeared, to the naked eye, to be but
one object. At the end of 159 years, Mars, Jupiter, and

---

[1] Bailly, *Astron. Ancienne*, p. 345. Desvignoles' original memoir
appears in *Mém. de l'Acad. de Berlin*, vol. iii. p. 166, and Kirch's, in
vol. v. p. 193.

[2] De Mailla, *Hist. Gén. de la Chine*, vol. i. p. 155.

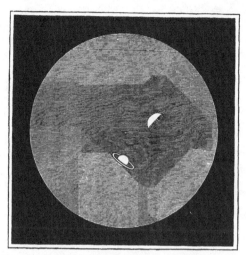

VENUS and SATURN, December 19, 1845.

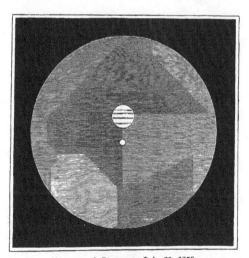

VENUS and JUPITER, July 20, 1859.

CONJUNCTIONS OF THE PLANETS.

Saturn return to nearly the same part of the heavens; this will next happen about the year 2000.

The following brief remarks on the different theories of the

Fig. 6.

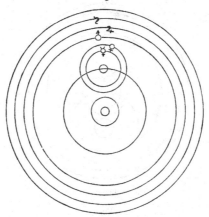

The Egyptian System.

Fig. 7.

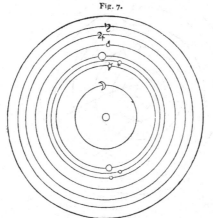

The Ptolemaic System.

planetary system which have at various times been current, will appropriately conclude this chapter.

The Egyptian system supposed that the Earth was the centre; and that round the Earth revolved the Moon, the Sun, and the

Fig. 8.

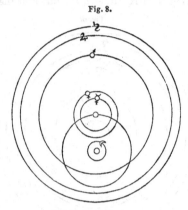

The Tychonic System.

Fig. 9.

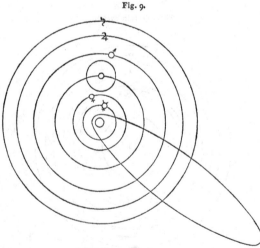

The Copernican System.

major planets *except Mercury and Venus*, which revolved by themselves round the Sun.

The Ptolemaic system also placed the Earth in the centre of the universe, but supposed that the Sun, Moon, planets, and stars, all moved round the Earth together in 24 hours.

The Tychonic system supposed (with the preceding) that the Sun revolved round the Earth as a centre, but that all the other planets moved round the Sun, as solar satellites.

The Pythagorean or *Copernican system,* (the one now known to be the true one) places the Sun in the centre; and supposes all the other planets to move round that centre in the order of the signs.

CHAPTER III.

VULCAN.

Le·Verrier's Investigation of Mercury. — Narrative of the Discovery of Vulcan. — Le Verrier's Interview with M. Lescarbault. — Approximate Elements of Vulcan. — Concluding Note.

BEFORE entering upon the story of the supposed discovery of a new planet to which this name has been given, a brief prefatory statement seems called for.

M. Le Verrier, the well-known astronomer of Paris, having conducted an investigation into the theory of the orbit of Mercury, was led to the conclusion that a certain error in the perihelion could only be accounted for by supposing the mass of Venus to be at least $\frac{1}{10}$ greater than was commonly imagined, or else that there existed some unknown planet, situated between Mercury and the Sun, capable of producing a disturbing action. Le Verrier offered no opinion on these hypotheses, but contented himself with laying them before the scientific world in the autumn of 1859.[1]

On this becoming public, a certain M. Lescarbault, a physician at Orgères,. in the department of Eure-et-Loire France, came forward and stated that on March 26 in that year (1859), he had observed the passage of an object across the Sun's disc which he thought might be a new planet, but which he did not like to announce as such until he had obtained a confirmatory observation; he then related the details, and Le Verrier determined to seek a personal interview.

---

[1] *Compt. Rend.*, vol. xlix. p. 379.

The following account of the meeting will be read with interest.

" On calling at the residence of the modest and unobtrusive medical practitioner, he refused to say who he was, but in the most abrupt manner, and in the most authoritative tone, began, ' It is then you, Sir, who pretend to have observed the intra-Mercurial planet, and who have committed the grave offence of keeping your observation secret for nine months. I warn you that I have come here with the intention of doing justice to your pretensions, and of demonstrating either that you have been dishonest or deceived. Tell me then, unequivocally, what you have seen?' The doctor then explained what he had witnessed, and entered into all the particulars regarding his discovery. On speaking of the rough method adopted to ascertain the period of the first contact, the astronomer inquired what chronometer he had been guided by, and was naturally enough somewhat surprised when the physician pulled out a huge old watch with only minute hands. It had been his faithful companion in his professional journeys, he said; but that would hardly be considered a satisfactory qualification for performing so delicate an experiment. The consequence was, that Le Verrier, evidently now beginning to conclude that the whole affair was an imposition or a delusion, exclaimed, with some warmth, ' What, with that old watch, showing only minutes, dare you talk of estimating seconds? My suspicions are already too well founded.' To this Lescarbault replied, that he had a pendulum by which he counted seconds. This was produced, and found to consist of an ivory ball attached to a silken thread, which, being hung on a nail in the wall, is made to oscillate, and is shown by the watch to beat very nearly seconds. Le Verrier is now puzzled to know how the number of seconds is ascertained, as there is nothing to mark them; but Lescarbault states that with him there is no difficulty whatever in this, as he is accustomed ' to feel pulses and count their pulsations,' and can with ease carry out the same principle with the pendulum. The telescope is next inspected, and pronounced satisfactory. The astronomer then asks for the original memorandum, which, after some searching, is found ' covered with grease and laudanum.' There is a mistake of 4 minutes on it when compared with the doctor's letter, detecting which, the *savant* declares that the observation has been falsified. An error in the watch regulated by sidereal time accounts for this. Le Verrier now wishes to know how the doctor managed to regulate his watch by sidereal time, and is shown the small telescope by which it is accomplished. Other questions are asked, to be satisfactorily answered. The doctor's rough drafts of attempts to ascertain the distance of the planet from the Sun ' from the period of four hours which it required to describe an entire diameter' of that luminary are produced, chalked on a board. Lescarbault's method, he being short of paper, is to make his calculations on a plank, and make way for fresh ones by planing them off. Not being a mathematician, it may be remarked he had not succeeded in ascertaining the distance of the planet from the Sun.

" The end of it all was, Le Verrier became perfectly satisfied that an intra-Mercurial planet had been really observed. He congratulated the medical practitioner upon his discovery, and left with the intention of making the facts thus obtained the subject of fresh calculations."[1]

In March or April, 1860, it was anticipated that the planet would again pass across the Sun, which was carefully scrutinised by different observers on several successive days, but no trace of it was obtained. However, this proves nothing, and we are prepared to regard the existence of this planet as a *fact*, to be fully demonstrated at some future time.

Meanwhile some approximate elements may be offered.

Longitude of ascending node	$= 12°$ $59'$
Inclination of orbit	$= 12°$ $10'$
Semi-axis-major ($\oplus=1$)	$= 0\cdot143$
Daily heliocentric motion	$= 18°$ $16'$
Period	$= 19$d. 17h.
Mean distance	$= 14{,}008{,}000$ miles.

[1] Epitomised from the *North British Review*, vol. xxxiii. pp. 1–20, August, 1860. A full account will also be found in *Cosmos*, vol. xvi. pp. 22–28, 1860; see also *Cosmos*, same vol., pp. 50–56.

Apparent diameter of ☉ as seen
  from Vulcan $= 3° \ 36'$
      ,,   ,, $(\oplus = 1)$ $= 6\cdot79$
Greatest possible elongation from
  the Sun $= 8°$

From the heliocentric position of the nodes, it appears that
transits can only occur in March and September or there-
abouts.

Instances are not wanting of observations of spots of a
planetary character passing across the Sun[1], and which may
turn out to have been transits of Vulcan ; but further specu-
lation is at present somewhat premature. It is, however,
right to add that M. Liais states he was watching the Sun, in
Brazil, at the same time Lescarbault professes to have seen the
black spot, and that he is positively certain nothing of the
kind was visible in his telescope, which was considerably more
powerful than the doctor's.

---

[1] Some will be found in *Month. Not.* R.A.S., vol. xx. p. 100.

CHAPTER IV.

MERCURY. ☿

*Difficulty attending Observations of Mercury.—Its Phases.—Determina-
tion of its Mass.—Acquaintance of the Ancients with Mercury.—
Copernicus and Mercury.—Astrological Connections.*

OWING to its proximity to the Sun, observations of the physical appearance of this planet are obtained with difficulty, and
even then are liable to much uncertainty. Schröter, who paid
considerable attention to Mercury, thought he had detected
traces of the existence of high mountains on its surface; his
observations, however, were not confirmed by Sir W. Herschel.
Mercury exhibits *phases* similar to those of the Moon. At its
maximum elongations, only half its disc is illuminated, but as
it approaches its superior conjunction, the breadth of the illuminated part increases, and its form becomes *gibbous*, and
ultimately circular in conjunction, when, however, the planet
is lost in the Sun's rays and invisible; on emerging, the gibbous form is still preserved, but the gibbosity is on the opposite side. The breadth of the illuminated part diminishes as
the planet draws near its greatest elongation, when it again
appears like a half moon, and continues to become more and
more crescented as it approaches the inferior conjunction;
having passed this, the crescent (now on the opposite side)
gradually increases until the planet again reaches its greatest
elongation.

As far as we know, Mercury is not attended by any satellites, and this renders the determination of its mass somewhat
difficult; it is effected, however, by means of Encke's comet.
The greatest elongation of Mercury not exceeding 29°, it can
never be seen but in the strong twilight, when it may be

detected by a keen eye, shining like a star of the 3rd magnitude, and of a pale rosy hue.   An old English writer, of the
name of Goad, in 1686, humorously termed this planet "a
squinting lacquey of the Sun, who seldom shows his head in
these parts, as if he was in debt."   The ancients were not only
acquainted with the existence of this planet, but were able to
approximate, with considerable accuracy, its period, and the
nature of its motions in the heavens.[1]   From the Greeks it
received the epithet ὁ στίλβων (the sparkling one).   Egyptian
observations have also been handed down to us in the *Almagest.*   So difficult is it to see Mercury in these northern
latitudes, that Copernicus, who died at the age of 70 years,
complained in his last moments that, much as he had tried,
he had never succeeded in detecting it, owing, according to
Gassendi, to the vapours prevailing near the horizon on the
banks of the Vistula, where that illustrious philosopher lived.
The best time to see Mercury is about $1\frac{3}{4}$ hours before sunrise
in autumn, and after sunset in spring.

   It occasionally happens that Mercury in the course of its
revolution around the Sun passes directly between the Earth
and that luminary, in which case the planet will be seen projected on the Sun's disc—in other words, a *transit* is taking
place.   The next recurrence of this phenomenon will take
place on the morning of Nov. 12, 1861.

   It is worthy of remark that in times gone by, when astrology
was a subject much in vogue, that Mercury was always looked
upon as a most malignant planet, and was stigmatised as a
sidus dolorosum.   From its extreme mobility, chemists adopted
it as the symbol for quicksilver.

---

[1] Pliny, *Hist. Nat.,* lib. ii. cap. 7; Cicero, *De Naturâ Deorum,* lib. ii.
cap. 20.

CHAPTER V.

VENUS. ♀

Phases of Venus resemble those of Mercury.— Galileo's Discovery of them.— Its Brilliancy.— Mountains suspected by Schröter.— Its apparent Motions.— The only Planet mentioned by Homer.— Most favourably placed for Observation once in 8 Years.— Suspected Satellite.— Transits.— Used for Nautical Observations.—Ancient Observations.

VENUS, when in a favourable position, is to us the most conspicuous member of the planetary system. Its apparent diameter differs considerably at different times : this is owing to the greater extent of variation of distance than occurs in the case of Mercury. Its phases are similar to those of Mercury, to which the reader is referred. The discovery of the phases of Venus is due to Galileo[1], who announced the fact to his friend Kepler in the following logograph or anagram : —

"Hæc immatura, a me, jam frustra, leguntur.—oy."

These things not ripe, as yet concealed from others, are read by me.

Which, when transposed, becomes —

"Cynthiæ figuras, æmulatur mater Amorum."

Venus imitates the phases of Cynthia [the Moon].

Little is known of the appearance of this planet's surface,

---

[1] It was one of the objections urged to Copernicus against his theory of the solar system, that if it were true, then the inferior planets ought to exhibit phases. He is said to have yielded assent, adding, "Se non e vèro, è ben trovato."

[2] *Opere di Galileo*, vol. ii. p. 42. Ed. Padua, 1744.

NEAR ITS GREATEST ELONGATION. (*Schröter.*)

NEAR ITS INFERIOR CONJUNCTION. (*Schröter.*)

VENUS.

owing to its dazzling brilliancy when viewed in a telescope; notwithstanding this, however, spots are occasionally visible. These spots are not supposed to be connected with the planet's surface, but rather to belong to its atmosphere. Schröter has suspected the existence of numerous high mountains on its surface. The rotation of Venus on its axis was discovered by D. Cassini in the year 1667.

The sidereal period of Venus is 225 days; that is to say, as viewed from the Sun it performs a revolution round the ecliptic in that time, but on account of the Earth itself being in motion, Venus is seen on the same side of the Sun for 290 days successively, when in that part of the orbit which is furthest from the Earth. The synodic revolution extends to 584 days, which is the average interval between two conjunctions, inferior or superior.

Venus is a morning star, rising before the Sun from inferior to superior conjunction; and an evening star, setting after the Sun, from superior to inferior conjunction. It was called "Lucifer" or "Hesperus" by the ancients[1] according to the circumstances under which it was visible. It was the sagacity of Pythagoras which discovered that Venus was sometimes a morning and sometimes an evening star.

It is somewhat remarkable that Venus is the only planet mentioned by Homer, who describes her as —

"Ἕσπερος ὃς κάλλιστος ἐν οὐρανῷ ἵσταται ἀστήρ."[2]

As the greatest elongation it ever attains amounts to about 47° 51′, it is only observable for 3 or 4 hours after sunset, and before sunrise. Once in 8 years, when at or near its greatest northerly latitude, about 5 weeks from inferior conjunction, and about ¼th illuminated, Venus shines with such brilliancy as frequently to be distinguished with the naked eye in the daytime, and to cast a very sensible shadow at night. It is uncertain whether or not this planet is attended by a satellite.

---

[1] Cicero, *De Nat. Deor.*, lib. ii. cap. 20.
[2] *Iliad*, lib. xxii. ver. 318.

Several observers of note during the last century[1], considered
they had obtained undoubted proofs of its existence; but
as more recent observations, with superior instruments, have
not confirmed the suspicion, we must at present look upon the
discovery as premature ; though it is difficult to imagine how
the observers in question could have been deceived.

As in the case of Mercury, transits of Venus across the Sun
occur from time to time, but they are very rare.   The next
will take place on Dec. 8, 1874.

To the mariner Venus is, owing to its rapid motion, a useful
auxiliary for taking lunar distances, or "lunars " as they are
shortly termed, when continuous bad weather may have prevented an observation of the Sun.

Many observations of Venus have been preserved in Ptolemy's *Almagest;* the most ancient is dated Oct. 12, 271 B.C.,
according to our reckoning, on which day Timocharis saw the
planet in conjunction with a star in the wing of Virgo.   It is
not improbable that Venus is referred to in Isaiah, xiv. 12.

---

[1] D. Cassini, in 1672 and 1686 ; Short, in 1740; Montaigne, in 1761.

CHAPTER VI.

THE EARTH. ⊕

"O let the Earth bless the LORD: yea let it praise Him and
magnify Him for ever."—*Benedicite.*

*Form of the Earth.—The Ecliptic.—The Equinoxes.—The Solstices.—
Diminution in the Obliquity.—The Earth's Orbit, an Ellipse.—
Motion of the Line of Apsides.—Familiar Proofs and Illustrations
of the Sphericity of the Earth.—Extracts from the Opinions of
Ancient Philosophers.*

THE form of the Earth is not strictly spherical, the polar
diameter being less than the equatorial by about $26\frac{1}{2}$ miles;
it is, in fact, like many, probably all, the planets, an oblate
spheroid. The great circle of the heavens *apparently* described
by the Sun every year, owing to our revolution round that
body, is called the *ecliptic*[1], and is usually employed by
astronomers as a fixed plane of reference. The Earth's equator
prolonged in the direction of the fixed stars, differs from the
equator of the heavens, which is inclined to the plane of the
ecliptic at an angle which in January 1, 1860, was equal to
23° 27′ 33″; and which angle is known as *the obliquity of the
ecliptic.* It is this inclination which gives rise to the vicissi-
tudes of the seasons during our annual revolution round the
Sun. The two points where the celestial equator is intersected
by the ecliptic, are called the *equinoxes*[2]; the points exactly

[1] "The line of eclipses."
[2] From *æquus*, equal, and *nox*, a night; because when the sun is at
these points, day and night are theoretically equal throughout the
world. In 1860 this occurs on March 19 and Sept. 22.

midway between these being the *solstices*.[1] It is from the
vernal (or spring) equinox, that right ascensions are measured
along the equator and longitudes along the ecliptic. The
obliquity of the ecliptic is now slowly decreasing at the rate
of 48″ in 100 years. It will not, however, always be on the
decrease, for before it can have attained $1\frac{1}{2}°$, the cause which
produces this diminution must act in a contrary direction,
and thus tend to increase the obliquity. Consequently, the
change of obliquity is a phenomenon in which we are concerned only as astronomers, since it can never become sufficiently great to produce any sensible alteration of the climate
on the Earth's surface. It is stated by Pliny, that the discovery of the obliquity of the ecliptic is due to a disciple of
Thales, named Anaximander, in the 6th century B.C. Laplace, however, says that he has found some Chinese observations, purporting to have been made not less than 1100 years
before the Christian era, by Tcheou-kong.[2] All the ancient
observations were made with gnomons or armillæ.

The path of the Earth round the Sun is an ellipse of small
eccentricity; in 1800 it amounted to 0·0167917, but is subject
to a small secular diminution not exceeding 0·000041 in a
period of 100 years. Supposing the change to go on continuously, the Earth's orbit must eventually become circular;
but we are enabled to prove, by the theory of attraction, that
this progressive diminution is only to proceed for a certain
interval of time; and though we are not yet in a position to
assign any very definite limits to the oscillations, yet we know
that after the lapse of some thousands of years, the eccentricity will become permanent for a time, and then increase
again; and that, unless some external cause of perturbation
arise, these variations must continue throughout all ages, within
certain not very distant, though as yet indeterminate limits.

---

[1] From *Sol*, the Sun, and *sistere*, to stand still; because the Sun,
when it has reached these points, has attained its greatest declination
N. or S., as the case may be. In 1860 this occurs June 20, 17h.
43m., and Dec. 21, 1h. 51m. G. M. T.

[2] *Conn. des Temps*, 1811.

The line of apsides is subject to an annual direct change of 11·29″ independent of the effects of precession; so that, allowing for the latter cause of disturbance, the annual movement of the apsides may be taken at about 1′. One important consequence of this motion of the major axis of the Earth's orbit, is the variation in the lengths of the seasons at different periods of time. In the year 4089 B.C., or, singularly enough, near the supposed epoch of the Creation, the longitude of the Sun's perigee coincided with the vernal equinox; so that summer and autumn were of equal length, but longer than winter and spring, which were also equal. In the year 1250 A.D. the perigee coincided with the winter solstice; spring was therefore equal to summer, and autumn to winter, the former being the longest. In the year 6589 A.D. the perigee will have completed half a revolution, and will then coincide with the autumnal equinox; summer will then be equal to autumn, and winter to spring; the former seasons however, being the shortest. In the year 11928 A.D. the perigee will have completed three fourths of a revolution, and will then coincide with the summer solstice; autumn will then be equal to winter, but longer than spring and summer, which will also be' equal. And finally in the year 17267 A.D. the cycle will be completed by the coincidence, for the second time since the creation of the world, of the solar perigee with the vernal equinox.

The spherical form of the Earth is so well known, that argument to prove it is now quite superfluous. It is known by the appearance presented by a ship in receding from the observer: first the hull disappears, then the lower parts of the rigging, and finally the topmasts: also by the shadow cast on the Moon during a lunar eclipse: by the varying appearances of the constellations as we proceed to the north or to the south: by the varying elevation of the Pole Star above the horizon: by the culmination of the heavenly bodies: by the length of shadows; by the duration of day and night, &c.

Among the ancients we find that Aristarchus of Samos, and Philolaus, maintained that not only did our globe revolve on its own axis, but that it revolved round the Sun in 12 months.[1] Nicetas of Syracuse is also mentioned as a supporter of this doctrine.[2] The Egyptians taught the revolution of Mercury around the Sun[3]; and Apollonius Pergæus assigned a similar motion to Mars, Jupiter, and Saturn. Hesiod states that the Earth is situated half way, exactly, between Heaven and 'Tartarus.[4] Lovers of mediæval English history may like to know that their ancestors 300 and 400 years ago termed the ecliptic the "thwart circle;" the meridian, the "noonsteede circle;" the equinoxial, "the girdle of the sky"; the zodiac, "the Bestiary" and "our Lady's waye." The origin of the division of the zodiac into animal constellations, if we may so speak of them, is lost in obscurity. Though commonly attributed to the Greeks, it now seems certain that the custom is much older than this; possibly being due to the ancient Hindoos or the Chinese, in whose behalf, however, a claim to prior knowledge is always put in, whenever we Europeans fancy we have made a discovery.

---

[1] Archimedes, *In Arenario;* Plutarch, *De Placit. Philos.*, lib. ii. cap. 24; Diog. Laërt. *In Philolao.*

[2] Cicero, *Acad. Quest.*, lib. ii. cap. 39.

[3] Macrobius, *Comment. in Somn. Scip.*, lib. i. cap. 19, and others.

[4]   "From the high heaven a brazen anvil cast,
    Nine days and nights in rapid whirls would last,
    And reach the Earth the tenth; whence strongly hurl'd,
    The *same the passage to th' infernal world.*"
                       HESIOD, *Theogonia,* ver. 721.

Fig. 55. *Plate* VII.

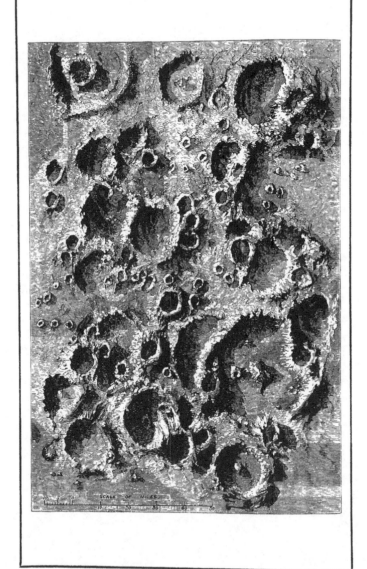

VIEW OF A PORTION OF THE MOON'S
SURFACE ON THE S.E. OF TYCHO.
DRAWN BY NASMYTH.

CHAPTER VII.

THE MOON. ☽

The Earth's only Satellite. — Its Phases. — Its Motion and their Complexity. — Libration. — Evection. — Variation. — Parallactic Inequality. — Annual Equation. — Secular Acceleration. — Diversified Character of the Moon's Surface. — Lunar Mountains, &c. — Anecdote related by Mr. Frend. — Lunar Atmosphere. — Researches of Schröter, Beer, and Mädler, Du Séjour, &c. — The Earthshine. — The Harvest Moon. — The Hunter's Moon. — Calorific Rays.

THE MOON is the Earth's only attendant during its annual journey round the Sun, and as such must always be an object of peculiar interest to us. To it we owe the phenomenon of the Tides, to say nothing of the other influences supposed

Fig. 10.

Probable appearance of the Earth as viewed from the Moon.

to be possessed by it. The phases of the Moon are similar in character to those of the inferior planets, which, having been already fully explained, will not be adverted to again. They were first treated of by Anaxagoras, who at the same time wrote of the Moon's rotatory motion round the Earth. The Moon's motions are very complicated, and for a long time greatly embarrassed astronomers; though they are, however, now pretty well known, and understood.

Speaking generally, we may say that the same hemisphere of the Moon is always turned towards us; and although this is, in the main, correct, yet there are certain small variations at the edge which it is necessary to notice. The Moon's axis, although nearly, is not exactly perpendicular to the plane of its orbit, deviating therefrom by an angle of 1° 30° 10·8″; owing to this, the two poles of the Moon lean alternately to and from the Earth. When the north pole leans towards the Earth, we see somewhat more of the region surrounding it, and somewhat less when it leans the contrary way; this is known as *libration in latitude*.[1] In order that, strictly speaking, the same hemisphere should be continually turned towards us, the time of the Moon's rotation on its axis must not only be equal to the time of its revolution in its orbit, but its angular velocity on its axis must, in every part of its course, exactly equal its angular velocity in its orbit. This, however, is not the case, for its angular velocity in its orbit is subject to a slight variation; in consequence of which a little more of its eastern or western edge is seen at one time than another; this is known as the *libration in longitude*, and was discovered by Hevelius, who published an account of it in the year 1647.[2] On account of the diurnal rotation of the Earth, we view the Moon under somewhat different circumstances, at its rising and at its setting, according to the latitude of the Earth in which we are placed. By thus viewing it in different positions, we see it under different aspects; this gives rise to another phenomenon, the *diurnal libration*. This periodical variation in the visible portion of the Moon's disc seems to have been first remarked by Galileo—a discovery very credit-

---

[1] *Librans*, swinging or oscillating.   [2] *Selenographia*.

able to him, when we consider the inefficient means with which he worked.

The following are the most important perturbations in the motion of our satellite :—

1. The *Evection,* " caused by a change in the eccentricity of its orbit, whereby its mean longitude is sometimes increased or diminished to the amount of 1° 20' ;" discovered by Hipparchus.

2. The *Variation,* which depends on the angular distance of the Moon from the Sun : and at certain times amounts to 37'. The discovery of this has usually been attributed to Tycho Brahe, but Sedillot and others claim the merit of its first recognition for Abùl Wefa, who flourished in the 9th century A.D.

3. The *Parallactic Inequality,* which arises from the difference in the influence of the Sun's attraction, when the Moon traverses that part of its orbit nearest the Sun, as compared with the attractive force exerted when in that part which is most distant.

4. The *Annual Equation,* which is the inequality in the Moon's motion arising from the eccentricity of the Earth's orbit, whereby the diurnal motion of our satellite is sometimes quicker and sometimes slower than its mean motion.

5. The *Secular Acceleration,* caused by a slow change in the eccentricity of the Earth's orbit, which has recently diminished the length of the Moon's revolution since the time of the earliest observations. This inequality was detected by Halley, in 1693, from a comparison of the periodic time of the Moon deduced from Chaldæan observations of eclipses, made at Babylon in the years 720 and 719 B.C., and the Arabian observations of the 8th and 9th centuries.

When examined with the naked eye the Moon is found to present an irregular mottled appearance, which the telescope shows to be owing to numerous mountains existing on its surface, as was discovered by Galileo. [1]   That such is the case

---

[1] An excellent list is given by Webb, *Celest. Objects,* pp. 67-104.

is known by the shadows cast by the high peaks on the surrounding plains, when the Sun shines obliquely; these shadows disappear, however, at the full phase, as the Sun then shines directly on the Moon's surface. When not near the full, that edge of the Moon turned from the Sun has a rough jagged appearance; this is caused by the Sun's light illuminating, first, the summits of the peaks; the surrounding valleys being still dark, which gives a disconnected form to the whole edge, causing the irregular appearance above mentioned.

Most of the lunar mountains have received names, chiefly those of men eminent in science, both ancient and modern; the heights of many have also been measured by the Prussian astronomer, Mädler: the most elevated has an altitude of 23,800 feet, and has received the appropriate name of Newton. Many of the shaded regions were called by former observers, seas; and although it has been found convenient to retain this designation, yet it is not thereby to be supposed that these localities are really considered to be masses of water, for there is every reason to believe that no liquids of any kind exist in the Moon. Amongst others we meet with the names *Oceanus Procellarum, Mare Nubium, Sinus Iridum, &c.*

Walled Plains, or cavities, are also to be found in many parts of the Moon, and are, doubtless, merely portions of the surface depressed much below the general level: *Plato* is a well-known instance of this kind. These cavities are usually of a circular form, but owing to their often being seen obliquely, they present an oval outline. There is one lunar mountain, *Aristarchus*, which has been thought by many eminent observers to be a volcano at times in active operation; although, however, the peculiar appearance often presented by this mountain, can otherwise be explained, yet there is the most conclusive evidence of the past, though probably not of the present, existence of volcanic agency on our satellite.

Since the axis of the Moon is very nearly perpendicular to the plane of the ecliptic, of course she has scarcely any change of seasons; but, what is still more remarkable, one half of the Moon has no darkness at all, then the other half has a fortnight

Figs. 56, 57.   *Plate* VIII.

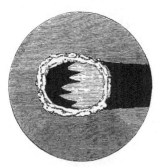

ARCHIMEDES.

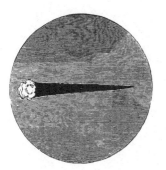

PICO.

LUNAR MOUNTAINS.

of light, and the same of darkness alternately. To the lunarians (if there be any) the Earth appears the largest orb in the universe; it appears to them 3 times the size of the Sun, and 13 times greater than the Moon does to us[1], exhibiting similar phases but in a reverse order. For when the Moon is full, the Earth is invisible to them, and when the Moon is new, they will see the Earth full. As seen from the Sun, the Moon never departs more than 18′ at its greatest elongation.

It is related by the late Mr. W. Frend, that early one morning a lady of his acquaintance noticed the thin crescent of the Moon, then approaching near its conjunction; and that in the evening of the following day she observed the opposite crescent in the west, soon after sunset. Thus having seen in the morning of one day, and in the evening of the next, the waning and the waxing Moon.

The question of a lunar atmosphere is at present rather an open one. Schröter considered that there was one, but he estimated the height at only 5376 feet[2]; and Laplace thought it to be more attenuated than the vacuum of an air-pump. At the Moon, refraction must be at least 1000 times less than at the Earth, and the horizontal parallax cannot exceed 1·7″.

MM. Beer and Mädler have concluded that the Moon has an atmosphere, but that it is small on account of the smallness of the planet's mass : and they also say, " It is possible that this weak envelope may sometimes through local causes, in some measure, dim or condense itself,"—an idea which, if proved, would go far to clear up many of the conflicting details of occultation phenomena. Auzout remarked, that if the Moon had an atmosphere, it must likewise have a twilight.[3] None was detected for about half a century, when Schröter obtained traces of one[4], so feeble, however, as only, according to calculation, to be caused by an atmosphere of the small extent previously referred to. Du Sèjour found that the Moon's atmosphere was

---

[1] Smyth's *Cycle.*, vol. i. p. 129.

[2] *Phil. Trans.*, vol. lxxxii. p. 354. 1792.

[3] *Mém. Acad. des Sciences*, vol. vii. p. 106.

[4] *Phil. Trans.*, vol. lxxxii. p. 337 *et seq.* 1792.

at least 1400 times rarer than common atmospheric air. The Moon viewed in a large telescope, under a magnifying power of 1000, will appear the same as it would if it were only 250 miles off.

For a few days, both before and after new Moon, an attentive observer may often detect the outline of the un-illuminated portion, without much difficulty; this lustre is the light reflected on the Moon by the Earth, "Earth shine" in fact; the French call it, *la lumière cendrée;* but it is popularly known in England as "the old Moon in the new Moon's arms." The ancient Latin appellation was *lumen incinerosum.* This light is stronger during the waning of the Moon than at any other time.[1]

The *Harvest Moon* is the name given to that full Moon which falls nearest to the autumnal equinox, as it is then that it rises almost at the same time on several successive evenings, at the same time making its first appearance at points of the horizon most distantly apart, and is therefore of great assistance to the farmer at that important time. Although this near coincidence in the several successive risings of the Moon takes place in every lunation when our satellite is in the signs Pisces and Aries, yet the phenomenon is only prominently noticeable when it is full in these signs, and this only occurs in August or September, and when the Sun is in Virgo or Libra. The least possible variation between the times of two successive risings in these latitudes, is about 15 m. and the greatest possible, about 1 h. 15 m., which takes place when the Moon is in Libra, and, at the same time, at or near its descending node. The following full Moon, in October, is termed the *Hunter's Moon.*

The whole heavens covered with full Moons would scarcely make daylight, and it has been supposed that the Moon's light is only $\frac{1}{800,000}$ that of the Sun.[2] Until recently it was considered to possess no calorific rays, but Melloni, by concentrating them in a lens 3 feet in diameter, succeeded in obtaining a

---

[1] Arago, *Pop. Ast.*, vol. ii. p. 300, Eng. ed.
[2] *Phil. Trans.*, vol. cxix. p. 27. 1829.

sensible elevation of temperature.   More recently, in 1856,
proofs of the heat in the Moon's rays were obtained by Professor Smyth on Teneriffe without much difficulty.[1]

---

[1] C. P. Smyth, *An Astronomer's Experiment, &c.*, p. 213.   Professor
Tyndall has informed the writer, that neither of these statements are
to be depended upon; that Melloni doubts his own observations, and
that Smyth's apparatus was not sufficiently delicate.   (*Note.*   May,
1861.)

CHAPTER VIII.

MARS. ♂

Its Position in the System.—Its Brilliancy.—Telescopic Appearance.—
Phases.—Explanation of its Ruddy Hue.—Axial Rotation.—Analogy
between Mars and the Earth.—Epitome of its apparent Motions.—
Early Observations.

MARS is the first planet whose orbit is exterior to that of
the Earth, and when in opposition to the Sun, is a very conspicuous object in the heavens, and of a fiery red colour; this,
however, only takes place once in two years, the planet's synodic period being 780 days. When in perigee and in perihelion
at the same time, as it is in opposition, Mars shines with a
brilliancy rivalling Jupiter. Such was the case in August,
1719, when the planet was only $2\frac{1}{2}°$ from perihelion; its brightness was then so considerable as to give rise to a sort of panic.[1]
In order to be seen to the greatest advantage it must not only
be in opposition, but also in perihelion at the same time; this
occurs only once in 8 years.

Telescopically examined, Mars is found to be covered with
numerous dusky patches; these are supposed to be the outlines
of continents and seas analogous to those in our own globe:
near the poles there are brilliant white spots, considered by
many to be masses of snow; this idea is somewhat strengthened
by the fact that they have been observed to diminish when
brought under the Sun's influence at the commencement of
the Martial summer; and increase again on the approach of
winter.

---

[1] De Zach, *Corr. Astronomique*, vol. ii. p. 293.

1858: June 3.

1858: June 14.

MARS.
DRAWN BY SECCHI.

Mars exhibits slight phases; when in opposition its disc is circular; at all other times it is gibbous. It is surrounded by a considerable atmosphere. No satellite has yet been detected, but analogy would lead us to suppose it has one, which has hitherto escaped observation. Mars is of a ruddy colour, due, it is conjectured by Sir J. Herschel, possibly to some peculiar geological structure.[1]

The axial rotation of Mars was detected by Hook and D. Cassini in 1666[2], but it is stated that Fontana, of Naples, suspected this rotation 23 years previously, or in 1643. Mars, though at times gibbous, is never "horned," and this shows that its orbit is exterior to the Earth's.

There is a greater analogy between the Earth and Mars than between any of the other planets. Their diurnal motion is nearly the same; the inclination of their equator, — upon which their seasons depend, — is nearly the same, nor are the lengths of their years very different when compared with those of Jupiter and the other planets exterior to it. The Earth, however, it would seem is the more favoured of the two, since water would not remain fluid, even at the Martial *equator*, and alcohol would freeze at its temperate zones. The inhabitants of Mars (and possibly there are such) receive but $\frac{4}{9}$ of the Sun's light and heat that we do, liable, however, to variations on account of the great eccentricity of its orbit. If the atmosphere of Mars is as dense as it is commonly supposed, they will probably seldom catch a glimpse of Mercury and Venus. The Earth and the Moon, however, will appear as a pair of planets alternately changing places with each other, with horned phases, never quite full, and never more than $\frac{1}{4}°$ from each other.[3]

After conjunction, when Mars first emerges from the Sun's rays, it rises some minutes before the Sun, and has a progressive easterly motion. The Earth's movement in the same direction being nearly double that of Mars, the consequence

<hr/>

[1] *Outlines of Ast.*, p. 339.
[2] *Phil. Trans.*, vol. i. pp. 239, 242. 1666.
[3] Smyth, *Cycle of Cel. Obj.*, vol. i. p. 147, abridged.

D

is, that that planet appears to recede from the Sun in a *westerly* direction, notwithstanding that its real motion is towards the east. This continues for nearly a year, when its angular distance from the Sun amounts to 137°; for a few days it then appears stationary. After that, its motion becomes retrograde, or westerly, and continues so until the planet is 180° distant from the Sun, or in opposition, and consequently in the meridian at midnight. Then its retrograde motion is the swiftest; it afterwards becomes slower, and ceases altogether when the planet is again at a distance of about 137° on the other side of the Sun. Its motion then again becomes progressive, and continues so, till, once more, the planet becomes lost in the solar rays, where the phenomena are renewed, but with a considerable difference in the extent and duration of the movements. The retrogradation commences or finishes when the planet is at a distance from the Sun which varies from 128° 44' to 146° 37', the arc described being from 10° 6' to 19° 35'; the duration of the retrograde motion in the former case is 60d. 18h., and in the latter 80d. 15h. The period in which all these changes take place, or the interval between one conjunction and one opposition, constitutes the synodic period, which we have already stated to be 780 days. Mars and the Earth come *nearly* to the same relative position every 23 years; but several centuries must elapse before precise coincidence occurs.[1]

The earliest recorded observation of Mars bears a date equivalent to January 17, 272 B.C.[2] The Jewish appellation of this planet signified "blazing," as also did the Greek (πυρόεις). The epithet seems to be a wide-spread one.

It was from observing the orbit of Mars, and the very varying difference of its diameter at different times, that Copernicus decided to adopt the Pythagorean system.[3]

[1] Smyth, *Cycle of Cel. Obj.*, vol. i. pp. 151, 152, abridged.
[2] Ptolemy, *Almagest*.   [3] *Hist. Ast.*, L. U. K., p. 43.

CHAPTER IX.

THE MINOR PLANETS.[1]

Sometimes called Ultra-Zodiacal Planets. — Table of them. — Summary of Facts. — Notes on Ceres. — Pallas. — Juno. — Vesta. — Olbers' Theory.

BETWEEN the orbits of Mars and Jupiter there is a wide interval, which, until the present century, was not known to be occupied by any planet. The researches of late years, as previously intimated in Chapter II., have led to the discovery of a numerous group of small bodies revolving round the Sun, which are known as the Minor Planets [2], and which have received names taken chiefly from the mythology of ancient Greece and Rome.[3]

The planets differ in some respects from the other members of the system, especially in point of size, the largest being probably not more than 200 or 300 miles in diameter. Their orbits are also much more inclined, as a general rule, than the orbits of the older planets, whence they are sometimes termed the *ultra-zodiacal planets.*

The following is a list of these planets, together with the chief elements of their orbits:—

---

[1] The use of symbols has been discontinued, except for the four early ones, as follows: Ceres ♀, Pallas ⚴, Juno ⚵, Vesta ⚶.

[2] The old name of *asteroids*, proposed by Sir W. Herschel, has nearly fallen into disuse. Nothing could be more inappropriate than such a designation; *planetoids* would have been better. However, *minor planets* is preferable to either.

[3] See Appendix I.

No.	Name	Discovered			λ		π		Ω	
		on	by	at	°	′	°	′	°	′
1	Ceres . . .	1801. Jan. 1 .	Piazzi	Palermo .	346	38	149	26	80	49
2	Pallas . .	1802. March 28	Olbers . . .	Bremen .	318	17	122	10	172	39
3	Juno . . .	1804. Sept. 1 .	Harding . . .	Lilienthal .	206	17	54	4	171	0
4	Vesta . . .	1807. March 29	Olbers . . .	Bremen .	2	25	250	20	103	25
5	Astræa . .	1845. Dec. 8 .	Hencke . . .	Driesen .	80	56	134	35	141	24
6	Hebe . . .	1847. July 1 .	Hencke . . .	Driesen .	15	4	15	12	138	36
7	Iris . . .	— Aug. 13 .	Hind	London . .	114	59	41	29	259	47
8	Flora . . .	— Oct. 18 .	Hind	London . .	68	48	32	54	110	17
9	Metis . . .	1848. April 25 .	Graham . . .	Markree .	209	3	71	9	68	32
10	Hygeia . .	1849. April 12 .	De Gasparis .	Naples . .	354	47	227	47	287	38
11	Parthenope	1850. May 11 .	De Gasparis .	Naples . .	283	56	316	10	125	3
12	Victoria . .	— Sept. 13 .	Hind	London . .	7	42	301	39	235	34
13	Egeria . .	— Nov. 2 .	De Gasparis .	Naples . .	11	24	119	31	43	19
14	Irene . . .	1851. May 19 .	Hind	London . .	63	39	179	26	86	40
15	Eunomia .	— July 29 .	De Gasparis .	Naples . .	149	54	27	47	293	55
16	Psyche .	1852. March 17	De Gasparis .	Naples . .	313	1	12	30	150	32
17	Thetis . .	— April 17 .	Luther . . .	Bilk . . .	210	1	259	22	125	27
18	Melpomene	— June 24 .	Hind	London . .	95	6	15	14	150	1
19	Fortuna . .	— Aug. 22 .	Hind	London . .	150	1	30	22	211	30
20	Massilia . .	— Sept. 19 .	De Gasparis . .	Naples . .	318	35	98	36	206	42
21	Lutetia . .	— Nov. 15 .	Goldschmidt .	Paris . .	41	23	327	2	80	27
22	Calliope . .	— Nov. 16 .	Hind	London . .	76	59	58	7	66	36
23	Thalia . .	— Dec. 15 .	Hind	London . .	173	39	123	11	67	55
24	Themis . .	1853. April 5 .	De Gasparis . .	Naples . .	130	4	139	7	36	9
25	Phocea . .	— April 6 .	Chacornac . .	Marseilles .	75	18	302	54	214	4
26	Proserpine .	— May 5 .	Luther . . .	Bilk . . .	181	21	235	17	45	53
27	Euterpe . .	— Nov. 8 .	Hind	London . .	260	43	87	39	93	44
28	Bellona . .	1854. March 1 .	Luther . . .	Bilk . . .	94	6	122	24	144	38
29	Amphitrite .	— March 1 .	Marth . . .	London . .	293	11	56	39	356	26
30	Urania . .	— July 22 .	Hind	London . .	19	30	31	23	308	13
31	Euphrosyne	— Sept. 1 .	Ferguson . . .	Washington	53	49	93	51	31	25
32	Pomona .	— Oct. 26 .	Goldschmidt .	Paris . .	57	34	194	22	220	48
33	Polyhymnia	— Oct. 28 .	Chacornac . .	Paris . .	23	5	340	41	9	14
34	Circe . . .	1855. April 6 .	Chacornac . .	Paris . .	210	3	149	19	184	51
35	Leucothea .	— April 19 .	Luther . . .	Bilk . . .	89	34	198	37	356	9

ι	ε	μ	Period.	Semi-Axis, Major.	Diameter.	App. opp. Star Mag.	Epoch Berlin M. T.	Calculator.
° ′		″	Years.		Miles			
10 36	0·08024	771·30	4·600	2·7660	227	7·7	1859. Sept. 6·5 .	Wolfers.
34 42	0·23969	769·64	4·610	2·7700	172	7·9	— Aug. 11 .	Galle.
13 3	0·25590	813·44	4·362	2·6687	112	8·7	— April 26 .	Bremiker.
7 8	0·09012	978·22	3·627	2·3607	228	6·6	— Oct. 5 .	Encke.
5 19	0·18999	857·95	4·136	2·5775	61	10·0	1850. Jan. 0. .	Zech.
14 46	0·20115	939·37	3·777	2·4254	100	8·6	1859. Sept. 30 .	Luther.
5 27	0·23125	962·51	3·686	2·3862	96	8·6	1860. Feb. 9 .	Schubert.
5 53	0·15670	1086·33	3·266	2·2014	60	8·9	1848. Jan. 1 .	Brünnow.
5 36	0·12320	962·61	3·686	2·3862	76	8·9	1859. April 28·5	Wolfers.
3 47	0·10056	634·85	5·589	3·1494	111	9·8	1851. Sept. 17 .	Zech.
4 36	0·09888	923·78	3·841	2·4526	62	9·5	1858. June 27 .	Luther.
8 23	0·21890	994·83	3·567	2·3344	41	9·6	1851. Jan. 0. .	Brünnow.
16 32	0·08775	858·42	4·133	2·5756	73	9·9	1858. Sept. 26 .	Günther.
9 7	0·16525	851·49	4·167	2·5895	68	9·7	1857. Nov. 5 .	Bruhns.
11 44	0·18601	825·80	4·297	2·6429	12	9·1	1854. Jan. 0 .	Schubert.
3 4	0·13575	708·80	5·006	2·9263	93	10·1	— July 14 .	Klinkerfues.
5 35	0·12686	911·98	3·890	2·4737	52	9·9	1856. April 4 .	Schönfeld.
10 9	0·21723	1019·97	3·479	2·2958	54	9·5	1854. Jan. 0. .	Schubert.
1 32	0·15792	930·16	3·815	2·4414	61	9·7	1858. March 9 .	Powalky.
0 41	0·14383	948·77	3·740	2·4093	68	9·3	1859. Aug. 2 .	Günther.
3 5	0·16204	933·56	3·801	2·4354	40	10·3	1853. Jan. 2. .	Lesser.
13 44	0·10361	715·11	4·962	2·9091	96	10·6	— Jan. 0. .	Hornstein.
10 13	0·23521	835·29	4·253	2·6250	42	11·0	1854. Jan. 0 .	Schubert.
0 48	0·11701	637·09	5·570	3·1420	36	11·6	1858. April 14 .	Krüger.
21 34	0·25335	952·93	3·723	2·4023	31	10·5	— Dec. 23 .	Günther.
3 35	0·08752	819·68	4·329	2·6556	47	10·7	1857. March 20	Hoek.
1 35	0·17290	986·63	3·596	2·3473	39	9·9	1859. Jan. 14 .	Günther.
9 21	0·15039	766·14	4·631	2·7784	59	9·8	1857. Dec. 15 .	Bruhns.
6 7	0·07238	868·87	4·084	2·5548	83	9·1	1859. July 9. .	Günther.
2 5	0·12718	976·07	3·635	2·3642	51	10·1	1858. Oct. 9 .	Günther.
26 25	0·21601	632·80	5·607	3·1561	50	11·6	1855. Jan. 0 .	Winnecke.
5 29	0·08240	852·86	4·160	2·5831	35	11·0	— Jan. 5. .	Lesser.
1 56	0·33769	731·83	4·848	2·8646	38	11·4	— Jan. 0. .	Pape.
5 26	0·10961	806·98	4·397	2·6389	29	11·5	— June 23 .	Powalky.
8 12	0·22251	688·01	5·157	2·9850	25	12·1	1858. Dec. 1·25	Schubert.

No.	Name	Discovered on	by	at	λ	π	Ω
36	Atalanta	1855. Oct. 5	Goldschmidt	Paris	36 19	42 22	359 8
37	Fides	— Oct. 5	Luther	Bilk	42 34	66 4	8 9
38	Leda	1856. Jan. 12	Chacornac	Paris	112 56	100 44	296 27
39	Lætitia	— Feb. 8	Chacornac	Paris	146 43	2 7	157 19
40	Harmonia	— March 31	Luther	Bilk	213 54	0 55	93 30
41	Daphne	— May 22	Goldschmidt	Paris	202 28	230 21	180 5
42	Isis	— May 23	Pogson	Oxford	276 59	317 57	84 27
43	Ariadne	1857. April 15	Pogson	Oxford	224 5	277 14	264 29
44	Nysa	— May 27	Goldschmidt	Paris	278 9	111 37	131 1
45	Eugenia	— June 28	Goldschmidt	Paris	294 .34	228 51	148 5
46	Hestia	— Aug. 16.	Pogson	Oxford	87 48	354 25	181 30
47	Pseudo· Daphne	— Sept. 9	Goldschmidt	Paris	330 53	294 57	194 54
48	Aglaia	— Sept. 15	Luther	Bilk	11 17	314 29	4 29
49	Doris	— Sept. 19	Goldschmidt	Paris	16 2	77 37	185 14
50	Pales	— Sept. 19	Goldschmidt	Paris	31 25	32 50	290 29
51	Virginia	— Oct. 4	Ferguson	Washington	31 41	10 0	173 32
52	Nemausa	1858. Jan. 22	Laurent	Nismes	172 45	190 12	175 37
53	Europa	— Feb. 6.	Goldschmidt	Paris	136 25	102 12	129 57
54	Calypso	— April 4	Luther	Bilk	162 13	91 32	144 15
55	Alexandra	— Sept. 10	Goldschmidt	Paris	324 1	293 39	313 50
56	Pandora	— Sept. 10.	Searle	Albany,U.S.	16 7	10 9	10 55
57	Mnemosyne	1859. Sept. 22	Luther	Bilk	324 10	54 1	200 8
58	Concordia	1860. March 24	Luther	Bilk	183 14	183 38	161 19
59	Danaë	— Sept. 9	Goldschmidt	Châtillon	345 41	340 8	334 18
60		— Sept. 12	Chacornac	Paris	9 52	18 55	170 18
61	Erato	— Sept. 14	Förster	Berlin	18 8	31 53	126 21
62	Titania	— Sept. 14	Ferguson	Washington	14 50	96 49	192 3
63	Ausonia	1861. Feb. 10	De Gasparis	Naples	180 14	268 7	338 3
64	Angelina	— March 4	Tempel	Marseilles	170 42	126 28	311 2
65	Maximiliana	— March 8	Tempel	Marseilles	194 15	254 37	159 9
66	Maia	— April 9	H. P. Tuttle	Cambridge,	183 21	43 54	8 11
67	Asia	— April 17	Pogson	Madras,U.S.	241 22	294 23	202 0
68	Hesperia	— April 29	Schiaparelli	Milan	163 8	125 42	186 51
69	Leto	— April 29	Luther	Bilk	240 24	346 15	44 49
70	Panopea	— May 5	Goldschmidt	Châtillon	253 11	299 3	48 21

ι	ε	μ	Period.	Semi-Axis, Major.	Diameter.	App. opp. Star Mag.	Epoch. Berlin M. T.	Calculator.
° '	'	''	Years.		Miles.			
18 42	0·29788	778·60	4·557	2·7847	20	12·9	1856. Jan. 0 . .	Förster.
3 7	0·17489	826·17	4·295	2·6422	41	10·5	— Jan. 0 .	Rümker.
6 58	0·15552	782·32	4·535	2·7399	29	10·9	— Jan. 0. .	Allé.
10 21	0·11081	769·20	4·613	2·7710	87	9·3	— Jan. 1. .	Allé.
4 15	0·04608	1038·90	3·415	2·2679		9·1	— June 2 .	Powalky.
15 48	0·20249	954·11	3·719	2·4003		10·2	— June 0·5 .	Pape.
8 35	0·22566	930·89	3·812	2·4401		10·6	— July 1 . .	Seeling.
3 27	0·16756	1084·52	3·272	2·2038		10·0	1857. April 17 .	Weiss.
3 41	0·14933	940·08	3·774	2·4242		10·4	1858. Jan 0 . .	Powalky.
6 34	0·08200	792·78	4·476	2·7159		11·0	— Jan. 0. .	Löwy.
2 17	0·16184	888·12	3·995	2·5178		12·4	1859. Jan. 0. .	Karlinsky.
7 56	0·22702	852·80	4·152	2·5835		10·1	1857. Sept. 13 .	Luther.
5 0	0·12788	724·77	4·896	2·8831		11·5	1855. Jan. 8·0 ·	Powalky.
6 29	0·07580	648·67	5·470	3·1044		11·4	1858. Feb. 3 .	Powalky.
3 8	0·23783	654·47	5·421	3·0861		10·9	— Feb. 23 .	Powalky.
2 47	0·28695	823·14	4·310	2·6486		12·3	— Jan. 0 .	Förster.
10 14	0·06285	967·64	3·667	2·3779		10·4	— March 2·5	Förster.
7 24	0·00450	650·11	5·458	3·0999		10·7	— Jan. 0. .	Hornstein.
5 7	0·21263	841·39	4·217	2·6102		11·5	— April 10·5	Oeltzen.
11 47	0·19941	796·39	4·553	2·7076		11·2	— Sept. 20·5	Schultz.
7 20	0·13895	769·96	4·608	2·7692		11·1	— Nov. 4·5 .	Möller.
15 4	0·10752	631·73	5·616	3·1597		10·0	1859. Nov. 0 .	Auwers.
5 2	0·04103	800·71	4·431	2·6979		11·6	1860. April 10. G.	Seeling.
18 17	0·16308	691·58	5·131	2·9746		10·5	— Sept. 29·0	Luther.
8 36	0·11883	793·56	4·472	2·7147		10·0	— Oct. 25. G.	Ellis.
2 12	0·16678	642·56	5·522	3·1242		11·7	— Oct. 23·5	Schmidt, jr.
3 33	0·19194	951·39	3·729	2·4049		11·0	— Oct. 0. W.	Ferguson.
5 45	0·12732	956·00	3·712	2·3972		9·5	1861. March 16.	Tietjen.
1 19	0·12482	809·50	4·385	2·6783		10·0	— April 0 ·	Förster.
3 29	0·14070	553·14	6·413	3·4523		10·5	— March 18·5	Schmidt, jr.
3 4	0·15422	820·71	4·322	2·6539		13·0	— May 16·3. W	A. Hall.
5 48	0·14408	958·41	3·702	2·3931		11·5	— April 21 .	Tietjen.
8 27	0·17552	641·15	5·534	3·1287		11·0	— May 15·4 ·	Schiaparelli
7 58	0·18566	767·64	4·622	2·7748		11·0	— June 1 ·	Seeling.
11 14	0·22378	813·22	4·363	2·6701		10·5	— June 0 ·	Förster.

It is needless to give any detailed account of each, but a short summary may not be out of place.

The nearest to the Sun is *Flora*, which revolves round that luminary in 1193 days or 3¼ years, at a mean distance of 209,819,000 miles.

The most distant is *Maximiliana*, whose period is 2343 days or 6·4 years, and whose mean distance is 329,000,000 miles.

The least eccentric orbit is that of *Europa*, in which ε amounts to only 0·004.

The most eccentric orbit is that of *Polyhymnia*, in which ε amounts to 0·337.

The least inclined orbit is that of *Massilia*, in which ι amounts to 0° 41′.

The most inclined orbit is that of *Pallas*, in which ι amounts to 34° 42′.

The brightest planet is *Vesta*.

The faintest, *Atalanta*.

The largest planet is *Pallas*, whose diameter, according to Lamont, is 670 miles: other estimations make it less.

The smallest is not yet determined.

Under favourable circumstances *Ceres* has been seen with the naked eye, being then of the brightness of a star of the 7th magnitude; it is, however, more usually that of an 8th. The light is somewhat of a red tinge, and some observers have remarked a haziness surrounding the planet, which is attributed to the density and extent of its atmosphere. Sir W. Herschel once fancied he had detected two satellites accompanying Ceres; but analogy would lead us to infer that its mass would be insufficient to retain satellites around it. *Pallas*, when nearest the Earth in opposition, shines as a full 7th magnitude, with a decided yellowish light. Traces of an atmosphere have also been observed. *Juno* usually shines as an 8th magnitude star, and is of a reddish hue. *Vesta* appears at times as bright as a 6th magnitude star, and may then constantly be seen without optical aid, as was the case in the autumn of 1858. The light of Vesta is usually considered to be a pure white, but Hind considers it a pale yellow.[1]

---

[1] *Sol. Syst.* p. 85.

Sir J. Herschel remarks : " A man placed on one of the minor planets, would spring with ease 60 feet, and sustain in his descent no greater shock than he does on the Earth from leaping a yard. On such planets giants might exist; and those enormous animals which, on Earth, require the buoyant powers of water to counteract their weight, might there be denizens of the land." [1] But of such speculations there is no end.

Respecting the past history, so to speak, of the minor planets we can say but little. The hypothesis of Olbers seems to be the most satisfactory that has yet been propounded. That astronomer on calculating the elements of the orbit of Pallas, was forcibly struck with the close coincidence he found to exist between the mean distance of that planet and Ceres. He then surmised that they were the fragments of some large planet which had, in remote antiquity, by some great catastrophe, been shivered to pieces, the two small planets being amongst the fragments. When this theory was started it was certainly a bold one, but the discoveries of late years have materially strengthened it, and there can be now but little doubt that this theory truly points out the origin of these celestial wanderers.

The circumstances which led originally to search being made for planetary bodies, in the space intervening between Mars and Jupiter, were these. In the year 1800, 6 astronomers, of whom Baron De Zach was one, assembled at Lilienthal, and there resolved to establish a society of 24 practical observers, to examine all the telescopic stars in the whole zodiac, which was to be divided into 24 zones, each containing one hour of right ascension, for the express purpose of fishing for undiscovered planets.[2] They elected Schröter president of the association, and the Baron was unanimously chosen their perpetual secretary. Such determined organisation was ere long rewarded by the discovery of 4 planets, but as no more seemed to be forthcoming, the search was relinquished in 1816. Thus was afforded

---

[1] *Outlines of Ast.*, p. 352.   [2] See above, p. 19.

" a striking instance of those anticipations by which sagacity
sometimes outstrips its age and country." [1]

---

[1] In the Table, page 53, the column headed "Diameter," is derived
from the results of the photometric experiments of Prof. Stampfer of
Vienna, an account of which will be found in Bruhns's *De Planetis
Minoribus*, Berlin, 1856. Recent discoveries, not included in the table
on pages 54, 55, will be found in another part of the volume.

Plate X.

Fig. 60.

JUPITER, OCTOBER 25, 1856.

DRAWN BY DE LA RUE.

CHAPTER X.

JUPITER. ♃

Its great Brilliancy. — Its Belts. — Their Physical Nature. — Discovered by Zuppi. — Jupiter's Axial Rotation. — Its Spots. — Illustration of its immense Size. — Its apparent Motions. — Astrological Influences. — Its Satellites. — Sometimes visible to the naked Eye. — Singular Circumstance connected with the Three Interior ones. — Instances of all the Satellites being invisible. — Observation of the Satellites for ascertaining the Longitude. — Eclipses of the Satellites.

JUPITER is the largest planet of the system, and when in and near opposition shines with a brilliancy inferior only to Venus. A friend of the writer has informed him that on October 16, 1857, the planet's lustre was so considerable as to throw a sensible shadow. When telescopically examined, the surface of Jupiter is found to be marked with a series of dusky streaks, commonly known as the " belts." These belts vary greatly in form, size, and number from time to time. Occasionally only two or three broad ones are seen, at other times as many as eight, ten, or even a dozen narrow ones. Their physical nature is not well understood, but they are usually considered to be masses of cloud, acted upon in a manner in some way analogous to our terrestrial trade-winds. If this is the true hypothesis, as probably it is, then the planet has wind, rain, water and clouds, and is consequently fitted for the existence of animal and vegetable life. The existence of the belts was first detected by Zuppi, at Naples, in May 1730[1], though a claim has been put in on behalf of Torricelli.[2]

---

[1] Riccioli, *Almag. Nov.*, vol. i. p. 486.
[2] Moll., *Journ. Roy. Inst.*, vol. i. p. 494, May 1831.

The rotation of Jupiter on its axis is performed in about 10 hours; it consequently happens that not only is its form spheroidal, but the compression (owing to the great bulk of the planet) is so very considerable that it can be detected without difficulty in a telescope. The axial rotation of Jupiter was discovered by D. Cassini in 1665, and its spheroidal form by the same observer in 1691. Spots have from time to time been seen on Jupiter, as was the case in the year 1857.[1] As a proof of the immense size of Jupiter, it may be stated that a line carried from the Earth to the Moon would not reach round Jupiter's equator.

Seen from the Earth the apparent motion of Jupiter is sometimes retrograde. The length of the retrograde arc varies from 9° 51′ to 9° 59′; its duration varying from 116d. 18h. to 122d. 12h. The retrograde motion begins or ends, as the case may be, when the planet is at a distance from the Sun which varies from 113° 35′ to 116° 42′.[2]

In days gone by, Jupiter has not been without its astrological influences. It was supposed to be the cause of storms and tempests, and to have power over the prosperity of the vegetable kingdom. Pliny thought that lightning, amongst other things, owed its origin to Jupiter. An old MS. Almanac for 1386 states, that " Jubit es hote and moyste, and doos weel til al thynges, and noyes nothing."

Jupiter is attended by 4 satellites[3], all discovered by Galileo, at Padua, on January 8, 1610.[4] They shine with the brilliancy of stars of the 7th magnitude, or perhaps rather less, but owing to their proximity to their primary, are usually invisible to the naked eye, though several instances to the contrary are on record. Two of them were

---

[1] *Month. Not.* R.A.S., vol. xviii. p. 6.

[2] We may here observe that, as a general rule, the further a superior planet is from the Sun, the less will be the extent of its arc of retrogradation, but the greater will be the time occupied in describing it.

[3] Named respectively *Io, Europa, Ganymede, Callisto.* These names are not in general use.

[4] *Sidereus Nuncius; Opere di Galileo,* vol. ii. p. 15 *et seq.* Ed. Padua, 1744.

1857 : November 27.　(*Dawes.*)

1858 : November 18.　(*Lassell.*)

JUPITER.

Plate XII.

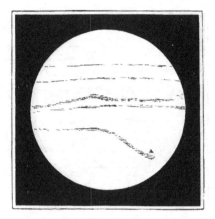

1860: March 12. (*Jacob.*)

1860: April 9. (*Baxendell.*)

JUPITER.

THE SATELLITES OF JUPITER.

	Discoverer.	Mean Distance.		Sidereal Period.		Mean Apparent Diameter.	Diameter.	Diameter seem from ♃.	Apparent Diam. of ♃ seen from Sat.	Max. Elong. ♃ in ☉.	Mass ♃=1 (Laplace).	Density.		Appar. Star Mag.
		Radii of ♃	Miles.	d. h. m.	d.		Miles.	′ ″	° ′	′ ″		⊕=1.	Water =1.	
1. Io	Galileo,	6·05	278,542	1 18 28	1·77	1·02	2440	38 11	19 49	2 15	0·000017	0·020	0·114	7
2. Europa . .	Padua,	9·62	444,904	3 13 4	3·55	0·91	2192	17 35	12 25	3 35	0·000023	0·030	0·171	7
3. Ganymede .	Jan. 7th,	15·35	706,714	7 3 43	7·15	1·49	3579	18 0	7 47	5 46	0·000088	0·069	0·396	6
4. Callisto . .	1610.	26·99	1,242,619	19 16 32	16·69	1·27	3062	8 46	4 25	9 45	0·000043	0·039	0·222	7

The duration of an eclipse of the 1st is 2h. 2cm.
"   "   "   2nd 2 56
"   "   "   3rd 3 43
"   "   "   4th 4 56
(*Nautical Almanac*, 1835.)

The eccentricity of the orbits of 1 and 2 is = 0: of 3 and 4 it is "small and variable."

thus seen by some of the officers of H. M. S. Ajax, when in Kingstown Harbour, Dublin, on January 15, 1860.[1]

The Jovian satellites move in orbits nearly circular, and between the motions of the first three, two singular relations exist. " *The mean sidereal motion of the first added to twice that of the third, is constantly equal to three times that of the second; and the sidereal longitude of the first, minus three*

Fig. 11.

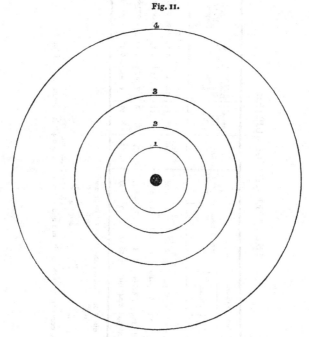

Plan of the Jovian System.[2]

times that of the second, plus twice that of the third, is always equal to 180°. From this it follows, that for a very long period of time, the 3 first satellites cannot all be eclipsed

---

[1] *Month. Not.* R.A.S., vol. xx. p. 212.
[2] All the plans are drawn to scale.

August 26.

August 26 (two hours later).

August 27.

September 9.

The 3rd SATELLITE in 1855. (*Secchi.*)

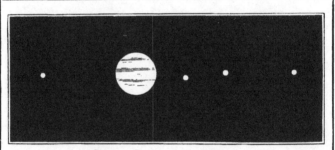

GENERAL TELESCOPIC APPEARANCE.

The 4th SATELLITE, 1849, Feb. 12.
(*Dawes.*)

The 3rd SATELLITE, 1860, Jan. 31.
(*Dawes.*)

THE SATELLITES OF JUPITER.

at the same time; for in the simultaneous eclipses of the second and third the first will always be in conjunction with Jupiter, and *vice versâ*.[1] Making use of his own tables, Wargentin has calculated that simultaneous eclipses of the 3 satellites cannot take place before the lapse of 1,317,900 years[2], and an alteration of only 0·33″ in the annual motion of the second satellite would suffice to render the phenomenon for ever impossible.

It occasionally, though rarely, happens that all 4 satellites are for a short time invisible, being either in front of, or behind, the planet. Such was the case, according to Molyneux[3], on Nov. 12, 1681 (o. s.) A similar occurrence was noticed by Sir W. Herschel on May 23, 1802, by Wallis on April 15, 1826, and more recently by Dawes and Griesbach, on September 27, 1843. On this occasion the planet was seen apparently deprived of its *comites* for 30 m.

Soon after the discovery of the satellites, it suggested itself to Galileo that they might be made use of for ascertaining the longitude.[4] He offered the secret to the Spaniards, *then* the great navigators of the world. His offer was refused, but soon after it was accepted by Holland.

The eclipse phenomena visible on Jupiter are on a great scale; for in consequence of the small inclination of the orbits of the satellites to the planet's equator, and also the small inclination of the latter to the ecliptic, all the satellites, the fourth excepted, are eclipsed some time in every revolution; so that an inhabitant of Jupiter witnesses during his year 4500 eclipses of the Moon, and about the same number of the Sun. The eclipses and transits of Jupiter's satellites form a very interesting series of phenomena, which a moderate sized telescope will enable the reader to observe; a complete table of them being inserted every year in the *Nautical*

---

[1] Laplace has demonstrated by the theory of gravitation, that if this relation be once approximately begun, it will *always* last.

[2] *Acta. Soc. Upsal.* 1743, p. 41.

[3] *Optics*, p. 271.

[4] *Opere di Galileo*, vol. ii. p. 439, Ed. Padua, 1744.

Almanac. " The discovery of 4 bodies revolving round a primary, exhibited a beautiful illustration of the Moon's revolution round the Earth, and furnished a most favourable argument in favour of the Copernican theory. The announcement of this fact pointed out also the long vista of similar discoveries which have continued from time to time down to the present day to enrich the solar system, and to shed a lustre on the science of astronomy."

It was from observations of these satellites that Römer discovered the progressive transmission of light.[1]

---

[1] *Opere di Galileo*, vol. ii. p. 33, Ed. Padua, 1744.

Plate XIV.

Fig. 72.

SATURN, MARCH 27 and 29, 1856.

DRAWN BY DE LA RUE.

65

CHAPTER XI.

SATURN. ♄

Interest attaching to it.—Its Atmosphere.—Historical Notices.—Observations of Galileo.—The Logograph sent by him to Kepler.—Huyghens' Discovery of the Ring.—His Logograph.—The Bisection of the Outer Ring.—Discovery of the Interior Transparent Ring.—Facts relating to the Rings.—Appearances presented by them under different circumstances.—Observations of W. Herschel, Schröter, and Harding.—Measurements of W. Struve.—Bessel's Investigations.—Observations of Maraldi in 1714.—Appearance of the Rings when viewed from the Planet.—The Satellites of Saturn.—Table of them.—Particulars relating to Titan.—Prediction of Mr. Harris, F.R.S., on Telescopes.—The Uranography of Saturn.—Ancient Observations.

SATURN is undoubtedly the most interesting member of the planetary system, not only from its being accompanied by 8 satellites, but especially on account of the system of rings by which it is surrounded. Belts are also occasionally noticed, but they are far more indistinct than those of Jupiter, though doubtless due to the same physical cause, whatever that may be. Spots are rare. Sir W. Herschel considered this planet to be surrounded by a very dense atmosphere; an idea that has been fully confirmed by subsequent observers.

When this planet was first telescopically examined by Galileo, he noticed that it presented a very oval outline, which he conjectured was owing to a larger planet having on each side of it two smaller ones. He added, that with telescopes of superior power, the planet did not appear triple, but exhibited an oblong form, somewhat like the shape of an olive.[1]

---

[1] *Opere di Galileo*, vol. ii. p. 41, Ed. Padua, 1744.

Continuing his observations, the illustrious astronomer was not long in noticing that the two (supposed) bodies gradually decreased in size, though still in the same position as regards their primary[1], until they finally disappeared altogether. Galileo's amazement at this was unbounded, and his third letter to Welser, in which he expresses his feelings on the subject, is still extant.   He remarks: —

" What is to be said concerning so strange a metamorphosis ?   Are the two lesser stars consumed after the manner of the solar spots?   Have they vanished or suddenly fled? Has Saturn, perhaps, devoured his own children?   Or were the appearances indeed illusion or fraud, with which the glasses have so long deceived me, as well as many others to whom I have shown them?   Now, perhaps, is the time come to revive the well-nigh withered hopes of those who, guided by more profound contemplations, have discovered the fallacy of the new observations, and demonstrated the utter impossibility of their existence.   I do not know what to say in a case so surprising, so unlooked for, and so novel.   The shortness of the time, the unexpected nature of the event, the weakness of my understanding, and the fear of being mistaken, have greatly confounded me." [2]

The original discovery was announced to Kepler in the following logograph : [3] —

smaismrmilmepoetalevmibvnenvgttaviras;

which being transposed, becomes —

altissimvm planetam tergeminvm observavi;

" I have observed the most distant planet to be threefold."

As time wore on, more correct ideas were obtained of the phenomenon, which gradually came to be looked upon as due to the existence of two ansæ or handles to the planet, though the cause of their disappearance from time to time was yet inexplicable.   It was not till after the lapse of nearly 50 years

---

[1] *Opere di Galileo*, vol. ii. p. 46, Padua Ed. 1744.
[2] Ibid., p. 152.   [3] Ibid., p. 40.

that the true cause of the appearance seen by Galileo and others became known. Huyghens was the discoverer, and he intimated his discovery in the following logograph [1] : —

aaaaaaa ccccc d eeeee g h iiiiiii llll mm nnnnnnnnn oooo pp q rr s ttttt uuuuu;

which letters, when placed in their proper order, give —

annulo cingitur, tenui plano, nusquam cohaerente, ad eclipticam inclinato.

" The planet is surrounded by a slender flat ring, everywhere distinct from its surface, and inclined to the ecliptic." [2]

The duplicity of the ring was detected on October 13, 1665, by two English observers named Ball, at Minehead, in Devonshire.[3]

The observations of modern times have brought to light many peculiarities never even before suspected. It is now known that the planet is surrounded by more than two rings, though they are only to be detected by powerful telescopes. Quetelet, in December 1823, Kater, on several occasions in December 1825[4], and Encke, in 1838, noticed an appearance which seemed to indicate that the outermost was itself a double one[5], and on September 7, 1843, Lassell and Dawes, unaware apparently of Encke's observations, saw with a 9-foot reflector, what they considered to be a division in the outer ring.[6]

We have yet another discovery to refer to, of still greater

---

[1] *De Saturni Luna Observatio Nova.* Hagæ, 1656. Followed in 1659 by detailed particulars in the *Systema Saturnium.*

[2] Maurice (*Indian Antiquities*) gives an engraving of *Sani*, the Saturn of the Hindùs, from an image in an ancient pagoda. A circle is formed around him by the intertwining of two serpents; whence the writer infers that, by some means or other, the existence of Saturn's ring may have been known in remote ages.

[3] *Phil. Trans.* vol. i. p. 152. 1666.

[4] *Mem.* R.A.S., vol. iv. p. 383 *et seq.*

[5] *Trans. Berl. Acad.*, 1838.   [6] *Month. Not.* R.A.S., vol. vi. p. 12.

interest. It is the existence of an interior, light-reflecting, and yet transparent ring of extreme tenuity. In 1838, Dr. Galle, of Berlin, noticed a phenomenon which he described as a gradual shading off of the innermost ring, then known, towards the surface of the planet to a faint, but still perceptible, annulus.[1] Little attention was paid to these observations of Galle until the close of the year 1850, when the same appearance was seen by Mr. Dawes in England, and Professor Bond in America; the former observer having drawn Lassell's attention to the subject; and their joint observations were published nearly simultaneously with those of Bond.[2] It was not, however, until 1852 that the transparent nature of the ring was fully ascertained; when further details and some elaborate drawings were published by Mr. Dawes, Captain Jacob, and Mr. Lassell.

Having said this much on the history of these discoveries, we must point out some of the facts connected with the rings. The true form of the rings is no doubt circular, or nearly so; but as we always see them foreshortened, they appear more or less oval, when the Earth is above or below the plane of the rings, but when we are in the plane, if visible at all, they appear projected on the ball as a single black line. In the true position of the rings, during Saturn's revolution round the Sun, there is no change: they remain continually parallel to themselves.

The plane of the rings is inclined 28° to the ecliptic, and intersects it at present in the 18° of Virgo and Pisces; the former point being called the place of the ascending node, because the Earth there ascends from beneath the plane of the rings to their northern side; and the latter, of the descending node, for the contrary reason. In the course of the next 100 years these points will have advanced about $1\frac{1}{4}$° on the ecliptic, in the order of the signs.[3]

Whether viewed from the Earth or from the Sun, the pheno-

---

[1] *Trans. Berl. Acad.*, 1838.

[2] Vide *Month. Not.* R.A.S., vol. xi. pp. 18–27, where the observations of the three will be found.

[3] Hind, *Illust. Lond. Ast.*, p. 66.

mena seen in connection with the rings are pretty much the
same, but the motion of the former in its orbit, the inclination
of which differs somewhat from that of Saturn, gives rise to
certain phases in the rings which would not be witnessed by

Fig. 12.

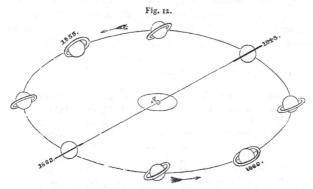

Phases of Saturn's Rings.

an observer placed in the Sun. Thus it usually happens
that there are two, if not three, disappearances about the
time of the planet's arrival at the nodes. The plane of the
ring may not pass through the Earth and Sun at the same
times, but they may become invisible under both conditions,
because its edge only will be directed towards us. It is also
invisible when the Earth and the Sun are on opposite sides of
its plane—a state of things that may continue a few weeks;
in this case we have the dark surface turned towards our
globe. In very powerful telescopes it has been found that the
disappearance of the ring is complete under the latter condition; it has, however, been perceived as a faint broken line
of a dusky colour, not only when the Sun is in its plane, but
likewise when its edge is directed to the Earth.

By a careful examination of the ring, Sir W. Herschel
ascertained that it revolves round the ball in 10h. 32m. 15s.,
or in about the same time as the latter rotates round its axis,
and in the same direction.[1]   MM. Schröter and Harding

---

[1] *Phil. Trans.*, vol. lxxx. p. 480. 1790.

found that the ring was not mathematically concentric with
the ball [1] ; but that the planet lay to the west of the ring, a
condition which by oscillations of much complexity tends to
maintain the stability of the system. This announcement
induced Professor Struve rigidly to investigate the matter, and
he found as the results of careful measurements executed with
the great Dorpat refractor, that the dimension of the right
vacancy was 11·288″ but of the left, only 11·073." [2]

The following are Struve's measurements, reduced to the
mean distance of the planet : —

	″	Miles.
Outer diameter of exterior ring .	40·095 =	176418
Inner diameter of ,, ,, . .	35·289 =	155272
Breadth ,, ,, . .	2·403 =	10573
Outer diameter of interior ring . .	84·475 =	151690
Inner ,, ,, .	26·668 =	117339
Breadth ,, ,, .	3·903 =	17176
Interval between the two . . .	0·408 =	1791
Distance of ring from ball . . .	4·339 =	19090
Equatorial diameter of ball . .	17·60	

The measure of De La Rue[3], Main[4], and Jacob[5] are appended
for comparison : —

	De La Rue.	Main.	Jacob.
	″	″	″
Outer diameter of exterior ring . .	39·83	39·73	39·99
Inner ,, ,, . .	35·33		35·82
Breadth ,, ,, . .	2·25		2·08
Outer diameter of interior (middle) ring	33·45		34·85
Inner ,, ,, ,,	26·91	27·65	26·27
Breadth ,, ,, ,,	3·27		4·29
Interval between the two . . .	0·94		0·48
Distance of ring from ball . . .	4·62	5·07	4·16
Equatorial diameter of planet . .	17·66	17·50	17·94

[1] *Conn. des Temps*, 1808.   [2] *Mem.* R.A.S., vol. iii. p. 300.
[3] *Month. Not.* R.A.S., vol. xvi. p. 43.
[4] Ibid., p. 35.   [5] Ibid., pp. 124.

Bessel of Königsberg has entered upon some investigations to determine the mass of the rings, by ascertaining the perturbing effect on the 6th satellite, Titan. He estimates it at $\frac{1}{118}$ of the mass of the planet.[1] The thickness of the rings being too minute for measurement, no precise determination can be arrived at as to their density; if, however, we assume it as approximately equal to that of the planet, as is probably the case, it will follow that the thickness is about 138 miles, a quantity which tolerably well coincides with the estimations of several observers. Supposing this correct, at the mean distance of the planet, the rings would only subtend an angle of about 0·03″; it will then be inferred that the ring will at stated times become wholly invisible even in the most powerful telescopes. Sir W. Herschel surmised that the ring is not flat, but that the inner edge was spherical or hyperbolical.[2] Maraldi, in October 1714, noticed that just previously to the actual passage of the Earth through the plane of the ring, and whilst the ansæ were decreasing, the eastern one appeared a little larger than the other for three or four nights, and yet it vanished first.[3] This induced him to suspect that it was not bounded by parallel lines, and also was endued with an axial rotation. J. Cassini and Messier observed certain phenomena of a confirmatory character. Bessel, however, has shown the fallacy of this idea.[4]

" Seen from Saturn, the rings must appear as vast and inconceivably splendid luminous arches, stretched across the heavens from horizon to horizon, while the satellites adding their variety of phase, increase the radiance of the scene. In the planet's course round the Sun, the zone assumes a beautiful variety of oval forms, from its being seen obliquely; gradually contracting from a certain elliptical form to an almost imperceptible line, or, according to the telescope's power, to entire disparition; then expanding again, till it assumes its maximum

---

[1] *Conn. des Temps*, 1838.
[2] *Phil. Trans.*, vol. xcii. p. 463. 1802.
[3] *Mém. Acad. des Sciences*, 1715, p. 12.
[4] *Conn. des Temps*, 1838.

of ellipticity.   The intersection of the ring and the ecliptic is
in 173° 32′ and 352° 32′; consequently, the ring wanes when
Saturn is near either of those points; on the contrary, it is
seen to the greatest advantage when he is in 77° 31′ or 257° 31′,
according to the Earth's position; thus the ring is most open
when the planet is in the 19th degree of Gemini, or of Sagittarius; and least so when in the 19th degree of Pisces, or
of Virgo.   In other words, between each of Saturn's revolutions, that is, every 15 years, the Earth being in the
plane of its ring, we only see its thin edge; therefore it
becomes invisible except in the mightiest telescopes.   When
the Sun illumines the side of the ring opposite to that on
which the Earth is, it will also be invisible to us, except that
with good optical aid, its existence may be detected by a
narrow shadow cast across the body of the planet."

De La Rue has recently suspected the existence of mountains
on the ring.[1]   Lassell and Jacob have confirmed this.[2]

Some years since O. Struve propounded a theory, to the
effect that the ring is contracting upon the ball, at a rate
which at no distant period may bring it in contact with the
globe.   Struve supports his idea by deductions from the observations of Picard, Campana, D. Cassini, and others, made in
the 17th century; if these observations can be relied upon,
possibly there may be some foundation for the surmise, but as
they cannot, no opinion can be safely expressed in its favour.

Saturn is accompanied by 8 satellites, 7 of which move
in orbits whose planes coincide nearly with that of the
planet's equator, and therefore with the plane of the rings;
the orbit of the remaining and most distant satellite is inclined
about 12° to the aforesaid planes.   The consequence of this
coincidence in the orbits of the first 7 satellites, is that they
are always invisible to the inhabitants of both hemispheres,
when not under eclipse in their primary's shadow.   They
move with greatly varying rapidity; Mimas traverses its
orbit at the rate of 16′ (of arc) per minute, so that in 2
minutes it moves over a space equal to the apparent diameter

THE SATELLITES OF SATURN.

	Order of Dis.	Discoverer	Mean Distance		Sidereal Period		Diameter		Diam. seen from ♄	App. Diam. of ♄ seen from Satel.	Max. Elong. ♄ in ☉	Mass ♄=1.	App. Star Mag.
			Rad. of ♄	Miles	d. h. m.	d.	Mean App.	Miles					
1. Mimas . .	7	Sir W. Herschel. 1789, Sept. 17 .	3·36	129,746	0 22 37	0·94	?	1000	?	17·0	0 33		17
2. Enceladus .	6	„ 1789, Aug. 28 .	4·31	166,430	1 8 53	1·37	?	?	?	13·3	0 43		15
3. Tethys .	5	J. D. Cassini. 1684, March .	5·34	206,204	1 21 18	1·88	?	?	?	10·7	0 53		13
4. Dione . . .	4	„ 1684, March .	6·84	264,126	2 17 41	2·73	?	?	?	8·4	1 8		12
5. Rhea . . .	3	„ 1672, Dec. 23 .	9·55	358,773	4 12 25	4·51	?	1200	?	6·0	1 35		10
6. Titan . . .	1	C. Huyghens. 1655, March 25	22·14	854,936	15 22 41	15·94	0·75	3300	25·0	2·6	3 41		8
7. Hyperion . .	8	W. Bond and Lassell. 1848, Sept. 19 .	28·00	1,086,220	21 7 7	21·29	?	?	?	2·0	4 40		17
8. Iapetus . .	2	J. D. Cassini. 1671, Oct. . .	64·36	2,485,526	79 7 53	79·33	?	1800	?	0·9	10 43		9

E

of the Moon. Titan is visible with comparatively trifling optical assistance, but the others are all invisible except with

Fig. 13.

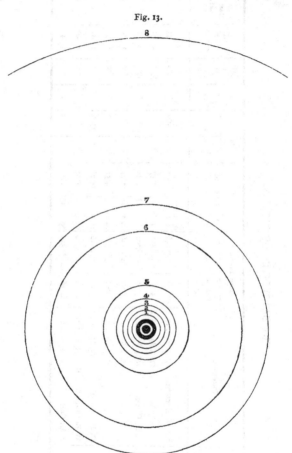

Plan of the Saturnian System.

telescopes of great power. An inspection of the table will show that the periodic times of the 3rd and 4th are respectively double those of the 1st and 2nd. The line of apsides of Titan completes a revolution in 718 years, and the nodes in 3600 years. When Huyghens discovered this satellite in 1655, he was imprudent enough to predict that there were no others. The danger of prediction in matters of this kind is well illustrated in the case of Mr. John Harris, F.R.S. That learned gentleman published a book in 1729, in which he says: " 'Tis highly probable that there may be more than 5 moons revolving round this remote planet [the number of satellites which Saturn was then known to possess]; but their distance is so great as that they have hitherto escaped our eyes, and perhaps may continue to do so for ever; *for I do not think that our telescopes will be much further improved*"!!

The following elements are by Capt. Jacob.[1]

	1 Jan. 1857. λ		π		Ω		ι		φ	ε	Semi-axis maj. a	Daily Hel. Mot. μ
	°	'	°	'	°	'	°	'			''	°
Mimas . . .	210	±										381·94
Enceladus . .	301	55										262·73
Tethys . . .	281	42	169	7	167	37	28	10		0·01086	42·60	190·69
Dione	115	30	145	4	167	37	28	10		0·00310	54·85	131·53
Rhea	288	43	185	0	167	19	28	8		0·00080	76·13	79·69
Titan	299	42	257	6	167	58	27	36		0·027937	176·90	22·57
Hyperion . .												
Iapetus . . .	78	9	349	20	143	1	18	37		0·28443	514·96	4·53

From Saturn, the Sun only appears 3' in diameter, and the greatest elongation of the planets are Mercury, 2° 19': Venus, 4° 21': Earth, 6° 1': Mars, 9° 11': Jupiter, 33° 3': so that a Saturnian, can see but Mars and Jupiter; and the former only with difficulty. Saturn, on account of its slow dreary pace, was chosen by the alchemists as the symbol for lead.

The earliest observation of Saturn is that recorded to have been made by Ptolemy, bearing a date corresponding to March 1, 228 B.C. according to our reckoning.[2] An occultation of this planet by the Moon is recorded to have been observed by Thius, at Athens, 513 A.D.

---

[1] *Month. Not.* R.A.S., vol. xviii. p. 1.   [2] *Almagest*, lib. xi.

CHAPTER XII.

URANUS. ♅

Circumstances connected with its Discovery by Sir W. Herschel. — Other
Names which it has received. — Just perceptible to the naked eye. —
Amount of Light received from the Sun. — Position of its Axis. —
Attended by 8 Satellites. — Peculiarities connected with them.

THIS planet was discovered by Sir W. Herschel on March 13,
1781, whilst he was engaged in scrutinising some small stars
in Gemini. He observed one of them which seemed to have a
more sensible diameter than the others, and to be less luminous.
The application of high magnifying powers rendered these
peculiarities more perceptible : he therefore made some very
careful observations, and found that it was moving at the rate
of $2\frac{1}{4}''$ per hour. He then announced to the Royal Society
the discovery of a new comet, so little was a planet expected.[1]
Maskelyne found it, and soon suspected its true nature. On
proper inquiries being made, it was found that Flamsteed had
seen it 3 times, Mayer, once, and Le Monnier, 11 times previously to the epoch of Herschel's discovery; all of whom
had been ignorant of its real nature.

A brisk discussion took place on the name the new planet
was to have. Herschel himself desired to call it the *Georgium*
Sidus in compliment to his friend and patron, our most excellent Sovereign King George III.; some of the foreign
astronomers, amongst whom was Laplace, insisted that it
ought to bear the name of its discoverer; Bode proposed

---

[1] Herschel's original observations appear in the *Phil. Trans.* vol.
lxxiii. p. 1, *et seq.* 1783.

Uranus as the mythological father of Saturn; and this name finally triumphed, though for a long course of years it was frequently known as the *Georgian Planet.*

Uranus is just perceptible to the naked eye[1]; but owing to its small size and great distance, no marks have yet been noticed on its surface, whereby the time of rotation can be ascertained.

It has been calculated that the amount of light received by Uranus from the Sun is equal to about the quantity which would be afforded by 300 full Moons. The inhabitants of Uranus can see Saturn, and perhaps Jupiter, but none of the planets included within the orbit of the latter.

The axis of Uranus falls in the plane of its orbit, in consequence of which " the Sun turns in a spiral form round the whole planet, so that even the 2 poles sometimes have that luminary in their zenith."[2] Sir W. Herschel fancied that Uranus was surrounded by a pair of rings at right angles to each other; the idea, however, like others of the same observer, was wholly imaginary.

Uranus is attended by 8 satellites, but such is their extreme minuteness that they are invisible except in the most powerful telescopes; there are probably not more than a dozen instruments that will show them. They revolve in circular orbits, nearly perpendicular to that of the Earth, and in a *retrograde* direction, or from east to west; a circumstance which does not obtain in the case of any member of the solar system yet known, except the satellite of Neptune.

It is found that the satellites and all small stars lose much of their lustre in the neighbourhood of the planet, and disappear when within a few seconds of it. That it is not due to any atmosphere belonging to the planet is known by the fact,

---

[1] It is a somewhat singular fact that the Burmese mention eight planets: the Sun, Moon, Mercury, Venus, Mars, Jupiter, Saturn, and *Rahu*, which latter is invisible. "An admirer of Oriental literature," says Buchanan, "would here discover the Georgium Sidus, and strip the illustrious Herschel of his recent honours."

[2] Sir W. Herschel, quoted in Smyth's *Cycle*, vol. i. p. 205.

THE SATELLITES OF URANUS.

	Order of Discovery.	Discoverer.	Mean Distance. Radii of ♅.	Mean Distance. Miles.	Sidereal Period. d. h. m.	Sidereal Period. d.	Apparent Star Magnitude.	Max. Elong.
1. Ariel .	7	Lassell, 1847. Sept. 14	7·44	128,340	2 12 28	2·52		" 12
2. Umbriel	8	O. Struve. 1847. Oct. 18 .	10·37	198,882	4 3 27	4·14		15
3.	3	Sir W. Herschel. 1790. Jan. 18 .	13·12	226,520	5 21 25	5·89		18
4. Titania	1	,, ,, 1787. Jan. 11 .	17·01	293,422	8 16 55	8·71		33
5.	6	,, ,, 1794. Mar. 26	19·85	342,412	10 23 3	10·96		38
6. Oberon .	2	,, ,, 1787. Jan. 11 .	22·75	392,507	13 11 6	13·46		44
7.	4	,, ,, 1790. Feb. 9 .	45·51	785,047	38 1 48	38·08		88
8.	5	,, ,, 1794. Feb. 28 .	91·01	1,569,922	107 16 39	107·69		177

Inclination . . . . . $79°±$

Eccentricity . . . . Small.

Direction of motion . —

that the phenomenon happens equally in whatever part of its orbit the satellite may be. Sir W. Herschel surmised that it was an effect of contrast, and this is probably the true explanation: at least, as yet we know no better.

Hind gives the following[1] (calculated for a direct motion):

					Ω	ι
Titania	.	.	.	.	165·25	100·34
Oberon	.		.	.	165·28	100·34

Fig. 14

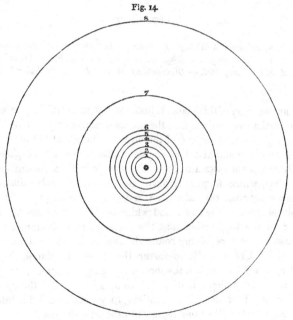

Plan of the Uranian System.

---

[1] *Month. Not.* R.A.S., vol. xv. p. 48.

CHAPTER XIII.

NEPTUNE. ♆

*Circumstances which led to its Discovery.— Investigations of Adams and
Le Verrier.— Suspected Ring.— Discovery of a Satellite by Lassell.—
A Second suspected.— Observations of the Planet by Lalande in
1795.*

THE discovery of this planet is justly esteemed one of the greatest
triumphs ever recorded in the annals of astronomy.   The following is a brief résumé of the circumstances that led to it:—
From a long and attentive course of observation on the planet
Uranus, it was ascertained that it was subject to some disturbing
agency, which frequently gave rise to a considerable discordance between the calculated and the observed path of the
planet amongst the stars, and which could only be accounted
for on the supposition that there was another, though as yet
unseen, planet revolving round the Sun in an orbit exterior to
that of Uranus.   To discover the suspected planet, MM.
Adams and Le Verrier, the former in 1843, the latter in 1845,
commenced independently, an examination of the theory of
Uranus.   From their investigations, they were both led to infer
the existence of the unseen planet in the constellation Aquarius.
M. Le Verrier having informed Dr. Galle, of the Royal Observatory, Berlin, of the result of his labours, that gentleman was
fortunate enough to discover the planet on the night of Sept. 23,
1846, very near the position assigned by theory.   Mr. Adams,
in July, 1846, likewise placed the results of *his* investigations
in the hands of Mr. Airy and the Rev. Professor Challis, the
Director of the Cambridge Observatory, the latter of whom

on July 29, commenced a diligent search for the planet with the then recently-erected Northumberland Equatorial. An object, afterwards found to be the anxiously-sought-for planet, was observed on the 4th and 12th of August, but its true nature was not detected till Sept. 29. The priority of the discovery unquestionably rests with Mr. Adams, he having commenced his researches more than 2 years before Le Verrier did. After a good deal of discussion, the name ultimately agreed upon for the newly discovered body was "Neptune." "Such is a brief history of this most brilliant discovery, the grandest of which astronomy can boast, and one that is destined to a perpetual record in the annals of science — an astonishing proof of the power of the human intellect." [1]

Fig. 15.

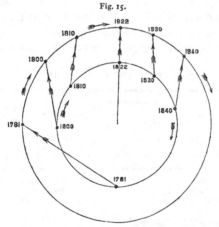

Illustration of the Perturbation of Uranus by Neptune.

The accompanying diagram shows the paths of Uranus and Neptune from 1781 to 1840, and will illustrate the direction of the perturbing action of the latter planet, on the former.

From 1781 to 1822 it will be evident, from the direction of

---

[1] As far back as October 25, 1800, Lalande and Burckhardt came to the conclusion that there existed an unseen planet beyond Uranus. (*Year Book of Facts*, 1852.)

the arrows, that Neptune tended to draw Uranus in advance of its place, computed independent of exterior perturbation.

In 1822 the two planets were in conjunction, and the only effect of Neptune was to draw Uranus further from the Sun, without altering its longitude.

From 1822 to 1830 the effect of Neptune was to destroy the excess of longitude accumulated from 1781, and after 1830, the error in longitude changed its sign, and for some years subsequently, Uranus was *retarded* by Neptune.

Owing to its great distance from us, nothing is as yet known of the physical appearance of Neptune, though it was at one time suspected that the planet was surrounded by a luminous ring, as in the case of Saturn. A satellite was detected by Mr. Lassell, of Starfield, Lancashire, within a few weeks of the discovery of the planet; and Professor G. P. Bond, of Cambridge, U.S., has obtained tolerably conclusive evidence of the existence of a second, though this is not fully confirmed.

THE SATELLITE OF NEPTUNE.

	Discoverer	Mean Distance.		Sidereal Period.		Apparent Star Magnitude.	Max. Elong.
		Radii of $\Psi=1$.	Miles.	d. h. m.	d.		′ ′
1.	Lassell. 1846, Oct. 10	12·00	225,000	5 21 8	5·87	14	18

Inclination   .   .   .   .   . $29° \pm$
Eccentricity   .   .   .   .   Hardly appreciable.
Direction of Motion   .   .   .   .   − (?)

Hind gives the following[1] :

Epoch 1852, Nov. 0, G. M. T.

°   ′

Mean anomaly   243   32
Peri-neptunium   177   30
Ω   175   40
ι   151   0
Eccentricity   .   .   .   6 $5 = 0 \cdot 1059748$
Period   .   .   .   .   5·8769d.

---

[1] *Month. Not.* R.A.S., vol. xv. p. 47.

Mr. Lassell's satellite revolves round the primary at a distance about equal to that of the Moon from the Earth, and shines with the brightness of a star of the 14th magnitude.

Shortly after the discovery of Neptune, it was found that it had been seen on two occasions, in 1795, by Lalande, whose observations have materially assisted astronomers in obtaining a more correct determination of the planet's orbit than they would otherwise have been able to have done, if they had been compelled to depend on modern observations only.

Owing to its immense distance the inhabitants of Neptune can only see the planets Saturn and Uranus, but none of the planets which revolve within the orbit of the former. Though deprived of a view of the principal members of the solar system, the Neptunian astronomers are well circumstanced for inspecting comets, and are also able to take, probably with considerable success, observations on stellar parallax, seeing they are in possession of a base line of 5,700,000,000 miles, or 30 times the length of that to which we are confined.

BOOK II.

ECLIPSES AND THEIR ASSOCIATED PHENOMENA.

———•———

CHAPTER I.

GENERAL OUTLINES.

Definitions.—*Position of the Moon's Orbit as regards the Earth's.*—*Consequences resulting from their being inclined.*—*Retrograde Motion of the Nodes of the Moon's Orbit.*—*Coincidences of 223 Synodical Periods with 19 Synodical Revolutions of the Node.*—*Known as the Saros.*—*Statement of Diogenes Laërtius.*—*Illustration of the Use of the Saros.*—*Number of Eclipses which can occur.*—*Solar Eclipses more frequent than Lunar ones.*—*Duration of Annular and Total Eclipses of the Sun.*

THE class of phenomena we are about to describe are those produced by the interposition of celestial objects; for we know well that inasmuch as many of the heavenly bodies are constantly in motion, it follows that the direction of lines drawn from one to another will vary from time to time; and it must occasionally happen that three will come into the same line. "When one of the extremes of the series of 3 bodies, which thus assume a common direction, is the Sun, the intermediate body deprives the other body, either wholly or partially, of the light which it habitually receives. When one of the extremes is the Earth, the intermediate body intercepts, wholly or partially, the other extreme body from the view of observers situate at places on the Earth which are in the common line of direction, and the intermediate body is seen

---

[1] See Appendix II.

to pass over the other extreme body, as it enters upon or leaves the common line of direction. The phenomena resulting from such contingencies of position and direction are variously denominated '*Eclipses*' '*Transits*' and '*Occultations*' according to the relative apparent magnitudes of the interposing and obscured bodies, and according to the circumstances which attend them." We shall proceed to consider the several phenomena in detail, beginning with Eclipses.

It must be premised, that the Moon's orbit does not lie in exactly the same plane as the Earth's, but is inclined thereto at an angle of about 5° 8' 48". The two points where its

Fig. 16.

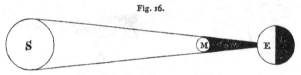

Theory of a Total Eclipse of the Sun.

path intersects the ecliptic are called the *nodes*, and the imaginary line joining these points is termed the *line of nodes*. When the Moon is crossing the ecliptic from south to north, it is passing its *ascending node* (☊); the opposite point of

Fig. 17.

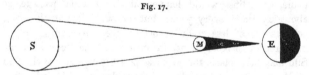

Theory of an Annular Eclipse of the Sun.

its orbit being the *descending node* (☋). If the Moon should happen to pass through either node at or near the time of conjunction, or New Moon, it will necessarily come between the Earth and the Sun, and the 3 bodies will be in the same straight line; it will therefore follow that, to certain parts of the Earth, the Sun's disc will be obscured, wholly or partially as the case may be: this is an *Eclipse of the Sun*. In the figures above, S, represents the Sun, E, the Earth, and M, the Moon.

The Earth and the Moon, being opaque bodies, must cast a shadow into space, though of course, owing to its larger size, the Earth's shadow is much the largest. If the Moon should happen to pass through either node, at or near the time of opposition or Full Moon, it will be again, as before, in the same straight line; only the Moon, in the present case, will be involved in the shadow of the Earth, and therefore will be deprived of the Sun's light: this is an *Eclipse of the Moon.*

Fig. 18.

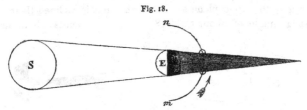

Theory of an Eclipse of the Moon.

In Fig. 18, S, represents the Sun: E, the Earth, and *m n*, the orbit of the Moon: that the Moon becomes involved in the Earth's shadow in passing from *m* to *n* is obvious.

If the orbits of the Earth and the Moon were in the same plane, an eclipse would happen at every nodal passage, or about 25 times every year; but as such is not the case, eclipses are of less frequent occurrence. According to the most recent investigations, in order that an eclipse of the Sun may take place, the greatest possible distance of the Sun or Moon, from the *true place* of the nodes of the Moon's orbit, is 18° 36', whilst the latitude of the latter body must not exceed 1° 34' 52". If, however, it be less than 1° 23' 15", an eclipse must take place, though between these limits the occurrence of the eclipse at any one station is doubtful, and depends upon the horizontal parallaxes and semi-diameters of the two bodies at the moment of conjunction. In order that a lunar eclipse may take place, the remark we have just made will equally hold good, provided only that 12° 24', 63' 45" and 51' 57", be substituted for the numbers given above.

The nodes of the Moon's orbit are not stationary, but have

an annual retrograde motion of about 19°, so that a complete revolution round the ecliptic is accomplished in 18y. 219d., nearly. The Moon performs a revolution with respect to the node in 27d. 5h. 6m. This is termed a " *synodical revolution of the node*," and must not be confounded with the " *synodical revolution of the Moon*." It is shorter than the latter, because the retrograde motion of the node upon the ecliptic brings the Moon in contact with it, before she comes again into conjunction or opposition as the case may be. We must now refer to a singular effect produced by the retrocession of the nodes on the ecliptic. The Moon's synodical period, or the time she occupies in passing from one conjunction or opposition to another, is 29d. 12h. 14m. 2·87s.; 223 of these periods amount to 6585·32 days, but 19 synodical revolutions of the node are completed in 6585·78 days : the near coincidence of these two periods produces this obvious result, that eclipses recur in almost, though not quite, the same regular order after the completion of 19 synodical revolutions of the Moon's node.

It was probably a knowledge of this fact that enabled the ancient astronomers to predict the occurrence of a great eclipse, since it is quite certain they did so, in more than one instance, before the nature of eclipses was fully understood : this cycle was known to the Chaldæans as the *Saros*. Diogenes Laërtius records 373 solar and 832 lunar eclipses observed in Egypt; and although his testimony is, generally, not of any great value, yet it is very singular that this is just the proportion of solar and lunar eclipses visible above a given horizon within a certain period of time (1200—1300 years),—a coincidence which cannot be accidental.[1]

From what we have just said, it might be imagined that a correct list of eclipses for 18 years would be sufficient for all purposes of calculation; as by adding the ecliptic period as many times as required, the period of an eclipse might be known at any distance of time. This would be correct if any eclipse appeared under precisely the same cir-

---

[1] *Hist. of Ast.*, L.U.K. p. 15.

cumstances as its corresponding eclipse in the preceding or
following period: but such is not the case. An eclipse of
the Moon, which in the year 565 was of 6 digits[1], was in
the year 583 of 7 digits, and in 601 of nearly 8. In 908
the eclipse became total, and it remained so for about 12
periods, or until the year 1088: this eclipse continued to
diminish until the commencement of the 15th century, when
it totally disappeared in the year 1413. In a similar
manner, an eclipse of the Sun, which appeared at the North
Pole in June, 1295, proceeded more southerly at each
period. On June 27, 1367, it made its first appearance in
the north of Europe; in 1439, it was visible all over Europe;
at its 19th appearance in 1601, it was central in London;
on May 5, 1818, it was visible at London, and was again
nearly central at that place on May 15, 1836. At its
39th appearance, August 10, 1980, the Moon's shadow
will have passed the equator; and, as the eclipse will take
place near midnight, will be invisible in Europe, Africa, and
Asia. At every subsequent period the eclipse will go more
and more towards the south, until, finally, at its 78th appearance on September 30, 2665, it will go off at the south
pole of the Earth, and disappear altogether.

In the 18-year eclipse period, there usually happen 70
eclipses, of which 41 are solar, and 29 are lunar. In any
one year the greatest number that can occur is 7, and the
least, 2; in the former case 5 of them *may* be solar, and 2
lunar; in the latter both must be solar. Under no circumstances can there be more than 3 lunar eclipses in 1 year,
and in some years there are none at all. Though eclipses of
the Sun are more numerous than those of the Moon in the
proportion of 41 to 29 (say 3 to 2), yet at any given place
there are more lunar eclipses visible than solar; because, whilst

---

[1] A digit is the $\frac{1}{12}$ part of the surface of the Sun or Moon; and of
course an eclipse of 6 digits will be understood to be one in which $\frac{1}{2}$ the
disc of the luminary is hidden. In the case of a lunar eclipse, when
the magnitude is said to exceed 12 digits, it means that the Earth's
shadow extends itself so many digits beyond the Moon's surface.

the former are visible through an entire hemisphere, the visibility of the latter is confined to a narrow strip of the Earth, seldom more than 140 miles or so in breadth.

The reason is this: an eclipse of the Moon is visible to an entire hemisphere, on account of the large size of the Earth's shadow, and the proximity of the 2 bodies: on the other hand, an eclipse of the Sun is only visible to a limited portion of the Earth's surface, on account of the comparative small size of the Moon, and its great distance from the Sun. In a solar eclipse, the Moon's shadow traverses the Earth at the rate of 1830 miles an hour, or rather more than half a mile per second.

Counting from first to last, a solar eclipse may last 4h. 30m., but the interval of time during which the Sun will be centrally eclipsed is very limited. The duration of the total obscuration is greatest when the Moon is in perigee and the Sun in apogee; for the apparent diameter of the Moon being then the greatest possible, while that of the Sun is the least possible, the excess of the former over that of the latter, upon which the totality depends, has then attained its maximum. Now the perigean diameter of the Moon = 33′ 31″; the apogean diameter of the Sun = 31′ 30″.

$$\Delta: 33' \ 31'' - 31' \ 30'' = 2' \ 1''.$$

This, then, is the arc which, under these circumstances, the Moon describes during the continuance of the total phase. Taking into consideration the rapid motion of the Moon, it will then be seen that, under the most favourable circumstances, the Sun cannot remain totally eclipsed for more than a few minutes.

The duration of the obscuration in a total eclipse of the Sun varies, *cæteris paribus*, with the latitude of the place of observation, being greatest under the equator. Du Sèjour found that, under the most favourable circumstances, the greatest possible duration of the total phase, at the equator, was 7m. 58s., and that at the latitude of *Paris*, it was 6m. 10s.[1]

---

[1] *Mém. Acad. des Sciences*, 1777, p. 318.

The duration of an annular eclipse is greatest when the Moon is in apogee and the Sun in perigee, for then the apparent diameter of the Sun is the greatest, whilst that of the Moon is the least possible, and consequently the excess of the former over the latter, upon which the annulus depends, is then at a maximum.

The perigean diameter of the Sun $= 32'\ 35''$. The apogean diameter of the Moon $= 29'\ 22''$.

$$\Delta:\ 32'\ 35'' - 29'\ 22'' = 3'\ 13''.$$

This, then, is the arc described by the Moon during the continuance of the angular phase. Du Sèjour found by calculation, that the greatest possible duration of an annular eclipse, under the equator, was 12m. 24s.[1], and that at the latitude of *Paris* it was 9m. 56s.[2]

It may be desirable just briefly to point out the reasons why the greatest possible duration of an annular eclipse exceeds that of a total one. They are 2 in number: 1st. Because the excess of the perigean diameter of the Sun over the apogean diameter of the Moon ($= 3'\ 13''$) is greater than the excess of the perigean diameter of the Moon over the apogean diameter of the Sun ($= 2'\ 1''$): 2nd. Because the motion of the Moon in apogee is much slower than it is in perigee.

From the above remarks it will be readily understood that though so many solar eclipses happen from time to time, yet the occurrence of an annular or total one, at any particular locality, is a very rare phenomenon. Thus, according to Halley, no total eclipse had been observed at London between March 20, 1140, and April 22, 1715, though at different times during that interval the shadow of the Moon had frequently passed over other parts of Great Britain.[3] There will not be another visible in the Metropolis till 1916.

---

[1] *Mém. Acad. des Sciences,* 1777, p. 317.

[2] Ibid., p. 316.

[3] *Phil. Trans.,* vol. xxix. p. 245. 1715.

91

CHAPTER II.

ECLIPSES OF THE SUN.

Grandeur of a Total Eclipse of the Sun. — How regarded in Ancient Times.—Effects of the Progress of Science.—Chief Phenomena seen in connexion with Total Eclipses. — Change of the Colour of the Sky. — The Obscurity which prevails. — Effect noticed by Piola. — Physical Explanation.—Baily's Beads.—Extract from Baily's original Memoir. —Probably due to Irradiation.—Supposed to have been first noticed by Halley in 1715. — His Description. — The Corona. — Hypothesis advanced to explain its Origin. — Probably caused by an Atmosphere around the Sun. — Remarks by Grant. — First alluded to by Philostratus. — Then by Plutarch.— Corona visible during Annular Eclipses. — The Red Flames.—Remarks by Dawes. — Physical Cause unknown. —First mentioned by Stannyan.—Note by Flamsteed.— Observations of Vassenius.—Aspect presented by the Moon.—Remarks by Arago.

A TOTAL eclipse of the Sun is a most imposing spectacle, especially when viewed from the summit of a lofty mountain. Words can but inadequately describe the grandeur and magnificence of the scene. On all sides, indications are afforded that something unusual is taking place. At the moment of totality the darkness is sometimes so intense that the brighter stars and planets are seen, birds go to roost, flowers close, and the face of nature assumes an unearthly cadaverous hue; not the least striking thing is the sudden and frequently considerable fall that takes place in the temperature of the atmosphere, as the time of the greatest obscuration draws near.

"During the early history of Mankind, a total eclipse of the Sun was invariably regarded with a feeling of indescribable terror, as an indication of the anger of the offended Deity, or the presage of some impending calamity; and various instances are on record of the (supposed) extraordinary effects

produced by so unusual an event. In a more advanced state
of society, when science had begun to diffuse her genial
influence over the human mind, these vain apprehensions
gave place to juster and more ennobling views of nature; and
eclipses generally came to be looked upon as necessary consequences flowing from the uniform operation of fixed laws,
and differing from the ordinary phenomena of nature, only in
their less frequent occurrence. To astronomers they have in
all ages proved valuable in the highest degree, as tests of
great delicacy for ascertaining the accuracy of their calculations
relative to the place of the Moon, and hence deducing a
further improvement of the intricate theory of her movements.
In modern times, when the physical constitution of the celestial bodies has attracted the attention of many eminent
astronomers, observations of eclipses have disclosed several
interesting facts, which have thrown considerable light on
some important points of inquiry respecting the Sun and
Moon." [1]

Among the Hindùs a singular custom exists. When during
a solar eclipse, the black disc of our satellite is seen advancing
over the Sun, the natives believe that the jaws of some
monster are gradually eating it up. They then commence
beating gongs, and rending the air with the most discordant
screams of terror and shouts of vengeance. For a time, their
efforts are productive of no good result—the eclipse still pro-
gresses. At length, however, the terrific uproar has the
desired effect on the voracious monster; it appears to pause,
and then, like a fish that has nearly swallowed a bait and then
rejects it, it gradually disgorges the fiery mouthful. When
the Sun is quite clear of the great dragon's mouth, a shout of
joy is raised, and the poor natives disperse, extremely self-
satisfied on account of their having (as they suppose) so successfully relieved their deity from his late perils. Times
have now happily altered. We no longer look on a total
eclipse of the Sun as a dire calamity, but merely as one of the

[1] Grant, *Hist. Phys. Ast.*, p. 359.

ordinary effects resulting from the due working of those laws
by which a Divine Being wills to govern the universe.

An eclipse of the Sun may be either *partial, annular,* or
total : it is partial when only a portion of the Moon's disc
intervenes between the Sun and the observer on the Earth ;
annular, when the Moon's apparent diameter is less than the
Sun's, so that when the former is projected on the latter, it is
not sufficiently large completely to cover it, — an annulus, or
ring of the Sun, being left unobscured ; and total when the
Moon's apparent diameter is greater than the Sun's, which is,
therefore, wholly obscured. In an annular eclipse, when the
centre of the Sun and Moon exactly coincide, it is said to be
central and annular—the Sun appearing, for an instant only,
as a brilliant ring of light around the dark body of the Moon.

We shall now proceed to describe the principal phenomena
which are usually witnessed in connexion with solar eclipses.
Not the least remarkable is the almost invariable change of
colour which the sky undergoes. Halley, in his account of
the eclipse of 1715, says : " When the eclipse was about 10
digits (that is when about $\frac{5}{6}$ of the solar diameter was
immersed), the face and colour of the sky began to change
from a perfect serene azure blue to a more dusky livid colour,
intermixed with a tinge of purple, and grew darker and
darker till the total immersion of the Sun." [1]

It has also been found that whilst the sky changes, colour
during the progress of an eclipse, similar effects are also
produced upon terrestrial objects. This seems to have been
noticed as far back as 840 A.D.[2] Kepler mentions that during
the solar eclipse which happened in the autumn of 1590, the
reapers in Styria noticed that everything had a yellow tinge.[3]
Similar effects have also been described in modern times.[4]

The darkness which prevails during a total eclipse of the
Sun is not usually so considerable as might be expected. It
is, however, subject to much variation. Ferrer, speaking of

---

[1] *Phil. Trans.,* vol. xxix. p. 247. 1715.
[2] *Ad Vitellionem Paralipomena,* p. 294.   [3] Ibid., p. 303.
[4] *Mem.* R.A.S., vol. xv. xxii ; *Annuaire,* 1846, p. 291, &c.

the eclipse of 1806, says, " that at the time of total obscuration as much light remained as was equal to that afforded by
a full Moon."[1] In general it has been found that the darkness
is sufficiently great to prevent persons from reading, though
exceptions to this rule have been known. The faint illumination which exists at the moment of the totality is due to
light reflected from those regions of the atmosphere which are
still exposed to the direct rays of the Sun. The corona
(presently to be described) also, no doubt, assists in producing
the effect. The degree of obscuration will also vary according as to whether the observer is or is not deeply immersed in
the lunar shadow, a fact first pointed out by Halley.[2]

In the case of the eclipse of 1842, it was remarked by
Piola at Lodi, and by O. Struve at Lipesk, that although
the obscurity was such that stars of magnitude 2 and 3
ought to have been visible, yet only those of magnitude 1
were actually seen.[3] M. Belli explains this curious fact
by reference to a physiological principle. He remarks that
during the short interval of total obscuration, the eye has
not sufficient time to recover from the dazzling effect of
the Sun's rays, and consequently is unable to take advantage
of the obscurity which actually prevails.[4] The suddenness
with which the light succeeds the darkness after a total
eclipse of the Sun is well known. Halley suggested 2 explanations of the phenomenon. 1st. That previously to the
total obscuration, the pupil of the eye might be very much
contracted by viewing the Sun, and, consequently the organ
of vision would be less likely to suffer by the effulgence of
the light than at the instant of emersion, when the pupil had
again expanded. 2nd. That, as the eastern margin of the
Moon, at which the Sun disappeared, had been exposed for a
fortnight to the direct action of the solar rays, the heat

---

[1] *Trans. Amer. Phil. Soc.*, vol. vi. p. 266.

[2] *Phil. Trans.*, vol. xxix. p. 250. 1715.

[3] *Giorn. dell' Ist. Lomb.*, vol. iv. p. 341; *Bibliothèque Universelle de Genève*, vol. xliv. p. 368.

[4] *Giorn. dell' Ist. Lomb.*, vol. iv. p. 341.

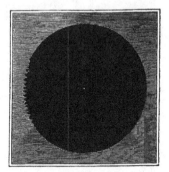

BAILY'S BEADS.

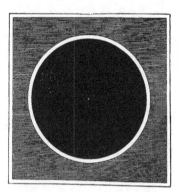

An ANNULAR ECLIPSE of the SUN.

SOLAR ECLIPSE PHENOMENA.

generated during this period might cause vapours to ascend in the lunar atmosphere, which, by their interposition between the Sun and the Earth, would have the effect of tempering the effulgence of the solar rays passing through them. On the other hand, the western margin of the Moon, at which the Sun reappeared, had just experienced a night of equal length, during which the vapours suspended in the lunar atmosphere had been undergoing a course of precipitation upon the Moon's surface under a process of cooling. In this case, therefore, the solar rays would meet with less obstruction in passing through the lunar atmosphere, and, consequently, it was reasonable to suppose they would produce a more intense effect.[1] The second hypothesis requires us to suppose the existence of a lunar atmosphere, which modern observation tends to prove does not exist. The first is doubtless the true explanation.

When the disc of the Moon advancing over that of the Sun has reduced the latter to a thin crescent, it is usually noticed that immediately before the beginning and after the end of complete obscuration, the crescent appears as a band of brilliant points, separated by dark spaces so as to give it the appearance of a string of beads. No satisfactory explanation has yet been given to account for this phenomenon, though the most probable hypothesis of their origin, is that which refers it to the effect of irradiation. It is often stated that the effects observed are due to the projection of some of the mountains of our satellite upon the solar disc. This explanation by no means satisfies us, though we are not in a position to offer a better one.

These phenomena are generally known as *Baily's beads*, having received their name from the late Mr. Francis Baily, who was the first to describe them in detail.[2] His original memoir was published about 25 years ago, and from it we make the following quotation :—

---

[1] *Phil. Trans.*, vol. xxix. p. 248. 1715.
[2] They were *noticed* long before his time.

"When [previous to the totality] the cusps of the Sun were about 40° asunder, a row of lucid points, like a string of bright beads, irregular in size and distance from each other, *suddenly* formed round that part of the circumference of the Moon that was about to enter, or which might be considered as having just entered, on the Sun's disc. Its formation indeed was so rapid, that it presented the appearance of having been caused by the ignition of a train of gunpowder. This I intended to note as the correct time of the formation of the annulus, expecting every moment to see the thread of light completed round the Moon, and attributing this serrated appearance of the Moon's limb (as others had done before me) to the lunar mountains, although the remaining portion of the Moon's circumference was comparatively smooth and circular, as seen through the telescope. My surprise, however, was great on finding that these luminous points, as well as the dark intervening spaces, increased in magnitude, some of the contiguous ones appearing to run into each other like drops of water; for the rapidity of the change was so great, and the singularity of the appearance so fascinating and attractive, that the mind was for the moment distracted and lost in the contemplation of the scene, so as to be unable to attend to every minute occurrence. Finally, as the Moon pursued her course, these dark intervening spaces (which, at their origin, had the appearance of lunar mountains in high relief, and which still continued attached to the Sun's border) were stretched out into long, black, thick parallel lines, forming the limbs of the Sun and Moon; when, all at once, they *suddenly* gave way, and left the circumferences of the Sun and Moon in those points, as in the rest, comparatively smooth and circular, and the Moon perceptibly advanced on the face of the Sun." [1]

Mr. Baily then goes on to describe the appearances which he saw after the total obscuration; they were, however, substantially the same as those recorded above.

The earliest account of the phenomenon of the beads is contained in Halley's Memoir on the total eclipse of 1715. He says: "About 2 minutes before the total immersion, the remaining part of the Sun was reduced to a very fine horn, whose extremities seemed to lose their acuteness, and *to become round like stars;* and, for the space of about a quarter of a minute, a small piece of the southern horn of the eclipse seemed to be *cut off from the rest* by a good interval, and

---

[1] *Mem.* R.A.S., vol. x. p. 6.

appeared like an oblong star rounded at both ends."[1] The first annular eclipse in which it appears that any beads were seen, was that of Feb. 18, 1736–7, observed by Maclaurin.[2]

One of the most interesting appearances seen during a total eclipse of the Sun is the *corona*, or halo of light which surrounds the Moon. It usually appears 3 or 4 seconds previous to the total extinction of the Sun's light, and continues visible for about the same interval of time after its re-appearance. In general, it may be compared to the *Gloria* commonly painted around the heads of the Virgin Mary, the Apostles, &c. Different explanations have been advanced to account for this phenomenon : Kepler thought it due to the presence of an atmosphere round the Moon[3] : La Hire suggested that it might be produced by the reflection of the solar rays from the inequalities of the Moon's surface, contiguous to the edge of her disc, combined with their subsequent passage through the Earth's atmosphere.[4] Professor B. Powell, some years ago, conducted a series of experiments which tended strongly to support the idea that refraction was the cause[5] : on the whole, however, the most probable theory is that which ascribes it to the presence of *an atmosphere about the Sun.* "Its round figure, its nebulous structure, and its gradually diminishing density onwards, are all favourable to the supposition of its being due to an elastic fluid encompassing the solar orb, and gravitating everywhere towards its centre. It is true that precisely similar results would ensue from the existence of an atmosphere about the Moon; but, in fact, there is no reason to suppose that the Moon possesses an atmosphere capable of producing an appreciable effect. On the other hand, the hypothesis of a solar atmosphere is not only warranted by

---

[1] *Phil. Trans.*, vol. xxix. p. 248. 1715.
[2] Ibid., vol. xl. p. 177. 1737.
[3] *Ad Vitell. Paralipom.*, p. 302; *Epit. Astron.*, p. 839.
[4] *Mém. Acad. des Sciences*, 1715, p. 161 *et seq.*
[5] *Mem. R.A.S.*, vol. xvi. p. 301 *et seq.*

the analogy of the other bodies of the planetary system, but is also supported by evidence of a positive nature, derived from observations on the physical constitution of the Sun. The changes presented by that body when viewed in a telescope, can only be consistently accounted for, by the supposition of two dissimilar envelopes of matter suspended in a transparent atmosphere at different altitudes above its surface." [1] Delisle conjectured that the luminous ring might be occasioned by the diffraction of the solar rays which pass near the Moon's edge.[2] Sir David Brewster has shown that this theory, though ingenious, is not tenable [3]; so we shall not refer to the subject further.

The earliest historical allusion to the corona is made by Philostratus. He mentions that the death of the Emperor Domitian had been announced previously by a total eclipse of the Sun. " In the heavens there appeared a prodigy of this nature. A certain *corona*, resembling the *Iris*, surrounded the orb of the Sun and obscured his light " [4] (*i.e.* it appeared coincidently with the total obscuration of his light.) Plutarch is still more precise in his allusion. Speaking of a total eclipse of the Sun which had recently happened, he endeavours to show why the darkness arising from such phenomena is not so profound as that of night. He begins by assuming, as the basis of his reasoning, that the Earth greatly exceeds the Moon in size, and, after citing some authorities, he goes on to say : — " Whence it happens that the Earth, on account of its magnitude, entirely conceals the Sun from our sight. . . . But even although the Moon should at any time *hide the whole of the Sun*, still the eclipse is deficient in diameter, as well as amplitude, for a peculiar effulgence is seen around the circumference, which does not allow a deep and very intense obscurity." (ἀλλὰ περιφαίνεται τις αὐγὴ περὶ τὴν ἴτον, οὐκ ἐῶσα βαθεῖαν γίνεσθαι τὴν σκιὰν καὶ ἄκρατον.) [5] The luminous ring seems to have been noticed by Clavius during the eclipse of April 9, 1567 : he thought it

---

[1] Grant, *Hist. Phys. Ast.*, p. 389.
[2] *Mém. Acad. des Sciences*, 1715, p. 166 *et seq.*
[3] *Edin. Encyc.*, art. Astronomy.   [4] *Vita Apollon. Tyan.*
[5] Plut. *Opera Mor. et Phil.*, vol. ix. p. 682.

was merely the uncovered margin of the Sun's disc; but Kepler showed that this was impossible.

There are one or two well-authenticated instances of the corona being visible during partial eclipses of the Sun. In 1842, M. D'Hombre Firmas, at Alais, which was contiguous to, though not actually in the path of the shadow, states that, "every one remarked the circle of pale light which encompassed the Moon when she almost covered the Sun." [1] Several observers of this eclipse noticed that the ring first appeared brighter on the side of the solar disc which was first covered by the Moon, but that previously to the close of the total phase, it was brightest at the part where the Sun was about to reappear.[2]

Not the least beautiful phenomena seen during a total solar eclipse, are the "Red Flames," which become visible around the margin of the Moon's disc immediately after the commencement of the total phase. Mr. Dawes, who saw them in July, 1851, has so minutely described them, that we cannot do better than quote his remarks in his own words. He says:—

"Throughout the whole of the quadrant from north to east there was no visible protuberance, the corona being uniform and uninterrupted. Between the east and south points, and at an angle of about 170° from the north point, appeared a large red prominence of a very regular conical form. When first seen it might be about 1½' in altitude from the edge of the Moon, but its length diminished as the Moon advanced.

"The position of this protuberance may be inaccurate to a few degrees, being more hastily noticed than the others. It was of a deep rose colour, and rather paler near the middle than at the edges.

"Proceeding southward, at about 145° from the north point, commenced a low ridge of red prominences, resembling in outline the tops of a very irregular range of hills. The highest of these probably did not exceed 40″. This ridge extended through 50° or 55°, and reached, therefore, to about 197° from the north point, its base being throughout formed by the sharply-defined edge of the Moon. The irregularities at the top of the ridge seemed to be permanent, but they certainly appeared to undulate from the west towards the east; probably an atmospheric phenomenon, as the wind was in the west.

"At about 220° commenced another low ridge of the same character,

[1] *Annuaire*, 1846, p. 339.   [2] *Mem.* R.A.S., vol. xv. p. 16.

and extending to about 250°, less elevated than the other, and also less irregular in outline, except that at about 225° a very remarkable protuberance rose from it to an altitude of $1\frac{1}{2}'$, or more. The tint of the low ridge was a rather pale pink; the colour of the more elevated prominence was decidedly deeper, and its brightness much more vivid. In form it resembled a *dog's tusk*, the convex side being northwards, and the concave to the south. The apex was somewhat acute. This protuberance, and the low ridge connected with it, were observed and estimated in height towards the end of the totality.

"A small double-pointed prominence was noticed at about 255°, and another low one with a broad base at about 263°. These were also of the rose-coloured tint, but rather paler than the large one at 225°.

"Almost directly preceding, or at 270°, appeared a bluntly triangular pink body, *suspended*, as it were, in the corona. This was separated from the Moon's edge when first seen, and the separation increased as the Moon advanced. It had the appearance of a large conical protuberance, whose base was hidden by some intervening soft and ill-defined substance, like the upper part of a conical mountain, the lower portion of which was obscured by clouds or thick mist. I think the apex of this object must have been at least 1' in altitude from the Moon's limb when first seen, and more than $1\frac{1}{2}'$ towards the end of total obscuration. Its colour was pink, and I thought it paler in the middle.

"To the north of this, at about 280° or 285°, appeared the most wonderful phenomenon of the whole. A red protuberance, of vivid brightness and very deep tint, arose to a height of, perhaps, $1\frac{1}{2}'$ when first seen, and increased in length to 2', or more, as the Moon's progress revealed it more completely. In shape it somewhat resembled a *Turkish cimeter*, the northern edge being convex, and the southern concave. Towards the apex it bent suddenly to the south, or upwards, as seen in the telescope. Its northern edge was well defined, and of a deeper colour than the rest, especially towards its base. I should call it a *rich carmine*. The southern edge was less distinctly defined, and decidedly paler. It gave me the impression of a somewhat conical protuberance, partly hidden on its southern side by some intervening substance of a soft or flocculent character. The apex of this protuberance was paler than the base, and of a purplish tinge, and it certainly had a flickering motion. Its base was, from first to last, sharply bounded by the edge of the Moon. To my great astonishment, this marvellous object *continued visible for about 5 seconds*, as nearly as I could judge, *after the Sun began to reappear*, which took place many degrees to the south of the situation it occupied on the Moon's circumference. It then rapidly faded away, *but it did not vanish instantaneously*. From its extraordinary size, curious form, deep colour, and vivid brightness, this pro-

tuberance absorbed much of my attention; and I am, therefore, unable
to state precisely what changes occurred in the other phenomena towards
the end of the total obscuration.

"The arc, from about 283° to the north point, was entirely free from
prominences, and also from any roseate tint."

It is difficult satisfactorily to assign any cause for these rose-
coloured emanations; but that their origin is to be referred to
the Sun rather than to the Moon, is, however, undoubted.
In the present state of our knowledge on the point, we can
only suspect that they are of a gaseous nature, and are, in
some manner or other, connected with the Sun's atmosphere.[1]

The earliest recorded account of the red flames is by
Captain Stannyan, who observed them at Berne, during the
total eclipse of 1706. He writes to Flamsteed:—

"That the Sun was totally darkened there for 4½ minutes of time;
that a fixed star and a planet appeared very bright; and *that his getting
out of his eclipse was preceded by a blood-red streak of light from its left
limb, which continued not longer than 6 or 7 seconds of time;* then part
of the Sun's disc appeared all of a sudden, as bright as Venus was ever
seen in the night; nay, brighter; and in that very instant gave a light
and shadow to things as strong as the Moon uses to do." [2]

On this communication Flamsteed remarks:—

"The captain is the first man I ever heard of, that took notice of a
red streak preceding the emersion of the Sun's body from a total eclipse.
And I take notice of it to you [the Royal Society], because it infers that
the Moon has an atmosphere; and its short continuance, if only 6 or 7
seconds' time, tells us that its height was not more than 5 or 6 hundredths
part of her diameter." [3]

The red flames were seen by Halley and Louville, in 1715,
and afterwards by Vassenius, at Göttenberg, who says:—

"But what seemed in the highest degree worthy, not merely of
observation, but also of the attention of the illustrious Royal Society,
were some reddish spots which appeared in the lunar atmosphere

---

[1] See Walker, *Essay on the Sun.* London, 1860. An able disserta-
tion on this question will be found in Grant's *History of Physical Astronomy*, p. 393 *et seq.*

[2] *Phil. Trans.*, vol. xxv. p. 2240. 1706.   [3] Ibid., p. 2241.

without the periphery of the Moon's disc, amounting to 3 or 4 in number, one of which was larger than the other, and occupied a situation about midway between the south and west. These spots seemed in each instance to be composed of 3 smaller parts or cloudy patches of unequal length, having a certain degree of obliquity to the periphery of the Moon. Having directed the attention of my companion to the phenomenon, who had the eyes of a lynx, I drew a sketch of its aspect; but while he, not being accustomed to the use of the telescope, was unable to find the Moon; I, again with great delight, perceived the same spot, or, if you choose, rather the invariable cloud occupying its former situation in the atmosphere near the Moon's periphery."[1]

The red flames have also been noticed in annular eclipses, as in that of 1737, observed by Maclaurin, which appears to be the earliest in which the phenomenon was seen[2], and in partial eclipses, of which that of 1605, observed by Kepler, is probably the first.[3]

The aspect presented by the Moon during eclipses of the Sun is frequently very singular. Kepler has stated that the Moon's surface is occasionally distinguishable by a ruddy hue which it possesses.[4] Mr. Baily, in his account of the annular eclipse of 1836, states, that " previous to the formation of the ring, the face of the Moon was perfectly black ; but on looking at it through the telescope, during the annulus, the circumference was tinged with a *reddish purple colour*, which extended over the whole disc, but increased in density of colour, according to the proximity to the centre, so as to be in that part nearly black."[5] Vassenius, in 1733, and Ferrer, in 1806, are the only observers who state that they have seen the irregularities in the Moon's surface during a central eclipse, whether total or annular.[6] Arago and others tried to do so in 1842, but failed. This fact of the lunar inequalities, being sometimes seen, and at other times not seen, is doubtless owing to meteorological causes.

In 1842, Arago saw the dark contour of the Moon projected

[1] *Phil. Trans.*, vol. xxxviii. p. 135. 1733.
[2] Ibid., vol. xl. p. 181. 1737.   [3] *De Stellâ Novâ*, p. 116.
[4] *Epit. Astron.*, p. 895.   [5] *Mem.* R.A.S., vol. x. p. 17.
[6] *Phil. Trans.*, vol. xxxviii. p. 135. 1733 ; *Trans. Amer. Phil. Soc.*, vol. vi. p. 267.

upon the bright sky 40m. after the commencement of the eclipse. He ascribes the phenomenon to the projection of the Moon upon the solar atmosphere, the brightness of which, by an effect of contrast, rendered the outline of the Moon's dark limb discernible.[1] On several occasions attempts have been made to detect the Moon's shadow in the course of its passage over the surface of the Earth. This, though a difficult matter, was done by Airy in 1851; he failed in 1842, though Plana and Forbes seem to have noticed it. The difficulty arises from the immense velocity of the shadow. Halley has calculated that it travels at the rate of 59 geog. miles per minute.[2] The earliest historical record of the eclipse-shadow being seen occurs in Duillier's account of the eclipse of 1706.[3]

---

[1] *Annuaire*, 1846, p. 372.
[2] *Phil. Trans.*, vol. xxix. p. 260. 1715.
[3] *Phil. Trans.*, vol. xxv. p. 2243. 1706.

CHAPTER III.

THE TOTAL ECLIPSE OF THE SUN OF JULY 28, 1851.

Observations by Airy.—By Hind.—By Lassell.

ONE of the most important total eclipses of the Sun. that
has occurred within the last few years was that of July 28,
1851. Though not visible in England, it was seen to great
advantage in Sweden, to which country astronomers flocked in
great numbers. From the numerous observations that were
made, and subsequently published, we select some extracts
which will doubtless interest the reader. The following remarks are from the pen of Mr. G. B. Airy, the Astronomer
Royal, who observed it at Göttenberg : —

" The approach of the totality was accompanied with that indescri-
bably mysterious and gloomy appearance of the whole surrounding
prospect, which I have seen on a former occasion. A patch of clear
blue sky in the zenith became purple-black while I was gazing at it. I
took off the higher power with which I had scrutinised the Sun, and
put on the lowest power (magnifying about 34 times.) With this
I saw the mountains on the Moon perfectly well. I watched carefully the approach of the Moon's limb to that of the Sun, which my
graduated dark glass enabled me to see in great perfection : I saw both
limbs perfectly well defined to the last, and saw the line becoming
narrower, and the curves becoming sharper, without any distortion or
prolongation of the limbs. I saw the Moon's serrated limb advance up
to the Sun's, and the light of the Sun glimmering through the hollows
between the mountain peaks, and saw these glimmering spots, extinguished one after another, in extremely rapid succession, but without
any of the appearances which Mr. Baily has described I
have no means of ascertaining whether the darkness really was greater
in the eclipse of 1842. I am inclined to think, that in the wonderful,
and I may say appalling obscurity, I saw the grey granite hills, within
sight of Hvaläs, more distinctly than the darker country surrounding

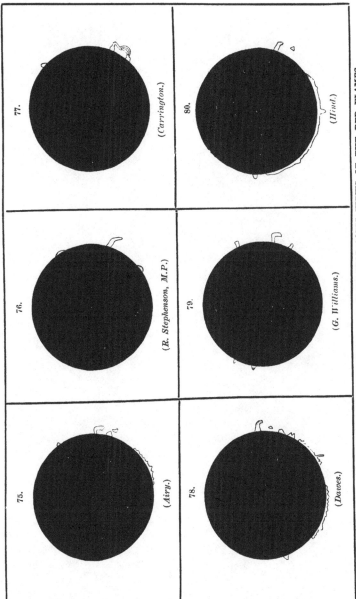

77.

(Carrington.)

80.

(Hind.)

76.

(R. Stephenson, M.P.)

79.

(G. Williams.)

75.

(Airy.)

78.

(Dawes.)

THE TOTAL ECLIPSE OF THE SUN OF JULY, 1851   TELESCOPIC VIEWS OF THE RED FLAMES.

the Superga. But whether, because in 1851 the sky was much less clouded than in 1842 (so that the transition was from a more luminous state of sky, to a darkness nearly equal in both cases), or from whatever cause, the suddenness of the darkness in 1851 appeared to me much more striking than in 1842. My friends who were on the upper rock, to which the path was very good, had great difficulty in descending. A candle had been lighted in a lantern, about a quarter of an hour before the totality; Mr. Haselgreen was unable to read the minutes of the chronometer's face, without having the lantern held close to the chronometer.

"The corona was far broader than that which I saw in 1842; roughly speaking, its breadth was a little less than the Moon's diameter, but its outline was very irregular. I did not remark any beams projecting from it, which deserved notice as much more conspicuous than the others; but the whole was beamy, radiated in structure, and terminated (though very indefinitely) in a way which reminded me of the ornament frequently placed round a mariner's compass. Its colour was white, and resembling that of *Venus*. I saw no flickering or unsteadiness of light. It was not separated from the Moon by any dark ring, nor had it any annular structure: it looked like a radiating luminous cloud behind the Moon . . . . The form of the prominences was most remarkable. One reminded me of a boomerang. Its colour, for at least two-thirds of its width, from the convexity to the concavity, was full lake red; the remainder was nearly white. The most brilliant part of it was the swell farthest from the Moon's limb; this was distinctly seen by my friends and myself with the naked eye. I did not measure its height; but judging generally by its proportion to the Moon's diameter, it must have been 3'. This estimation, perhaps, belongs to a later period of the eclipse. . . . . It was impossible to see the changes that took place in the prominences, without feeling the conviction that they belonged to the Sun, and not to the Moon.

"I again looked round, when I saw a scene of unexpected beauty. The southern part of the sky, as I have said, was covered with uniform white cloud; but in the northern part were detached clouds, upon a ground of clear sky. This clear sky was now strongly illuminated to the height of 30° or 35°, and through almost 90° of azimuth, with rosy red light shining through the intervals between the clouds. I went to the telescope, with the hope that I might be able to make the polarization-observation (which, as my apparatus was ready to my grasp, might have been done in 3 or 4 seconds), when I saw the *sierra*, or rugged line of projections, had arisen. This *sierra* was more brilliant than the other prominences, and its colour was nearly scarlet. The other prominences had perhaps increased in height, but no additional new ones had arisen. The appearance of this *sierra*, nearly in the

place where I expected the appearance of the Sun, warned me that I
ought not now to attempt any other physical observation. In a short
time, the white Sun burst forth, and the corona, and every other prominence, vanished.

"I withdrew from the telescope, and looked round. The country
seemed, though rapidly, yet half unwillingly, to be recovering its usual
cheerfulness. My eye, however, was caught by a duskiness in the
south-east, and I immediately perceived that it was the eclipse-shadow
in the air, travelling away in the direction of the shadow's path. For
at least 6 seconds, this shadow remained in sight, far more conspicuous
to the eye than I had anticipated."[1]

Mr. J. R. Hind watched the eclipse at Ravelsborg, near
Engelholm. He says:—

"The moment the Sun went out, the corona appeared; it was not
very bright, but this might arise from the interference of an extremely
light cloud of the *cirrus* class, which overspread the Sun at the time.
The corona was of the colour of tarnished silver, and its light seemed
to fluctuate considerably, though without any appearance of revolving.
Rays of light, the *aigrettes*, diverged from the Moon's limb in every
direction, and appeared to be shining through the light of the corona.
In the telescope many rose-coloured flames were noticed; one, far more
remarkable than the rest, on the western limb, could be distinguished
without any telescopic aid; it was curved near its extremity, and continued in view 4 *seconds after the Sun had disappeared,* i. e., after
the extinction of 'Baily's Beads,' which phenomena were very conspicuous in this eclipse, particularly before the commencement of the
totality. In this case, they were clearly to be attributed to the existence of many mountains and valleys along the Moon's edge, the Sun's
light shining through the valleys and between the mountain ridges, so
as to produce the appearance of luminous drops or beads, which continued visible some seconds. The colour of the 'flames' was a full rose
red at the borders, gradually fading off, towards the centres, to a very
pale pink. Along the southern limb of the Moon, for 40° or upwards,
there was a constant succession of very minute rose-coloured prominences, which appeared to be in a state of undulation, though without
undergoing any material change of form. An extremely fine line, of a
violet colour, separated these prominences from the dark limb of the
Moon. The surface of our satellite, during the total eclipse, was
purplish in the telescope; to the naked eye, it was by no means very

---

[1] *Mem.* R.A.S., vol. xxi. p. 8.

dark, but seemed to be faintly illuminated by a purplish grey light of uniform intensity, on every part of the surface.

"The aspect of nature during the total eclipse was grand beyond description. A diminution of light over the Earth was perceptible a quarter of an hour after the beginning of the eclipse; and about ten minutes before the extinction of the Sun, the gloom increased very perceptibly. The distant hills looked dull and misty, and the sea assumed a dusky appearance, like that it presents during rain; the daylight that remained had a yellowish tinge, and the azure blue of the sky deepened to a purplish violet hue, particularly towards the north. But notwithstanding these gradual changes, the observer could hardly be prepared for the wonderful spectacle that presented itself, when he withdrew his eye from the telescope, after the totality had come on, to gaze around him for a few seconds. The southern heavens were then of a uniform purple-grey colour, the only indications of the Sun's position being the luminous corona, the light of which contrasted strikingly with that of the surrounding sky. In the zenith and north of it, the heavens were of a purplish-violet, and appeared very near; while in the north-west and north-east, broad bands of yellowish crimson light, intensely bright, produced an effect which no person who witnessed it can ever forget. The crimson appeared to run over large portions of the sky in these directions, irrespective of the clouds. At higher altitudes, the predominant colour was purple. All nature seemed to be overshadowed by an unnatural gloom. The distant hills were hardly visible, the sea turned lurid red, and persons standing near the observer had a pale livid look, calculated to produce the most painful sensations. The darkness, if it can be so termed, had no resemblance to that of night. At various places within the shadow, the planets Venus, Mercury, and Mars, and the brighter stars of the first magnitude, were plainly seen during the total eclipse. Venus was distinctly seen at Copenhagen, though the eclipse was only partial in that city; and at Dantzic, she continued in view 10 minutes after the Sun had reappeared. Animals were frequently much affected. At Engelholm, a calf which commenced lowing violently as the gloom deepened, and lay down before the totality had commenced, went on feeding quietly enough, very soon after the return of daylight. Cocks crowed at Elsinborg, though the Sun was only hidden there 30 seconds, and the birds sought their resting-places, as if night had come on."[1]

Mr. W. Lassell, who was stationed near the Trollhätten Falls, thus describes the total obscuration : —

---

[1] *Sol. Syst.*, p. 71.

"I may attempt, but I cannot accomplish, an adequate description of the marvellous appearances, and their effect upon the mind, which were crowded into this small space of $3\frac{1}{3}$ minutes,—an interval which seemed to fly as if it were composed of seconds and not of minutes! Perhaps a naked eye observer would more fully grasp the awful effect of the sudden extinguishment of light,—the most overpowering of these appearances,—for, my eye being directed through the telescope, I must have been less, though sufficiently, struck with the unprecedented sensation of such instantaneous gloom. The amount of darkness may be appreciated from the fact that, on withdrawing my eye from the telescope, I could neither see the seconds hand of my watch, nor the paper sufficiently to write the time down; and was only able to do so by going to the candle, which I had by me burning on the table. Probably the suddenness of the gloom, not giving time for the expansion of the pupil of the eye, increased the sensation of apparent darkness; as I was obliged to repair close to the candle for the requisite light. After registering the time, I looked out for a few minutes with the naked eye over the landscape, north and south. The north was clear, and the line of horizon could be distinctly seen. The Sun, covered by the Moon, looking like a blue patch in the sky, had now the corona very symmetrically formed around it; but the Moon appeared to my unassisted eye to be not very round or smooth at its edge,— more as if one had rudely cut out with a knife on a board a circular disc of card,—the edges somewhat jagged and irregular in outline.

"The corona itself was perfectly concentric and radiating, some of the rays appearing in some parts of the circumference a *little* longer than in others; but the inequality was not great. I am unable to say whether the corona when *first* found was at all eccentric, for, as it is evident that any one observing with a telescope up to the moment of obscuration must have time to take off the dark glass before the corona can be seen, and as I had also to note the *time*, the centres of the Sun and Moon must have been pretty closely approximating before I again applied the eye to the telescope. It was indeed a great exercise of self-denial to spare the time from the exciting phenomena, which was necessary for accurately recording the duration of total darkness; but being inclined to think such record would be disregarded by many observers, I took my resolution to secure it."

The writer then proceeds to say that Venus was the only object visible to the naked eye. The corona he describes as "brilliant," and he considers that it afforded, speaking roughly, as much light as the Moon usually does when at its full.

"I had intended to direct my attention pointedly to the detection of the 'Red Flames,' which I had heard described as but faint phenomena. My surprise and astonishment may, therefore be well imagined when the view presented itself instantly to my eye as I am about to describe, or rather to attempt to give a notion of.

" In the middle of the field was the body of the Moon, rendered visible enough by the light of the corona around, attended by the apparent projections from behind the Moon of which I have attempted to sketch the positions. The effect upon my own mind of the awful grandeur of the spectacle I feel I cannot fully communicate. The prominences were of the most brilliant lake colour,—a splendid pink, quite defined and hard. They appeared to me to be not quiescent; but the Moon passing over them, and therefore exhibiting them in different phase, might convey an idea of motion. They are evidently to my senses belonging to the Sun and not at all to the Moon; for, especially on the western side of the Sun, I observed that the Moon passed over them, revealing successive portions of them as it advanced. In conformity with this observation also, I observed only the summit of *one*, on the eastern side, though my friends observing in adjoining rooms, had seen at least two: the time occupied by my noticing the time and observing with the naked eye not having allowed me to repair again to the telescope until the Moon had covered one, and three-fourths of the other. The point of the Sun's limit where the principal 'flame' appeared was (I judged) a few degrees south of the place where the cluster of spots was situated, and the flame which I observed on the eastern limb, was almost exactly where the eastern spot was situated. As, however, some prominences appeared adjacent to parts of the Sun's limit not usually traversed by spots, the attempt to trace a connection fails. The first burst of light from the emergent Sun was exactly in the place of the chief western flame, which it instantly extinguished . . . . . From the varying lengths of the red flames it is difficult to give an accurate estimation of their magnitude; but the extreme length of the largest, on the western limb, may have been about $2\frac{1}{2}'$. This estimation is rather rude, as I was so absorbed in contemplating their general phenomena that I had not time for exact measurement." [1]

---

[1] *Mem.* R.A.S., vol. xxi. p. 47.

CHAPTER IV.

THE ANNULAR ECLIPSE OF THE SUN OF MARCH 14-15, 1858.

Summary of Observations in England.

Of the different eclipses which have from time to time been
visible in England, few attracted such interest and attention
among all classes of society as that of March 14-15, 1858.
Though bad weather in most cases interrupted or altogether
prevented observations, yet many instructive features were
noticed, and therefore we feel that no apology is due for de-
voting a chapter to their consideration.

The line of central and annular eclipse passed across England from Lyme Regis, in Dorsetshire, to the Wash, between
Lincolnshire and Norfolk, traversing portions of the counties
of Somersetshire, Wiltshire, Berkshire, Oxfordshire, and
Northamptonshire. The following summary of the many and
various observations made, drawn up by Mr. Glaisher, will not
fail to be read with interest : —

"From returns received between Braemar and the Channel Islands,
from 30 to 40 in number, it is shown that the depression of temperature
during the eclipse was about $2\frac{1}{2}°$ at stations north of the line, and
nearly 3° at stations on and south of the line of central eclipse; that
at places where the usual diurnal increase had taken place in the morning the depression of temperature during the eclipse was greater; and
that at places where such increase had not taken place it was less than
the above numbers. Also that at places where the sky was uniformly
cloudy during the day the decrease in the readings of a black bulb
thermometer was less than 12°, while at places where the sky was
partially clear the depression was from 17° to 19°, and that, what
temperature soever the black bulb thermometer indicated in the morning, it fell during the eclipse to that of the temperature of the air at all
places.

" The humidity of the air was such that at places north of the line the wet bulb thermometer read 2·6° less; and on and near the line the depression was 3·2°, and south of it was 3·7° below the adjacent dry bulb thermometer.

" At some places the humidity of the air increased at the time of the greatest eclipse, but this was far from being universal.

" The sky was partially clear at some places on the east and south coasts, in the Channel Islands and north of Scotland, and it was for the most part overcast elsewhere. Near the southern extremity of the central line the sky was partially clear, and at its northern extremity near Peterborough the clouds were broken; at most intermediate places the sky was wholly overcast. The complete ring was seen at Charmouth, and neighbourhood near Lyme Regis, and at Peterborough, but so far as I can learn at no other places. My own station was on the calculated line of central eclipse, near Oundle, in Northamptonshire, and here I saw the Moon and Sun's apparent upper limb coincident, or very nearly so, and therefore that I was situated on or very near the northern limit of annularity, but distant from the centre line by 3 or 4 miles.

" It is very much to be regretted that the unfavourable weather precluded the witnessing the very beautiful attendant phenomena upon large solar eclipses. The time of year was unfavourable to all optical effects—whether of light and shade or colour, independently of the particular character of the day, which was more fatal still to their exhibition, for even where the Sun was visible their presence was only feebly indicated at a few parts of the country.

" At Oundle the weather for some time previous to the commencement of the eclipse was raw and ungenial for the time of year. The wind was gusty and the sky overcast, chiefly with cirro-stratus, and dark scud hurrying past the Sun's place from the north-west, the clouds occasionally giving way and allowing the Sun to be visible by glimpses. Shortly after 1 o'clock the sky became uniformly overcast, and a small steady rain set in for a considerable time.

" It was long before any sensible diminution of light took place. At 12·39 a gloom was for the first time perceptible to the north, and the crescent of the Sun shone out with a bright white light between breaks. At 0·43 the gloom was general, excepting around the Sun, which appeared the centre of a circle of light, and illuminated with fine effect some bold irregular masses of cumulus in its vicinity. At 0·45 the gloom increased, slight rain fell, and the wind rose, birds were heard chirping and calling. At 0·53. a severe storm might have been supposed impending, and numerous birds were flying homewards. The deepening of the gloom was gradual but very slow, and between 1h. and 1h. 1m. was at its greatest intensity; but even at this time the obscurity was not sufficient to require that any employment should be suspended. Messrs. Adams and Symons, situated 5 feet from a shed in an adjoining brick-field, spoke of the gloom as very intense for a period of 10 seconds, and sufficient to render it difficult to take the readings of the thermometer. A body of rooks rose from the ground at this moment, and flew homewards; a flock of starlings rose together, and various small birds flew wildly about; a hare was seen to run across a neighbouring field, as though it were daybreak; straw rustled, and the silence was peculiar and intense. The darkness and lull was that of an approaching thunder-storm. Directly after the greatest intensity the gloom was sensibly and instantaneously diminished, and the day was speedily restored to its ordinary appearance.

"After 0·50 the lark ceased to rise, and did not sing; at 1h. 10m. it rose again. The collected information tends to show that birds and animals, but particularly the former, were affected in some degree in most places; and that it is probable to suppose the gloom was referred by them to the approach of evening, and this not so much from the fact of the gloom as from the manner of its approach, without the accompanying signs of atmospheric disturbance which usher in a storm, and to which birds and animals are keenly sensitive.

"All over the country rooks seemed to have returned to their rookeries during the greatest obscuration; starlings were seen in many places taking flight, whole flocks of them together. At Oxford Dr. Collingwood remarked that a thrush commenced its evening song. At Grantham pigeons returned to their cote. At Ventnor Dr. Martin notes the fact that a fish confined in an aquarium, and ordinarily visible at evening only, was in full activity about the time of the greatest gloom. In Greenwich Park the birds were hushed and flew low from bush to bush, and at nearly all places the song of many birds was suspended during the darkness. At Camden Hill it was observed that the crocus closed about the same time, and at Teignmouth that its colour changed to that of the pink hepatica.

"The darkness was not sufficient at any place to prevent moderate-sized print being read at any convenient distance from the eye out of doors, but a difficulty was sometimes experienced in reading the instruments. At Grantham the darkness is described to have been about equal to the usual amount of light an hour before sunrise; near Oxford as about equal to that just after sunset on a cloudy day. The general impression communicated was that of an approaching thunder-storm. The sudden clearing up of the gloom after the greatest phase was likened by more than one observer to the gradual, but somewhat rapid withdrawal of a curtain from the window of a darkened room. The darkness is described to have been generally attended by a sensation of chilliness and moisture in the air. At Oxford the clouds surrounding the sun were beautifully tinted with red, which merged into purple as the obscuration increased. At Grantham as the eclipse progressed the light became of a decided grey cast, similar to that of early morning, but at the time of the greatest gloom it had a strong yellow tinge. At Teignmouth the diminution of light was very great; the sombre tints of the clouds became much deepened, and the remaining light thrown over the landscape was lurid and unnatural. At Greenwich the appearance of the landscape changed from a dull white to a leaden, and then to a slate-coloured hue; and as the darkness increased it had much the appearance of a November fog closing in on all sides. At Wakefield the tints of the clouds changed from the grey slate colour of clouds in a storm, and became of a purple hue. At Oundle, my own station, the clouds were one uniform leaden grey or slate colour, and quite in accordance with the general character of the day, nor could I perceive that the clouds appeared lower, or, in fact, that there was any very noticeable departure from the gloom we constantly experience during dull winter weather. Throughout the eclipse it occurred to me that the illuminating power of the Sun was much more than might have been supposed commensurate with the unobscured portion of the disc. When casual breaks permitted it to be visible the illuminated crescent up to the time of the greatest phase emitted beams of considerable brilliancy, which marked out a luminous track in the gloom, and were clearly and well defined in extent and figure. As the eclipse proceeded a decided change was to be observed in the colour of the Sun itself, which became of a pure silvery brightness, like that of Venus after inferior conjunction with the Sun. The absence of all colour in the light was remarkable, and at the time when the annulus was nearly formed it appeared like a line of silver wire. The departure from the usual amount of light we are accustomed to receive on an ordinary dull day during the greater part of the eclipse was so inconsiderable, that we might infer approximately the real amount of Sun our average daylight under a cloudy sky is equivalent to.

"As a photometric test during the eclipse, strips of photographic paper were exposed for equal intervals of time every 5 minutes. The result was a scale of tints which exhibited clearly the diminishing intensity of the light up to the period of greatest obscuration, and the rapid increase beyond. The range of tints is low owing to the cloudy state of the sky, but this does not interfere with the proportionate depths of tint; the time of greatest darkness is distinctly shown by the very feeble discolouration of the paper. The instruments used at Oundle were made specially for those observations, and were of a very delicate and accurate construction; the meteorological observations were made by Messrs. Adams and Symons.

"In conclusion, I beg sincerely to thank those gentlemen whose

returns have supplied the *data* for this investigation, of which we may
say, literally as well as figuratively, that it exhibits only the faint
outline of facts dimly visible through a screen of clouds. I think, however, it is reasonable to infer that the great paucity of effects and general
phenomena witnessed even in places where the Sun was visible, is due
to the conditions of the atmosphere, attributable alike to climate, time of
year, and unfavourable weather, and should by no means lessen our con-
fidence in previous accounts of the grandeur and beauty of the attendant
phenomena upon solar eclipses. Optical phenomena, we all know, are
dependent upon the medium through which we view them for the
nature and power of the effects produced."

Fig. 81.   *Plate* XVII.

The TOTAL ECLIPSE of the SUN of JULY, 1860:
TELESCOPIC VIEW OF THE CORONA AND RED FLAMES,
DRAWN BY FEILITZSCH.

CHAPTER V.

THE TOTAL ECLIPSE OF THE SUN OF JULY 18, 1860.

Prefatory Remarks on the "Himalaya Expedition" organised by the Astronomer Royal. — Extracts from his Observations. — Observations of the Red Flames by Bruhns. — Meteorological Observations by Lowe.

OF the total eclipses which have of late years been systematically observed, that of July 18, 1860, is by far the most interesting and important : it owes its interest to the agreeable circumstances connected with it, hereafter to be more fully spoken of, and its importance, to the very extensive and refined observations which were made by so many astronomers in America, Europe, and Africa.

Our limits wholly forbid our entering into any very lengthened statement: we shall therefore select from the published accounts of the observations, such portions as we deem most fitted to be placed on record in a work like the present, prefacing them by a brief epitome of the general circumstances attending "the Himalaya Expedition."

On Nov. 15, 1859, the Astronomer Royal, in an interview with the Duke of Somerset, First Lord of the Admiralty, drew his Grace's attention to the then approaching eclipse, at the same time suggesting the desirability of a ship being appropriated for the conveyance of observers to and from the coast of Spain. After the request was duly considered, Her Majesty's Government volunteered to place at the disposal of the Astronomer Royal and his friends, H.M.S. " Himalaya "; the offer was of course gratefully accepted, and in due course of time the expedition set forth.

Having received from the Admiralty an intimation of their willingness to supply the transport, Mr. Airy next applied to the Foreign Secretary, requesting him to memorialise the Spanish Government, so far to relax their Police and Customs Regulations, as to permit the free introduction into Spain of the various astronomical instruments required by the expeditionary party, — a request which the Spanish Government promptly expressed themselves willing to comply with.

In the meantime communications were received from Mr. C. Vignoles, C.E., Engineer-in-chief of the Tudela and Bilbao Railway, and from Mr. P. E. Sewell, C.E., who held a similar post in connexion with the Isabella Railway, extending southwards from Santander, offering their advice and assistance. With all these favourable concurrent circumstances before him, the Astronomer Royal had no difficulty in determining that on their arrival off Spain, the party should divide,—one portion landing at Bilbao, the other, in case the weather should prove unfavourable at Bilbao, proceeding on to try their chance at Santander. This arrangement was ultimately adopted, except that the Bilbao party were strongly recommended not to remain in that town for fear of a cloudy sky, but to pass overland through the Cantabrian Pyrenees into the Valley of the Ebro.

All the necessary preparations having been made, the "Himalaya" steamed majestically out of Plymouth Sound, on the morning of July 7, soon after 10 A.M. After passing Earne Head, England was soon lost sight of, and no more land seen. Cape Machicaco, on the Spanish coast, was sighted at 4 A.M. on July 9. From sunrise till noon on Sunday, July 8, a light gale prevailed, causing considerable discomfort to many of the passengers, so that when Divine Service was conducted by the Rev. W. R. Almond, but few of them were present. At 7 A.M., on the 9th, anchor was cast, outside the bar of Bilbao, and after breakfast the Bilbao party landed. It comprised the following persons :—The Astronomer Royal, and Mr. W. Airy ; the Rev. H. A. S. Atwood; Mr. J. Beck, and his assistant, Mr. W. Beck ; Mr. J. Bonomi ; Mr. W. De La Rue, and his assistants, Messrs. E. Beck, Beckley, Downes, and Reynolds;

Mr. F. Galton; Professor Grant; Mr. C. Gray; Capt. W. Jacob; Dr. M'Taggart; the Rev. J. S. Perowne; Mr. J. G. Perry; Professor Pole; Rev. C. Pritchard, and his assistants, Messrs. Fazel and Wright; Mr. R. Scott; M. O. Struve, and his assistants, MM. Oöm and Winnecke; the Rev. O. Vignoles, together with Mrs. G. B. Airy, Miss H. Airy, Mdlle. Struve, and Mrs. O. Vignoles. At noon, the "Himalaya" proceeded on her voyage to Santander, at which place, on July 10, she landed the following, forming party No. 2:—Mr. J. Buckingham, and his assistant, Mr. W. Wray; Mr. H.S. Ellis; Professor Fearnley; the Rev. H. A. Godwin; Mr. R. F. Heath, and his assistant, Mr. J. Turner; Mr. R. J. Hobbes; Mr. W. Lassell; M. Lindelöf; Professor Lindhagen; Mr. E. J. Lowe, and his assistants, the Rev. W. R. Almond, and Mr. S. Morley; Dr. Möller, Mr. Stanistreet, and Professor Swan. Both parties met with the most cordial reception alike from the English engineers employed out there, as well as from the "natives." What they and the other continental astronomers out there saw, will form the subject of the remainder of this chapter.

The Astronomer Royal, accompanied by the three above-named members of his family, stationed himself at the village of Pobes. From his memoir[1] we make the following extracts:—

"On the progress of the eclipse I have nothing to remark, except that I thought the singular darkening of the landscape, whose character is peculiar to an eclipse, to be sadder than usual. The cause of this peculiar character I conceive to be the diminution of light in the higher strata of the air. When the Sun is heavily clouded, still the upper atmosphere is brilliantly illuminated, and the diffused light which comes from it is agreeable to the eye. But when the Sun is partially eclipsed, the illumination of the atmosphere for many miles round is also diminished, and the eye is oppressed by the absence of the light which usually comes from it.

"I had a wax candle lighted in a lantern, as I have had at preceding total eclipses. Correcting the appreciations of my eye by reference to this, I found that the darkness of the approaching totality was much less striking than in the eclipses of 1842 and 1851. In my anxiety to lose nothing at the telescope I did not see the approach of the dark shadow

---

[1] *Month. Not.* R.A.S., vol. xxi. p. 9, *et seq.*

through the air; but, from what I afterwards saw of its retreat, I am sure it must have been very awful."

After describing the red flames [1], he says : —

"I may take this opportunity of stating, that the colour of these appearances was not identical with that which I saw in 1842 and 1851. The *quality* of the colour was precisely the same (full blush-red, or nearly lake), but it was diluted with white, and more diluted at the roots of the prominences close to the Moon's limb than in the most elevated points.

"About the middle of the totality I ceased for awhile my measures, in order to view the prospect with the naked eye. The general light appeared to me much greater than in the eclipses of 1842 and 1851, (one cloudy, the other hazy), perhaps 10 times as great; I believe I could have read a chronometer at the distance of 12 inches; nevertheless, it was not easy to walk where the ground was in the least uneven, and much attention to the footing was necessary. The outlines of the mountains were clear, but all distances were totally lost; they were in fact an undivided mass of black to within a small distance of the spectator. Above these, to the height perhaps of 6° or 8°, and especially remarkable on the north side, was a brilliant yellow or orange sky, without any trace of the lovely blush which I saw in 1851. Higher still, the sky was moderately dark, but not so dark as in former eclipses. The corona gave a considerable body, but I did not remark either by eye view or by telescope view anything annular in its·structure; it appeared to me to resemble, with some irregularities (as I stated in 1851), the ornament round a compass card. But the thing which struck me most was the great brilliancy of Jupiter and Procyon so near the Sun. It was impossible that they could have been seen at all, except under the circumstance of total absence of illumination on that part of the atmosphere through which the light passed. I returned to my measures, but I was soon surprised by the appearance of the scarlet sierra, announcing the approach of the Sun's limb. It disappointed me, for I had reckoned on a much longer time. All our party who were aware of the predicted duration fully believed that it must have been very erroneous. How the time passed I cannot tell. The Sun at length appeared, extinguishing the sierra, but the prominence and cloud remained visible, and my last measures were taken after reappearance. The prominences, &c. were then rapidly fading, and I quitted the teles-

---

[1] The description is less detailed than that given by some of the other observers, and therefore we pass it over. Mr. Airy nowhere makes mention of any "Baily Beads."

THE TOTAL ECLIPSE OF THE SUN JULY 1860

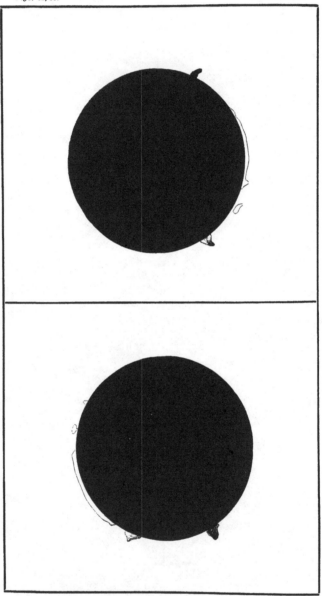

The TOTAL ECLIPSE of the SUN of JULY, 1860:
TELESCOPIC VIEWS OF THE RED FLAMES, DRAWN BY BRUHNS.

cope, not without the feeling that I had not done all that I had intended
or hoped to do.

"My companions saw 9 celestial objects, 8 of which were identified by means of Mr. Hind's chart, and the 9th by our knowledge of its position. They appeared to be (using Mr. Hind's numbers
of reference) 2 *Regulus*, 3 *Saturn*, 4 *Mercury*, 6 *Procyon*, 7 *Jupiter*,
8 *Venus*, 9 *Pollux* or 11 *Castor* (not quite certain), 13 *Capella* and
Arcturus.   *Sirius* and a *Orionis* were hidden by clouds."

The red flames were seen, and are described by many of the
observers; of the accounts which have come under our notice,
that drawn up by M. Bruhns, of Leipzic, seems the most
complete and methodical.[1]

He says : —

" Just before the totality there was visible, on the western border of
the Moon, only one protuberance and the corona ; but as the last rays of
the Sun disappeared, more protuberances started out on the eastern side,
and the corona shone forth with an intense white light, so lustrous in
fact as to dim the protuberances. I remarked that I saw them better
when a clear red glass was held before my eye.

"During the totality I sketched 4 drawings, and also measured
off the position-angles of the different protuberances, counting round
the circle from the north point through the east, &c.

"The figure marked [Fig. 82, Pl. XVIII.] was drawn during the first
minute of the totality. The first protuberance is the one already mentioned ; its position-angle was 35°, the length of its base $1\frac{1}{2}'$ or 2', and
its height about the same. The summit was somewhat curved, of an intense rose colour, but a little paler at the apex.

"The second protuberance situated at 60°, was completely separated from the Moon, there being between them an interval of $\frac{1}{2}'$. For
part of its extent it was parallel to the Moon's border, it then deviated
from it, and ended in a point. Its length was $1\frac{1}{2}'$ or 2', its height about
$\frac{1}{2}'$, and of a rose colour.

" The third protuberance, having a position-angle of 75°, resembled
a mountain, and had a base of $1\frac{1}{2}'$, and a height of fully $\frac{1}{2}'$. Extending
inwards for 50° from this protuberance was a narrow fringe, first of a
pale red, but a few seconds afterwards it came over a splendid rose
colour, and of a height of about $\frac{1}{2}'$, which soon narrowed as the Moon
passed over it, until at length it was quite covered.

" A fourth protuberance existed at 155°; its base was not more than

[1] *Ast. Nach.*, No. 1292, Jan. 22, 1861.

$\frac{3}{4}'$, but the height was as much as $1\frac{1}{2}'$. It had a hooked form with the curve trending northwards, and likewise of a rose colour.

"During no part of the totality were there any protuberances visible in the southern part of the Sun's disc.

"In the second minute the above-described protuberances became gradually smaller; the first, however, retained its magnitude and figure almost unchanged.[1]   The above-described unattached protuberance [No. 2] was reached by the Moon, and became gradually covered. By the end of the second minute the fringe was entirely covered, and at this juncture, on turning to examine the western border of the Moon, I perceived several protuberances, not previously visible.

"The protuberance situated at 260°, which I will call No. 5, had, at the beginning of the second [third ?] minute, only a base of $\frac{3}{4}'$, and about the same height, the colour being rose.

"Between 270° and 300° extended a second streak about $\frac{1}{4}'$ in height.

"A sixth protuberance was visible at 310°, having a base of 2', and a height of $2\frac{3}{4}'$.

"Lastly, I found at 340° a seventh protuberance, having a base of 1', and a height of $\frac{3}{4}'$, and of a rose colour, like all the preceding.

"On directing my attention to the first protuberance (the one at 35°), I fancied it had grown considerably larger. The sharp edge, first seen, had disappeared, and for a height of 3' or 4' flaming rays could be discerned, the colour (at the base a bright rose) was, at the top, hardly perceptible, but seemed to fade off and become merged in the corona.

"After I had observed these for about half a minute, without perceiving any alteration, I quitted the telescope to observe the corona and the sky for a short time with the naked eye. The black-looking Moon was surrounded by a crown of clear light of unequal breadth. Below [S.] it was considerably greater than above [N.]. I estimated that in the former case it was $\frac{3}{8}$°, in the latter about $\frac{1}{4}$°, and the general appearance of the thing gave me the idea that the Moon was eccentrically placed in the corona."

"The general form of the corona appeared circular, but on the eastern side a long ray shot out to a distance of about 1°; the breadth of its base was 3', but it tapered down to about $1\frac{1}{2}'$. During the 10 seconds that my attention was directed to it, neither the direction nor the length of the ray altered; its light was considerably feebler than that of the corona, which was of a glowing white, and seemed to coruscate or twinkle.

"With the naked eye I easily saw Venus and Jupiter, the former being much brighter than the latter. Although I knew whereabouts

---

[1] The sense in the original German, is here ambiguous; we translate literally.

Procyon, Castor, Pollux, Mercury, and Saturn were, yet in the few seconds available for seeking for them I failed to find them.

"My assistant, M. Auerbach, who observed the corona, and searched for the stars, during a longer period than I did, noticed in the south-western quadrant a curved ray about $\frac{1}{10}^\circ$ in length, which I in my hurry probably overlooked. He also saw Pollux, and another person saw Castor, but, as far as I am aware, no more than the above 4 objects were seen by any person in Tarragona.

"Towards the end of the 3rd minute of the totality, I again looked through the telescope, and made the drawing [Fig. 83, Pl. XVIII]. The western protuberances had altered considerably since the 2nd minute; the one at 35° had regained its original form and size, the flaming rays, previously spoken of, having disappeared. The protuberance in 340° had become much larger, the length of its base being now about 2′, and the height 1½′. The red streak extending from 270° to 300° had prolonged itself so as to take in the protuberance at 310° [No. 6], and had altogether now a length of 50°, its height having also become augmented from ¼′ to 1′, and its colour being an intense rose. The protuberance at 260° [No. 5] was now separated by about ¼′ from the Moon, its breadth being nearly 1′, and its height ½′. Finally at 240° a new and small protuberance had started into view, its base and height were both about ½′, and rose-coloured.

"As the end of the totality advanced so the protuberances became less distinct, the colour became brighter, and immediately after the 3rd minute of totality the protuberances at 240° and 260° disappeared; the fringe extended itself to a length of more than 90°, its height being 1½′, and embraced all the protuberances up to an angle of 35°. On the first appearance of the solar rays all suddenly vanished, with the exception of the first protuberance, which for some time afterwards remained visible in the thin red glass."

Meteorology found an able representative in Spain in the person of Mr. E. J. Lowe, who stationed himself at Fuente del Mar, near Santander, and who with his assistants, the Rev. W. R. Almond and Mr. S. Morley, during a period of 5 hours, on the memorable 18th of July, made upwards of 4000 observations, an abstract of which, in his own words, we now subjoin:—

"Commencing with underground temperature, a thermometer placed 6 inches below the surface of the ground ranged between 67·9° and 70·7°, i.e. 2·8°; at this depth the eclipse was not sensibly felt, whereas other thermometers, placed 4 inches, 2 inches, 1 inch, and ½ an inch below the surface, all exhibited in a very marked manner the effect of the eclipse. On the grass the temperature fell to 64° at 3h. 5m.; at ½ an inch below the surface, to 69° at 3h. 15m.; at 1 inch deep to 69·5° at 3h. 25m.; at 2 inches, to 71° at 3h. 55m.; at 4 inches, to 70·7° at 4h. 30m. P. M.

"The temperature on the grass was 77·5° at noon, rising to 91·7° at 1h. 50m., and then falling till 3h. 5m., and again rising to 85° at 4h. 10m., giving a range of 27·7°. At half an inch below the surface of the ground the temperature rose till 1h. 55m. P. M., when it was 78·5°, and then gradually fell to 69°, rising again to 74·7° at 4h. 30m. P. M., the range being 9·5° At 1 inch below the surface the temperature rose till 1h. 55m. to 76·2°, fell till 3h. 25m. to 69·5°, and rose till 4h. 55m. to 74·7°, the range being 6·7°. At 2 inches below the surface the temperature rose till 2h. 5m. to 74·4°, then fell till 3h. 55m. to 71·0°, and afterwards rose till 4h. 55m. to 73·7°, the range being 3·4; and at 4 inches below the surface the temperature rose till 2h. 50m. to 73°, then fell till 4h. 30m. to 70·7°, and again rose till 6h. P. M. to 73·2°, the range being 2·5°.

"The greatest cold on the ground occurred between 3h. and 3h. 5m. P. M.; ditto, ½ an inch below surface, 3h. 10m. and 3h. 15m. P. M.; ditto, 1 inch, 3·20, and 3h. 25m. P. M.; ditto, 2 inches, 3h. 50m., and 3h. 55m. P. M.; ditto, 4 inches, 4h. 25m. and 4h. 30m. P. M.

TABLE OF TEMPERATURES.

	Commencement of Eclipse.	Middle of Eclipse.	End of Eclipse.	Range During Eclipse.
	°	°	°	°
Of a blackened ball on grass . .	104·0	65·5	94·0	38·5
Of a blackened ball in vacuo . .	131·0	66·0	104·0	65·0
In sunshine at 2 feet above ground .	75·5	63·6	70·0	11·9
In sunshine 2 feet (wet bulb) . .	69·5	59·3	65·5	10·2
Diff. between dry and wet bulb at 2 ft.	6·0	4·4	4·5	1·6
In shade at 4 feet	70·0	64·7	71·0	6·3
In shade at 4 feet (wet bulb) . .	62·5	59·7	63·5	3·8
In shade at 3 feet	70·2	64·2	70·7	6·5
In shade at 2 feet	68·5	62·5	68·5	6·0
In shade at 1 foot	70·7	64·5	70·2	5·7

"The barometer rose from 1h. 40m. till 2h. 10m. 0·002 inch, then fell till 3h. 5m. 6·017 inch, and rose till end of eclipse, 0·009 inch.

"Intensity of photographic light, from salted papers conveyed, sensilised, in Marion's dark box, exposed for 10 seconds (with a scale of from 0 to 5°), at the commencement of the eclipse, $4\frac{1}{2}$° becoming 4° at 2h. 5m., 3° at 2h. 15m., 2° at 2h. 25m., 1° at 2h. 40m., o$\frac{3}{4}$ at 2h. 50m. 1° at 2h. 55m. (clear about Sun); o$\frac{1}{4}$° at 3h., 1° at 3h. 5m., 2° at 3h. 25m., $2\frac{1}{2}$° at 3h. 40m., 3° at 3h. 50m., and 4° at 4h. During totality a paper exposed for 1 minute gave o$\frac{3}{4}$°.

"The wind was N.W. and N.N.W. till 4h. 20m. then W.S.W., becoming S.W. at 4h. 25m., and south at 4h. 45m. The wind was brisk at the commencement of the eclipse, quite a calm during totality, and a gentle breeze afterwards. The distant prospect was very clear, except during totality, when the mountains disappeared, and only near objects were visible.

"The clouds, which were chiefly cumuli, diminished in amount till 1h. 50m., when only $\frac{4}{10}$ of the sky was overcast, then increased till 2h. 35m. with much cloud till 3h. 55m., then again diminished to $\frac{6}{10}$ at the termination of the eclipse, the range being $\frac{5 \cdot 5}{10}$ of the whole sky. Towards totality some of the cumuli became scud, which lasted from 2h. 5m. to 3h. 10m., giving the strongest impression that the change was due to the eclipse.

"The morning was fine, and from 12h. 45m. P. M. sunshine; at 1h. 25m. much open sky about the zenith; at 2h. 15m. a blackness about W. horizon, and slightly so in N. and S. ; at 2h. 30m. the hills dark, and the blue sky in N. and E., very pale in colour; at 2h. 35m., hills dark, with a blue haze among the more distant mountains; at 2h. 40m. horizon due W. pink; at 2h. 45m., clear sky, in N. pink; at 2h. 52m., splendid pink in W. horizon, warm purple in summits of mountains in S., clear sky, in N. deep lilac, and in E. very pale blue; at 2h. 57m. rapid change, the clear sky in N. deep marine blue with a red line.

"Before totality commenced the colours in the sky and in the hills were magnificent beyond all description ; the clear sky in N. assumed a deep indigo colour, while in the W. the horizon was pitch black (like night). In the E. the clear sky was very pale blue, with orange and red, like sunrise, and the hills in S. were very red ; on the shadow sweeping across, the deep blue in N. changed like magic to pale sunrise tints of orange and red, while the sunrise appearance in E. had changed to indigo. The colours increased in brilliancy near the horizon, overhead the sky was [of a] leaden [hue]. Some white houses at a little distance were brought nearer, and assumed a warm yellow tint; the darkness was great; thermometers could not be read. The countenances of men were of a livid pink. The Spaniards lay down, and their children screamed with fear. fowls hastened to roost, ducks clustered together, pigeons dashed against the sides of the houses, flowers closed (*Hibiscus Africanus* as early as 2h. 5m.); at 2h. 52m. cocks began to crow (ceasing at 2h. 57m., and recommencing at 3h. 5m.). As darkness came on, many butterflies, which were seen about, flew as if drunk, and at last disappeared; the air became very humid, so much so that the grass felt to one of the observers as if recently rained upon. So many facts have been noted and recorded that it is impossible to do more than give a brief statement of the leading features."

In the absence of the complete series of observations, which are not yet (May 1861) published, we will only sum up with one remark, namely, that it seems now to be conclusively decided that the red flames belong not to the Moon, but to the Sun.

CHAPTER VI.

HISTORICAL NOTICES.

Eclipses recorded in Ancient History.—Eclipse of 584 B.C.—*Eclipse of* 556 B.C.—*Eclipse of* 480 B.C.—*Eclipse of* 430 B.C.—*Eclipse of* 309 B.C.—*Allusions in old English Chronicles to Eclipses of the Sun.*

THE earliest eclipse on record is one given in the Chinese history the *Chou-king*, and which is supposed to refer to the solar eclipse of Oct. 13, 2127 B.C.[1]

One of the most celebrated eclipses of the Sun recorded in history is that which occurred in the year 584 B.C. It is so, not only on account of its having been predicted by Thales, who was the first ancient astronomer who gave the true explanation of the phenomena of eclipses, but because it seems to fix the precise date of an important event in ancient history. It appears that a war had been carried on for some years between the Lydians and Medes; and we are indebted to Herodotus for an account of the circumstances that led to its premature termination. He says: —

"In the sixth year, when they were carrying on the war with nearly equal success, on the occasion of an engagement, it happened that, in the heat of the battle, day was suddenly turned into night (συνήνεικε ὥστε τῆς μάχης συνεστεώσης τὴν ἡμέρην ἐαπίνης νύκτα γενέσθαι). This change of the day, Thales, the Milesian, had foretold to the Ionians, fixing beforehand, this year as the very period in which the change actually took place. The Lydians and Medes, seeing night succeeding in the place of day, desisted from fighting, and both showed a great anxiety to make peace." A peace was accordingly made and cemented by a marriage. "For, without strong necessity, agreements are not wont to remain firm."

[1] *Mem.* R.A.S., vol. xi. p. 47.

So adds the historian.[1] The exact date of this interesting event has long been disputed, and the solar eclipses of 609, 592, and particularly 584, have been fixed upon as the one mentioned by Herodotus, and it is only within the last few years that the point has been finally settled in favour of the last-mentioned eclipse, and that chiefly through the researches of the Astronomer Royal, who gives, as the date of the eclipse in question, May 28, 584 B.C.[2]

Another important ancient eclipse is that mentioned by Xenophon, in the *Anabasis*, as having led to the capture by the Persians of the Median city Larissa. In the retreat of the Greeks on the eastern side of the Tigris, not long after the seizure of their commanders, they crossed the river Zapetes, and also a ravine, and then came to the Tigris. At this place, according to Xenophon, there stood —

" A large deserted city called Larissa, formerly inhabited by the Medes; its wall was 25 feet thick, and 100 feet high; its circumference 2 parasangs; it was built of burnt brick on an understructure of stone 20 feet in height. When the Persians obtained the empire from the Medes, the king of the Persians besieged the city, but was unable by any means to take it; but a cloud covered the Sun and caused it to disappear completely, till [*i. e.* to such a degree, that] the inhabitants withdrew, and thus the city was taken." [3]

The historian then goes on to say, that the Greeks in continuing their march passed by another ruined city named Mespila. The minute description given by Xenophon has enabled Layard, Jones, and others, to identify Larissa with the modern Nimrùd, and Mespila with Mosul. It is plain that the phenomenon which the Greek author refers to as having led to the capture of the above-mentioned city, was no other than a total eclipse of the Sun. The Astronomer Royal has arrived at the conclusion that this eclipse occurred on May 19, ·556 B.C.

In the same year as that in which, according to the common account, the battle of Salamis was fought (479 B.C.) there occurred a phenomenon which is thus adverted to : —

---

[1] Herod. lib. i., cap. 74.
[2] *Phil. Trans.*, vol. cxliii. pp. 191—197. 1853. *Month. Not.* R.A.S., vol. xviii. p. 143, *et seq.*   [3] *Anab.*, lib. iii. cap. 4, § 7.

"With spring, the army [of the Persians] being ready, set out from Sardes on its march to Abydos; and as it was setting out, the Sun, leaving his place in the heavens, was invisible, (ὁ ἥλιος ἐκλιπὼν τὴν ἐκ τοῦ οὐρανοῦ ἕρδην ἀφανὴς ἦν,) when there was no cloud, but a perfectly clear sky; and instead of day it became night (ἀντὶ ἡμέρης τε νὺξ ἐγένε ο)."¹

This account, interpreted as a record of a total solar eclipse, has given great trouble to chronologers, and it is still a matter of uncertainty what eclipse is referred to. Mr. Airy "thinks it extremely probable" that the narrative relates to the total eclipse of the *Moon*, which happened 478 B.C. March, 13d. 15h. G. M. T.²

A total eclipse of the Sun, supposed to have been that of August 3, 430 B.C., would have seriously interfered with the Athenian expedition against the Lacedæmonians, had it not been for the intervention of Pericles, commander of the forces belonging to the former nation.

"The whole fleet was in readiness, and Pericles on board his own galley, when there happened an eclipse of the Sun. This sudden darkness was looked upon as an unfavourable omen, and threw them into the greatest consternation. Pericles observing that the pilot was much astonished and perplexed, took his cloak, and having covered his eyes with it, asked him if he found anything terrible in that, or considered it as a bad presage? Upon his answering in the negative, he said, 'Where is the difference, then, between this and the other, except that something bigger than my cloak causes the eclipse?'"³

Thucydides says:—

"The same summer, at the beginning of a new lunar month, (the only time at which it appears possible) the Sun was eclipsed after midday, and became full again after it had assumed a crescent form, and after some of the stars had shone out."⁴

An ancient eclipse, known by the name of Agathocles, has also been recently investigated by the Astronomer Royal, and previously by Mr. F. Baily. It took place on August 14, 309 B.C. The account of this eclipse, as given by the ancient writers, is

¹ Herod., lib. vii. cap. 37.
² *Phil. Trans.*, vol. cxliii. p. 197. 1853. See also Blakesley *in loco*.
³ Plutarch, *vita Periclis*. ⁴ Thucyd., lib. ii. cap. 28.

associated with an interesting historical event. Agathocles
having been closely blockaded in the harbour of Syracuse by
a Carthaginian fleet, took advantage of a temporary relaxation
in the blockade, occasioned by the enemy going in quest of a
relieving fleet, and quitting the harbour of Syracuse, he landed
on the neighbouring coast of Africa, at a point near what is now
called Cape Bon; and devastated the Carthaginian territories. It
is stated that the voyage to the African coast occupied 6 days,
and that an eclipse (which from the description was manifestly
total) occurred the 2nd day. Diodorus Siculus says that
the stars were seen[1], so that no doubt can exist as to the
totality of the eclipse. Mr. Baily, however, found that there
existed an irreconcilable difference between the calculated
path of the shadow and the historical statement, a space of
about 180 geographical miles appearing between the most
southerly position that can be assigned to the fleet of Agathocles and the northerly limit of the phase. " To obviate this
discordance, it is only necessary to suppose an error of about 3'
in the computed distances of the Sun and Moon at conjunction,
a very inconsiderable correction for a date anterior to the epoch
of the tables by more than 21 centuries.[2]

  In the writings of the early English chroniclers are to be found
numerous passages relating to total eclipses of the Sun. The
eclipse of August 2, 1133, was considered a presage of misfortune to Henry I. : it is thus referred to by William of
Malmesbury : — " The elements manifested their sorrow at this
great man's last departure. For the Sun on that day at the
6th hour shrouded his glorious face, as the poets say, in
hideous darkness, agitating the hearts of men by an eclipse ;
and on the 6th day of the week, early in the morning, there
was so great an earthquake that the ground appeared absolutely
to sink down ; an horrid noise being first heard beneath the
surface." [3] The same writer, speaking of the total eclipse of
March 20, 1140, says : — " During this year, in Lent, on the
13th of the calends of April, at the 9th hour of the 4th day of

[1] Diodor. Sic., lib. xx. cap. 1.   Justin., lib. xxii. cap. 6.
[2] *Phil. Trans.*, vol. cxliii. pp. 187–191. 1853.   [3] *Hist. Nov.*, lib. i.

the week, there was an eclipse, throughout England, as I have heard. With us, indeed, and with all our neighbours the obscuration of the Sun also was so remarkable, that persons sitting at table, as it then happened almost everywhere, for it was Lent, at first feared that Chaos was come again: afterwards learning the cause, they went out and beheld the stars around the Sun. It was thought and said by many, not untruly, that the king [Stephen] would not continue a year in the government." [1]

In this chapter and elsewhere, all dates before the Christian era are given according to the method of reckoning employed by astronomers, and not that of chronologists. Thus 584 B.C. astronomical reckoning, corresponds to 585 B.C. chronological reckoning. Sometimes astronomical dates of events happening before the birth of our SAVIOUR are given in this form : (—584), as being, in some sense, negative quantities.

In matters of this kind, due attention to these distinctions is of importance.

---

[1] *Hist. Nov.*, lib. ii. *Vide* also *Sax. Chron.*, Thorpe's Trans., p. 233. 8vo., London, 1861.

CHAPTER VII.

ECLIPSES OF THE MOON.

Lunar Eclipses of less interest than Solar ones.—Summary of Facts connected with them. — Peculiar Circumstances noticed during the Eclipse of March 19, 1848. — Observations of Forster. — Wargentin's Remarks on the Eclipse of May 18, 1761. — Kepler's Explanations of these Peculiarities being due to Metereological Causes. — Chaldæan Observations of Eclipses. — Other Ancient Eclipses. — Anecdote of Columbus.

An eclipse of the Moon, though inferior in importance to one of the Sun, is nevertheless not altogether devoid of interest; it is either partial or total, according to the extent to which our satellite is immersed in the Earth's shadow. In a total eclipse the Moon may be deprived of the Sun's light for 1h. 50m., and reckoning from the first to the last contact of the penumbra,

Fig. 19.

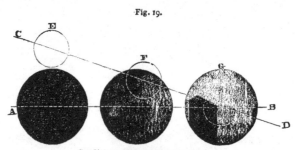

Conditions of Eclipses of the Moon.

the phenomenon may last 5h. 30m., but this is the outside limit. "Owing to the ecliptic limits of the Sun exceeding those of the Moon, there are more eclipses of the former luminary than of the latter, but on account of the comparatively small extent of the Earth's surface to which a solar eclipse is visible, the eclipses of the Moon are more frequently seen at any particular place than those of the Sun."

Fig. 19 is designed to illustrate the different conditions of eclipses of the Moon. A B is the ecliptic, C D the Moon's path. The 3 black circles are imaginary sections of the Earth's shadow, when in 3 successive positions in the ecliptic. If the conjunction in right ascension of the Earth and Moon occurs when the Moon is at E, it escapes eclipse; if the Moon is at F, it suffers a *partial* obscuration, but if the Moon is at or very near its node, at or very near G, it will be wholly involved in the Earth's shadow, and a *total* eclipse will be the result.

Even when most deeply immersed in the Earth's shadow, our satellite does not, except on rare occasions, wholly disappear, but may be generally detected in a telescope, and frequently with the naked eye, having a dull red, or coppery colour. This was exemplified in a very remarkable manner in the case of the eclipse of March 19, 1848, on which occasion the Moon, even when most deeply immersed, was so clearly seen that many persons doubted the reality of the eclipse.

Mr. Forster, who observed the eclipse at Bruges, writes as follows : —

"I wish to call your attention to the fact which I have clearly ascertained, that during the whole of the late eclipse of March 19, the shaded surface presented a luminosity quite unusual, probably about three times the intensity of the mean illumination of the eclipsed lunar disc. The light was of a deep red colour. During the totality of the eclipse, the light and dark places on the face of the Moon could be almost as well made out as on an ordinary dull moonlight night, and the deep red colour where the sky was clearer, was very remarkable from the contrasted whiteness of the stars. My observations were made with different telescopes; but all presented the same appearance, and the remarkable luminosity struck every one. The British Consul at Ghent, *who did not know there was an eclipse*, wrote to me for an explanation of the blood-red colour of the Moon at 9 o'clock." [1]

As a complement to this observation, we may quote one of

---

[1] *Month. Not.* R.A.S., vol. viii. p. 132.

Wargentin's. He was watching the total eclipse of May 18, 1761, and he says that 11m. after the commencement of the phase " the Moon's body *disappeared so completely, that not the slightest trace of any portion of the lunar disc could be discerned either with the naked eye, or with the telescope,* although the sky was clear, and the stars in the vicinity of the Moon were distinctly visible in the telescope." [1]

The red hue was long a phenomenon for which no explanation could be found; by some it was considered to be due to a light naturally inherent to the Moon's surface, but Kepler was the first to find out a more scientific explanation. He showed that the phenomenon was occasioned by the refraction of the Earth's atmosphere, which had the effect of turning the course of the solar rays passing through it, causing them to fall upon the Moon even when the Earth was actually interposed between them and the Sun. The deep red colour of the Moon's surface arises from the absorption of the blue rays of light in passing through the terrestrial atmosphere, in the same manner as the western sky is frequently seen to assume a ruddy hue, when illuminated in the evening by the solar rays. On account of the variable meteorological condition of our atmosphere, the quantity of light actually transmitted is liable to fluctuate considerably, and hence arises a corresponding variation in the appearances presented by the Moon's surface during her immersion in the Earth's shadow. If the portion of the atmosphere through which the solar rays have to pass is everywhere much saturated with vapour, the red rays will be almost totally absorbed, as well as the blue, and the illumination will be too feeble to render her surface visible : such was the case in the eclipse of June 15, 1620, observed by Kepler.[2] The eclipse of May 18, 1761, observed by Wargentin at Stockholm, and referred to above, furnishes a good example.

---

[1] *Phil. Trans.*, vol. li. p. 210. 1761. The original runs thus: "Tota luna, ita prorsus disparuerat, ut nullum ejus vestigium, vel nudis, vel armatis oculis, sensibile restaret, cœlo licet sereno, et stellis vicinis in tubo conspicuis."

[2] *Epist. Ast.*, p. 825.

THE MOON TOTALLY ECLIPSED, 1848, March 19.

THE MOON PARTIALLY ECLIPSED, 1860, February 6.

ECLIPSES OF THE MOON.

If, on the other hand, the region of the atmosphere through which the solar rays pass be everywhere very transparent, the red rays will be transmitted to the Moon in great abundance, and its surface will consequently be highly illuminated. Such was the case in the eclipse of March, 1848, already referred to. If moreover the region of the atmosphere through which the rays pass be saturated only in some parts, and not in others, it follows that some portions of the Moon's disc will be invisible whilst others will be more or less illuminated. Such an occurrence was seen by Kepler on Aug. 16, 1598.

The celebrated African explorers, the Landers, graphically describe what took place on the occasion of the eclipse of the Moon of Sept. 2, 1830. They say : — "The earlier part of the evening had been mild, serene, and remarkably pleasant. The Moon had arisen with uncommon lustre, and being at the full, her appearance was extremely delightful. It was the conclusion of the holidays, and many of the people were enjoying the delicious coolness of a serene night, and resting from the laborious exertions of the day ; but when the Moon became gradually obscured, fear overcame every one. As the eclipse increased, they became more terrified. All ran in great distress to inform their sovereign of the circumstance, for there was not a single cloud to cause so deep a shadow, and they could not comprehend the nature or meaning of an eclipse. Groups of men were blowing on trumpets, which produced a harsh and discordant sound ; some were employed in beating old drums ; others again were blowing on bullocks' horns. The diminished light, when the eclipse was complete, was just sufficient for us to distinguish the various groups of people, and contributed in no small degree to render the scene more imposing. If a European, a stranger to Africa, had been placed on a sudden in the midst of the terror-struck people, he would have imagined himself among a legion of demons, holding a revel over a fallen spirit."

It is to the Chaldæans that we owe the earliest recorded observations of lunar eclipses as mentioned by Ptolemy. The first of these took place in the 27th year of the era of Nabonasser, the first of the reign of Mardokempadius, on the 29th of the Egyptian month Thoth, answering to March 19, 720 B.C., according to our mode of reckoning. It appears to have been total at Babylon, the greatest phase occurring at about 9h. 30m. P.M. The second was a partial eclipse only; it happened at midnight on the 18th of the month Thoth, or on March 8, 719 B.C. The third took place in the same year, on the 15th of the month Phammuth, or Sept. 1, 719 B.C. The magnitude of the eclipse, according to Ptolemy, was 6 digits on the southern limb, and it lasted 3 hours, having commenced soon after the Moon rose at Babylon. An eclipse occurred in the 4th year of the 1st Olympiad, the 19th of the Peloponnesian war, answering to Aug. 27, 412 B.C., which produced very disastrous consequences to the Athenian army, owing to the obstinacy of their general Nicias.[1] Modern calculations show that it was total at Syracuse.

An eclipse of the Moon, which happened on March 1, 1504, proved of much service to Columbus. His fleet was in great straits, owing to the want of supplies, which the islanders of Jamaica refused to give. He accordingly threatened to deprive them of the Moon's light, as a punishment. " His threat was treated at first with indifference, but when the eclipse actually commenced, the barbarians vied with each other in the production of the necessary supplies for the Spanish fleet."

---

[1] Plutarch, *Vita Nicias.* Thucyd., lib. vii. cap. 50.

CHAPTER VIII.

SUGGESTIONS FOR OBSERVING ANNULAR ECLIPSES OF THE SUN.

PREVIOUS to the celebrated eclipse of 1858, the Astronomer Royal circulated a useful series of suggestions, which we here reproduce, in a slightly amended form, as applicable to other eclipses besides the one above referred to : —

" The following suggestions are offered as presenting grounds for consideration, which may tend to direct observers in deciding on the employment of the means which they may possess.

" I. *Observations not requiring Instruments.*

" 1. As the eclipse advances, it is desirable to obtain some notion or measure of the degree of darkness.

" 2. At what distance from the eye can a book or paper, exhibiting type of different sizes, be read?

" 3. Hold up a lighted candle nearly between the Sun and your eye. At how many sun-breadths' distance from the Sun can the flame be seen?

" 4. If you are in an elevated position, remark the changes of colour and appearance of the surrounding objects in the landscape.

" 5. If you see the spots of light formed by the intersecting shadows of the boughs of trees, remark whether they exhibit the luneform of the Sun.

" 6. When the annulus is formed, you will probably observe it with a darkened glass; but you are particularly requested to devote one instant (as early as possible) to the verification of this point, viz.: When the annular Sun is viewed with the naked eye, does it appear an annulus or a fully illuminated disc?

" II. *Optical, Astronomical, and Solar-physical Observations, requiring the use of Instruments.*

" 7. As the eclipse advances, estimate (on the image seen in the telescope) the comparative intensity of the Sun's light near the centre of his disc and near his limb.

" 8. For the more critical observations, it is desirable that the power of your telescope should be so low as to give you an easy view of the whole breadth of the Sun.

" 9. Remark irregularities on the Moon's limb.

" 10. As the cusps become very sharp, remark whether they are irregular. For this, and for all the observations near the annular phase, it is necessary that you be provided either with a graduated prismatic shade, or with a succession of shades of different intensity, and that you instantly select the shade which is most agreeable to your eye.

" 11. Remark whether the Sun's light extends beyond the intersection of the limbs of Sun and Moon, so as to make the Moon's limb visible beyond that intersection. For this purpose the bright parts of the Sun must be put out of the field of view, and the shade must be withdrawn.

" 12. As the annularity approaches and is formed, remark whether Baily's beads and strings are formed; whether first formed at points corresponding to large inequalities of the Moon's limb; whether they surround the Moon; how they form and break. Only an instant can be given to this observation. It is of the utmost importance to be assured that your vision at the instants immediately preceding, especially of the Moon's inequalities, is very distinct.

" 13. Remark, as one of the most important observations of the eclipse, whether any red flames are seen on the Sun's limb. For this purpose you must withdraw the shade, if you are on the annular track, the instant after formation of the annulus; if you are not on the annular track, as soon as the eye can bear the Sun. On the annular track the whole line must be rapidly scrutinised ; and when the ring breaks, the still illuminated part must be put out of the field, and the Moon's dark limb must be surveyed. At places not on the annular track, this plan (namely, to exclude the illuminated portion of the disc from the field, and to survey the Moon's dark limb) must be followed throughout. It seems not improbable that the best chance of seeing red flames will be obtained at places not on the annular track.

" 14. At the breaking up of the annulus, look for Baily's beads as before.

" 15. Do not attempt any record during or near the annularity. Endeavour to impress observations on your memory as well as you can. If you have an assistant seated at a table with a chronometer and writing-materials, you may give him signals for the register of time, but you must connect the phenomena with the time afterwards.

" 16. A good sextant-observer may obtain valuable observations for correction of the lunar tables, by measuring the intervals between the points of the bright cusps. The observations will require great nerve, and will be difficult, but where most difficult they will be most valuable.

" 17. It seems doubtful whether any valid photographic record can be made, on account of the extreme rapidity of the change of appearances.

" 18. If you have a doubly refracting prism, it will be desirable to make observations on the polarisation of the light from the Sun's limb. For this purpose, when the lune is narrow, place the prism so as to separate the 2 images transversely to the limb, and remark which image is brighter. Turn the prism 180° round the visual ray, and repeat the observation. Remark carefully the positions of the prism. The prism may be used with the naked eye or with the telescope, according to the amount of its angular separation of images.

" III. *Meteorological Observations.*

" 19. For change in intensity of solar radiation, observations

with the actinometer or the black-bulb thermometer should
be kept up during the eclipse.   The latter are most trustworthy when the bulb is enclosed in an exhausted glass
sphere.

" 20. The barometer should be observed from time to time,
but very frequent readings are unnecessary.

" 21. The thermometer should be frequently observed, and
the general feelings of cold should be noted.

" 22. Observations of humidity are very important.   They
should be made by the use of the wet-bulb thermometer.

" G. B. AIRY."

CHAPTER IX.

TRANSITS OF THE INFERIOR PLANETS.

Cause of the Phenomena. — Long Intervals between each Recurrence. — List of Transits of Mercury. — Of Venus. — Transit of Mercury of Nov. 7, 1631. — Predicted by Kepler. — Observed by Gassendi. — His Remarks. — Transit of Nov. 3, 1651. — Observed by Shakerley. — Number observed since this Date. — Transit of Nov. 9, 1848. — Observations of Dawes. — Of Forster. — Transit of Venus of Nov. 24, 1639. — Observed by Horrox and Crabtree. — Transit of 1761. — Of 1769. — Where observed. — Singular Phenomenon seen on both Occasions. — Explanatory Hypothesis.

WHEN an inferior planet is in superior conjunction, and "has a latitude or distance from the ecliptic less than the Sun's semi-diameter, it will be less distant from the Sun's centre than such semi-diameter, and will therefore be within the Sun's disc. In this case, the planet being between the Earth and the Sun, its dark hemisphere being turned towards the Earth, it will appear projected upon the Sun's disc as an intensely black round spot. The apparent motion of the planet being retrograde, it will appear to move across the disc of the Sun from east to west, in a line sensibly parallel to the ecliptic." Such a phenomenon is called a *transit,* and as it can only occur in the case of the inferior planets, or those which pass between the Earth and the Sun, it is limited to Vulcan, Mercury, and Venus. The observations of these planets are used in a manner which we cannot here explain, for the purpose of ascertaining the distance of the Earth from the Sun.

James Gregory (inventor of the reflecting telescope which bears his name) seems to have been the first to point out this application of transit observations.[1]

---

[1] *Optica Promota,* p. 130.

The transits of the inferior planets are phenomena of very rare occurrence, especially those of Venus, which occur only at intervals of $8,105\frac{1}{2}$; $8,121\frac{1}{2}$; $8,105\frac{1}{2}$, &c., years. Transits of Mercury usually happen at intervals of 13 and 7 years, and in the order, 13, 13, 13, 7, &c., this, however, is not altogether a correct expression of the intervals: owing to the considerable inclination of Mercury's orbit, it requires a period of about 217 years to bring the transits round in a complete cycle.

The following are the dates of the transits of Mercury and Venus from the beginning of the 19th century onwards: —

Mercury.				Venus.			
		d.	h.			d.	h.
1802	November .	8	20	1874	December .	8	16
1815	November .	11	14	1882	December .	6	14
1822	November .	4	14	2004	June . . .	7	21
1832	May . . .	5	0	2012	June . . .	5	13
1835	November .	7	7	2117	December .	10	15
1845	May . . .	8	8	2125	December .	8	3
1848	November .	7	1	2247	June . . .	11	8
1861	November .	11	19	2255	June . . .	8	16
1868	November .	4	18	2360	December .	12	13
1878	May . . .	6	6	2368	December .	10	2
1881	November .	7	2	2490 ?	June . . .	12	3
1891	May . . .	9	14	2498	June . . .	9	20
1894	November .	10	6	2603	December .	15	12
1901	November .	4		2611	December .	13	1
				2733 ?	June . . .	15	7
				2741	June . . .	12	23
				2846	December .	16	11
				2854	December .	14	0
				2984	June . . .	14	3

The transits of Mercury happen necessarily, from the heliocentric position of the nodes, always in May or November.

[1] Lalande, *Astronomie*, vol. ii. pp. 457 and 461.

RING ROUND MERCURY DURING ITS TRANSIT ACROSS THE SUN,
1799, May 7.

OCCULTATION OF JUPITER BY THE MOON:
1857, January 2.

TRANSIT AND OCCULTATION PHENOMENA.

When the transit occurs in May, the planet is passing through the *descending node*, and when in November, through the *ascending node.* The same remarks apply to the transits of Venus.

The first observed transit of Mercury occurred on November 7, 1631, and was previously predicted by Kepler[1], whose surmise was verified by Gassendi at Paris. The latter remarks: " The crafty god had sought to deceive astronomers by passing over the Sun a little earlier than was expected, and had drawn a veil of dark clouds over the Earth, in order to make his escape more effectual. But Apollo, acquainted with his knavish tricks from his infancy, would not allow him to pass altogether unnoticed. To be brief, I have been more fortunate than those hunters after Mercury, who have sought the cunning god in the Sun; I found him out, and saw him where no one else had hitherto seen him." [2] The next observation of a transit of Mercury was made by Jeremiah Shakerley, on November 3, 1651, who made a voyage to India to see the phenomenon, which his calculations told him would be invisible in England.[3] Since this date 27 transits have taken place, of which no less than 20 have been observed.[4]

The last transit of this planet happened on Nov. 8, 1848. Mr. W. R. Dawes, who observed it at Cranbrook in Kent, says:—" Nothing remarkable was noticed, till Mercury advanced on the Sun's disc to about three-quarters of its own diameter, when the cusps appeared much rounded off, giving a pear-shaped appearance to the planet. The *degree* of this deformity, however, *varied* with the steadiness and definition of the Sun's edge, being least where the definition was *best.* A few seconds before the complete entrance of the planet, the Sun's edge became much more steady, and the cusp sharper, though still occasionally a little thicker towards their points by the undulations. At the instant of their junction the

---

[1] *Admonitio ad Astronomos*, &c.   [2] *Opera Omnia*, vol. ii. p. 537.
[3] Wing, *Astronomia Britannica*, p. 312.
[4] Arago, *Pop. Ast.*, vol. i. p. 676. Eng. Ed.; Grant, *Hist. Phys. Ast.*, pp. 414–9.

definition was pretty good, and they formed the finest conceivable line, Mercury appearing at the same time *perfectly
round.* . . . No difference is recognised in the *Nautical
Almanac* between the polar and equatorial diameters of this
planet; yet my observations, both with the 5 feet achromatic
and the Gregorian, show a perceptible difference, and nearly
to the same amount. . . . The compression would appear
to be about $\frac{1}{29}$." [1] Forster observed the transit at Bruges.
He remarks the extreme *blackness* of the planet compared with
the spots: the intensities he estimates at 8 : 5. He also states,
that the planet had rather the appearance of a *globe* than of a
disc, and the difference of blackness between the planet and
the spots was less remarkable when he used a reflector with a
red shade. [2]

The first observed transit of Venus over the Sun took place
in the year 1639, and was seen by the Rev. Jeremiah Horrox,
of Hoole near Liverpool, who accidentally ascertained that the
phenomenon would take place, whilst engaged in computing
an ephemeris of the planet. Having fully satisfied himself
as to the correctness of his calculation, Horrox informed
his friend, William Crabtree, an enthusiastic astronomer, of
what was about to take place, and desired him to watch it.
The planet was seen on the Sun's disc on Nov. 24 (O. S.) by
both observers, though, owing to the clouds, Crabtree caught
but a single glimpse of it. Horrox was enabled to watch very
conveniently the transit of the planet, by adopting the ingenious
device of transmitting the image of the Sun through a telescope into a darkened room. [3] No other transit occurred till
June 5, 1761; this was observed in many parts of the world
for the purpose of ascertaining, in accordance with the suggestions of Halley, the parallax of the Sun, and thence of
finding the distance between the Earth and that luminary.
The results of the different observations have since been found
not so well to coincide as could be wished, and it is fortunate

[1] *Month. Not.* R.A.S., vol. ix. p. 21.   [2] Ibid., p. 4.
[3] Whatton, *Mem. of Horrox*, pp. 109–135.

that another transit on June 3, 1769, has afforded more satisfactory numbers.

Extensive preparations were made for observing the transit
of 1769, and the British Government dispatched a well
organised and equipped expedition to Tahiti, under the celebrated Captain James Cook, R.N.   Many of the continental
powers followed the example of Great Britain, and astronomers
were sent out to the most advantageous points for observation.
" The ingress of the planet on the Sun's disc was seen at
almost all the observatories of Europe; the egress at St. Petersburg, Pekin, Orenburg, Iakutsk, Manilla, Batavia, &c.; and
the complete duration of the transit at Cape Wardhus, Kola,
and Kajeneburg, in Lapland, at Tahiti, and Port Prince of
Wales, and St. Joseph in California." The observations were
all of the most trustworthy character.[1]

One phenomenon was seen in connexion with these transits
which requires a passing mention.   It was noticed on both
occasions, and by numerous observers, that the interior contact
of the planet with the Sun did not take place regularly at the
ingress, the planet appearing for some time, after it had entered
upon the solar disc, to be attached to the Sun's limb by a dark
ligament.   A similar phenomenon was noticed at the egress.
It was also found that even after the planet had wholly
separated from the Sun's limb, it did not acquire its round
form for several seconds.[2]   Lalande has suggested that irradiation is the cause of this phenomenon[3], and that is doubtless
the true explanation.

---

[1] See p. 3, *supra*.
[2] Vide *Phil. Trans.*, 1761, 1768, 1769, 1770; also *Mém. Acad. des Sciences*, for the same years.
[3] *Mém. Acad. des Sciences*, 1770, p. 409.

CHAPTER X.

OCCULTATIONS.

How caused. — Table annually given in the Nautical Almanac. — *Oc-
cultation by a young Moon. — Effect of the Horizontal Parallax. —
Projection of Stars on the Moon's Disc. — Occultation of Saturn,
May 8, 1859. — Occultation of Jupiter, January 2, 1857. — Historical
Notices.*

WHEN any celestial object is concealed by the interposition of
another, it is said to be *occulted*, and the phenomenon is called
an *occultation*.   Strictly speaking, an eclipse of the Sun is an
occultation of that luminary by the Moon, but usage has given
to it the exceptional name of " eclipse."   The most important
phenomena of this kind are the occultations of the planets
and larger stars by the Moon, but the occultation of one planet
by another, on account of the rarity of such an occurrence
is exceedingly interesting.   Inasmuch as the Moon's apparent
diameter is about $\frac{1}{2}$°, it follows that all stars and planets
situated in a zone extending $\frac{1}{4}$° on each side of her path, will
necessarily be occulted during her monthly course through the
ecliptic.   The great brilliancy of the Moon entirely over-
powers the smaller stars, but the disappearance of the more
conspicuous ones are visible in a telescope, a table of which is
inserted every year in the *Nautical Almanac*.

It must be remembered, that the disappearance always takes
place at that limb of the Moon which is presented in the direction of its motion.   From the period of New Moon to that of
Full Moon, the Moon moves with the dark edge foremost,
and from Full Moon to New Moon with the illuminated edge
foremost ; during the former interval, therefore, the objects
occulted disappear at the dark edge, and reappear at the illuminated edge ; and during the latter period they disappear at

the illuminated, and reappear at the dark edge. If the occultation be watched when the Moon is not more than 2 or 3 days old, the disappearance is extremely striking, inasmuch as the object occulted seems to be suddenly extinguished at a point of the sky where there is nothing apparently to interfere with it. Wargentin relates that he saw on May 18, 1761, an occultation of a star by the Moon, during a total eclipse of the latter. He says that the star disappeared in the twinkling of an eye (oculi ictu citius). In consequence of the horizontal parallax, the Moon as seen in the northern hemisphere, follows a path different from that seen in the southern hemisphere; it happens, therefore, that stars which are occulted in certain latitudes are not occulted at all in others, and of those which are occulted the duration of invisibility, and the moment and place of disappearance and reappearance, are different.

We must not omit a passing allusion to a circumstance occasionally noticed by the observers of occultations: we mean the apparent projection of the star *within* the margin of the Moon's disc.

Admiral Smyth gives an instance, under the date of October 15, 1829. He says: —

"I saw Aldebaran approach the bright limb of the Moon very steadily; but from the haze no alteration in the redness of its colour was perceptible. It kept the same steady line *to about ¾ of a minute inside the lunar disc*, where it remained, as precisely as I could estimate, 2¼ seconds, when it suddenly vanished. In this there could be no mistake, because I clearly saw the bright line of the Moon *outside* the star, as did also Dr. Lee, who was with me." [2]

Sir T. Maclear saw the same thing happen to the same star on October 23, 1831: —

"Previous to the contact of the Moon and star nothing particular occurred; but at that moment, and when I might expect the star to immerge, it advanced upon the Moon's limb for about 3 seconds and to rather more than the star's apparent diameter, and then instantly disappeared." [3]

---

[1] *Phil. Trans.*, vol. li. p. 210. 1761.

[2] *Mem.* R.A.S., vol. iv. p. 642. Other observers, Maclear included, saw the projection, though F. Baily and others did *not* see it.

[3] Ibid., vol. v. p. 273.

" This phenomenon seems to be owing to the greater proportionate refrangibility of the white lunar light, than that of
the red light of the star, elevating her apparent disc at the
time and point of contact." [1]

An occultation of the planet Jupiter took place on January
2, 1857. A dark shadowy streak which appeared projected
on the planet, from the edge of the Moon, was seen by several
observers.

Mr. W. Simms, Sen. thus describes it : —

"The only remarkable appearance noticed by me during the emersion
was the very positive line by which the Moon's limb was marked upon
the planet; dark as the mark of a black-lead pencil close to the limb,
and gradually softened off as the distance increased." [2]

A representation of this appearance, from a drawing by
Lassell, is given in Fig. 87, Plate XX.

An occultation of the planet Saturn by the Moon took place
on May 8, 1859. Mr. Dawes thus describes it : —

"At the disappearance, the dark edge of the Moon was sharply
defined on the rings and ball of the planet, without the slightest distortion of their figure. There was no extension of light along the Moon's
limb. Even the satellites disappeared without the slightest warning,
and precisely at the edge which was faintly visible.

"At the reappearance I could not perceive any dark shading contiguous to the Moon's bright edge, such as was seen by myself and
several other observers on *Jupiter* on January 2, 1858 [Qy. 1857].
The dark belt south of the planet's equator was clearly defined up to
the very edge; and there was no distortion of any kind, either of the
rings or ball.

" The very pale greenish hue of *Saturn* contrasted strikingly with the
brilliant yellowish light of the Moon." [3]

Mr. W. Simms, Jun. *did* see a dark shading on the planet
contiguous to the Moon's bright edge; but, in 1857, he failed
to notice it.

The earliest phenomenon of this kind of which we have any
mention is an occultation of Mars by the Moon, mentioned by
Aristotle.[4] Kepler found that it occurred on the night of

---

[1] Smyth.   [2] *Month. Not.* R.A.S., vol. xvii. p. 81.
[3] *Month. Not.* R.A.S., vol. xix. p. 291.   [4] *De Cœlo*, lib. ii. cap. 12.

April 4, 357 B.C.[1] Instances are on record of one planet occulting another, but these are of very rare occurrence. Kepler states that he watched an occultation of Jupiter by Mars on January 9, 1591. He also mentions that Mœstlin witnessed an occultation of Mars by Venus on October 3, 1590. Mercury was occulted by Venus on May 17, 1737.[2] As these observations were made before the invention of the telescope, it is possible that the two planets were not actually in front of each other, but only so close as to have the appearance of being only one object: as was the case with Venus and Jupiter on July 21, 1859. Sometimes stars are occulted by planets, as in the case of the occultation of a star in Aquarius by Mars on October 1, 1672, mentioned by D. Cassini.

---

[1] *Ad. Vitell. Paralipom.*, p. 307.
[2] *Phil. Trans.*, vol. xl. p. 394. 1738.

148

BOOK III.

THE TIDES.

CHAPTER I.

"O ye seas and floods, bless ye the LORD : praise Him, and magnify Him for ever."—*Benedicite.*

Introduction. — Physical Cause of the Tides. — Attractive Force exercised by the Moon. — By the Sun. — Spring Tides. — Neap Tides. — Summary of the Principal Facts. — Priming and Lagging.

EVERY inhabitant of a maritime country like Great Britain is more or less familiar with the phenomena now under our consideration, but beyond knowing the general fact, that the Moon has something to do with the tides, their physical history is not so well understood as it ought to be.

The phenomena of the tides are very frequently attributed to the attraction of the Moon, whereby the waters of the ocean are drawn towards that side of the Earth on which our satellite happens to be situated; in fact, that it is high water when the Moon is on or near the meridian of the place of observation.

This, though to a certain extent true, by no means adequately represents the facts of the case, for high water is not only produced on that side of the Earth immediately under the Moon, but also on the opposite side at the same time. The two tides are therefore separated from each other by 180°, or by a space equal to half the circumference of the globe. The diurnal rotation of the Earth, causing every portion of its surface to pass successively under the tidal waves, in about 24h., it follows that there are everywhere 2 tides daily, with an interval of about 12 hours between each; whereas, if the common supposition were correct, there would be only 1.

Such being the observed facts, it now devolves on us to show that although the attraction of the Moon gives rise to the upper tide, some other explanation must be sought for, to account for the lower one: the solution is extremely simple; it is only necessary to bear in mind that not only does the Moon attract the upper mass of water, but also the solid globe itself, which is consequently compelled to recede from the waters beneath, leaving them heaped up by themselves.

Besides the influence of the Moon in elevating the waters of the ocean, that of the Sun is to some extent concerned, but it is much more feeble than that of the former, on account of the much greater distance of the solar globe. The mean distance of the Sun from the Earth exceeds that of the Moon 401 times, its attractive power would consequently be $(401)^2$ or 160,801 less; but inasmuch as the mass of the Sun exceeds that of the Moon in the ratio of 28,394,880 to 1, which is much greater than 160,801 to 1, it will naturally be said that surely the attraction exercised by the Sun exceeds that of the Moon in the same proportion that 28,394,880 exceeds 160,801. This, however, is not the case, and for a reason we will now state. It must be borne in mind that the tides are due solely to the *inequality* of the attraction exercised on different sides of the Earth, and that the greater that inequality is, the greater will be the resulting tide, and *vice versâ*. The mean distance of the Sun from the Earth is 12,023 diameters of the latter, and consequently the difference between its distance from the one side of the Earth and from the other will be only $\frac{1}{12023}$ of the whole distance, while in the case of the Moon, whose mean distance is only 30 terrestrial diameters, the differences between the distances towards one side and towards the other, reckoned from the Moon, will be $\frac{1}{30}$ of the whole distance. The inequality of the attraction (upon which the height of the tidal wave depends) is therefore much greater in the case of the Moon than of the Sun; the ratio, according to Newton, being 58 : 23, or $2\frac{1}{2}$: 1'.

We thus see that there are 2 kinds of tides, lunar and solar. When therefore the Sun, Moon, and Earth are in the same straight line with each other, that is to say, when it is either *New* or *Full Moon*, the two former bodies give rise to what are known as spring tides; but when the Moon is in quadrature, or 90° from the Sun, the solar wave diminishing the effect of the lunar wave, causes the tides to be *neap*.

It may be convenient for us to state a few general facts relating to the tides :—

1. On the day of New Moon, the Sun and Moon cross the meridian at the same time, *i. e.* at noon, and at an interval after their passage, varying according to the place of observation, but fixed and definite for each place, high water occurs. The water having reached its maximum height, begins to fall, and after a period of 6h. 12m. attains a minimum depression; it then rises for 6h. 12m., and reaches a second maximum; falls for another interval of 6h. 12m., and rises again during a 4th interval of 6h. 12m. It has therefore 2 maxima and 2 minima in a period of 24h. 48m., which is called a *tidal day*.

2. On the day of Full Moon, the Moon crosses the meridian 12h. after the Sun, *i. e.* at midnight, and the tidal phenomena are the same as in (1).

3. As time is reckoned by the apparent motion of the Sun, the solar tide always happens at the same hour at the same place, but the lunar tide, which is the greater, thereby giving a character to the whole, happens 48m. 44s. later every day; it therefore separates eastward from the solar tide, at that rate, and gradually becomes later and later, till at the period of the 1st and 3rd quarters of the Moon it happens at the

same time as the low water of the solar tide : then the elevation of the high, and the depression of the low water, will be the difference of the solar and the lunar tides, and the tide will be neap.

4. The difference in height between the high and low water is called the *range of the tide.*

5. The spring tides are highest, especially those which happen on the 2nd or 3rd day after the New, or Full Moon.

6. The neap tides are the lowest, especially those which occur on the 2nd or 3rd day after the Moon is in quadrature.

7. The interval of time from Noon to the time of high water at any particular place is the same on the days both of New and Full Moon, and is termed the *Establishment of the port.*

The reason why an interval of time elapses, between the Moon's meridian passage, and the time of high water, is, that the waters of the ocean have to overcome a certain peculiar effect of friction, which cannot immediately be accomplished; it thus happens that the lunar tidal wave is not found immediately under the Moon, but follows it at some distance. Similar results ensue in the case of the solar wave. The tidal wave is also affected in another way, by the continued action of both these luminaries, and at certain periods of the lunar month is either accelerated or retarded in a way we will now describe: "In the 1st and 3rd quarters of the Moon, the solar tide is westwards of the lunar one; and consequently the actual high water (which is the result of the combination of the 2 waves) will be to the westward of the place it would have been at if the Moon had acted alone, and the time of high water will therefore be accelerated. In the 2nd and 4th quarters, the general effect of the Sun is, for a similar reason, to produce a retardation in the time of high water. This effect, produced by the Sun and Moon combined, is called the *priming* and *lagging* of the tides. The highest spring tides occur when the Moon passes the meridian about 1½h. after the Sun; for then the maximum effect of the

2 bodies coincides." The 2 tides following one another are also subject to a variation, called the *diurnal inequality*, depending on the daily change in declination of the Sun and Moon; the laws which govern it are, however, very imperfectly known.

CHAPTER II.

Local Disturbing Influences. — Table of Tidal Ranges. — Influence of the Wind. — Experiment of Smeaton. — Tidal Phenomena in the Pacific Ocean. — Remarks by Beechey. — Velocity of the Great Terrestrial Tidal Wave. — Its Course round the Earth sketched by Johnston. — Effects of Tides at Bristol. — Instinct of Animals. — Tides extinguished in Rivers. — Historical Notices.

WE have hitherto been considering the tidal wave, on the supposition of the Earth being a perfect sphere uniformly covered with water, but inasmuch as this is not the case, it follows that the actual phenomena of the tides are widely different, and of a much more complicated character, owing to the irregular outline of the land, the rugged surface of the ocean bed, the action of winds, currents, friction, &c. More especially are these disturbing influences rendered manifest in the difference of the range of the tides at different places on the Earth's surface. For if the surface of our globe were entirely covered with water, the height of a solar tide would be 1 ft. $11\frac{1}{30}$ in., and of a lunar one 8 ft. $7\frac{5}{22}$ in. The configuration of the land, however, as we have already mentioned, usually gives rise to a much greater difference in the level, for instance, in deep estuaries or creeks, open in the direction of the tidal wave, and gradually converging inward, the range is very much greater than elsewhere, as at

	Feet.
Bristol Channel (off Chepstow)	70
Fundy Bay	60
Gallegos River (Patagonia)	46
Mouth of the Avon	42
St. Malo	40
Bristol	40
Milford Haven	36

On the other hand, where promontories or headlands jut out
into the sea, the tidal range is frequently diminished; thus :—

	Feet.
Wicklow	4
Weymouth	7
The Needles	9
Cape Clear	11

In very large open tracts of water, like the Atlantic or the
Pacific Oceans, and in narrow confined seas, like the Baltic,
the Mediterranean, &c., the elevation of the tidal wave is very
inconsiderable; thus :—

	Inches.
Toulon	12
Antium	14
Porto Rico (S. Juan)	18
South Pacific	20
St. Helena	36

The usual range of the tides, at any particular place, is also
affected by certain conditions of the atmosphere. At Brest,
for instance, an alteration of 1 inch in the barometric column
gives rise to a difference of 16 in. in the elevation of the high-
water mark; at Liverpool the difference is about 10 in., and
at the London Docks about 7 in.; when therefore the barometer
is low, an unusual high tide may be expected, and *vice versâ*.
The influence of the wind is frequently very considerable, so
much so that during a violent hurricane, Jan. 8, 1839, there
was no tide at all at Gainsborough on the river Trent, a circumstance never before recorded. Smeaton found experimen-
tally in a canal 4 miles long, the water level at one end was 4
inches higher than at the other, owing to the force of the wind
acting on the surface of the water.

The tides in the Pacific Ocean present great anomalies; we
extract the following remarks from a missionary work published some years ago :—

" It is, to the missionaries, a well-known fact that the tides
in Tahiti and the Society Islands are uniform throughout the
year, both as to the time of the ebb and flow, and the heights
of the rise and fall, it being high water invariably at noon and
midnight, and consequently the water is at its lowest point at

6 o'clock in the morning and evening. The rise is seldom more than 18 inches or 2 feet above low water mark. It must be observed that mostly once, and frequently twice in the year, a very heavy sea rolls over the reef, and bursts with great violence upon the shore. But the most remarkable feature in the periodically high sea is that it invariably comes from the W. or S.W., which is the opposite direction to that from which the wind blows. The eastern sides of the island are, I believe, never injured by these periodical inundations. I have been thus particular in my observations, for the purpose of calling the attention of scientific men to this remarkable phenomenon, as I believe it is restricted to the Tahitian and Society Island Groups in the South Pacific, and the Sandwich Islands in the North. I cannot, however, speak positively respecting the tides at the islands eastward of Tahiti ; but all the islands I have visited, in the same parallel of longitude southwards, and in those to the westward, in the same parallel of latitude, the same regularity is *not* observed, but the tides vary with the Moon, both as to the time and the height of the rise and fall, which is the case at Raratonga."

The late Admiral Beechey is the only person, that we are aware of, who has given any solution to the question, and he proposes, as a simile, a basin to represent the harbour, over the margin of which the sea breaks with considerable violence, thereby throwing in a larger body of water than the narrow channels can carry off in the same time, and consequently the tide rises, and as the wind abates the water subsides.

This explanation the writer above quoted objects to, and he brings forward several arguments, and states several facts, of which we select the following : —

1. That the undeviating regularity of the tide is so well understood by the natives that they distinguish the hours of the day by terms descriptive of its state, such as the following: " Where is the tide ? " instead of, as we should say, " What o'clock is it ? "

2. There are many days during the year when it is perfectly calm, and yet the tide rises and falls in the same way,

and very frequently there are higher tides in calms than during the prevalence of the trade wind.

3. The tides are equally regular on the west side of the island, where the trade wind does not reach, as on the east, from which point it blows.

4. The trade wind is most powerful from noon till 4 or 5 o'clock, during which time the water ebbs so fast that it reaches its lowest level by 6 o'clock, instead of in the morning as Admiral Beechey states, at which time it is again high water.

Admiral Beechey's explanation does not seem a satisfactory one, but we are not yet in possession of any other.

The velocity of the tidal wave is subject to much variation, and we are not yet in a position to lay down the laws which govern it; if the whole globe were uniformly covered, the velocity would be rather more than 1000 miles per hour ($7926 \times 3 \cdot 1415 \div 248$). It is probably, however, nowhere equal to this, unless perhaps in the Antarctic Ocean. The following table of velocities is given by Whewell[1] :—

	Miles.
In Latitude 60° S.	670
In the Atlantic	700
Azores to Cape Clear	500
Cape Clear to Duncansby Head	160
Buchan Ness to Sunderland	60
Scarborough to Cromer	35
North Foreland to London	30
London to Richmond	13

Concerning the general character of the great terrestrial tidal wave, we cannot do better than quote the following description by a well-known eminent geographer : —

" The Antarctic is the cradle of tides. It is here that the Sun and Moon have presided over their birth, and it is here, also, that they are, so to speak, to attend on the guidance of their own congenital tendencies. The luminaries continue to travel round the Earth (apparently) from east to west. The tides no longer follow them. The Atlantic, for example,

---

[1] *Phil. Trans.*, vol. cxxiii. p. 212. 1833.

opens to them a long, deep canal, running from north to south, and after the great tidal elevation has entered the mouth of this Atlantic canal, it moves continually northward; for the second 12 hours of its life it travels north from the Cape of Good Hope to Cape Horn, and at the end of the first 24 hours of its existence, has brought high water to Cape Blanco on the west of Africa, and Newfoundland on the American continent. Turning now round to the eastward, and at right angles to its original direction, this great tidal wave brings high water, during the morning of the 2nd day, to the western coasts of Ireland and England. Passing round the northern cape of Scotland, it reaches Aberdeen at noon, bringing high water also to the opposite coasts of Norway and Denmark. It has now been travelling precisely in the opposite direction to that of its genesis, and in the opposite direction, also, to the relative motion of the Sun and Moon. But its erratic course is not yet complete. It is now travelling from the northern mouth of the German Ocean, southwards. At midnight of the 2nd day it is at the mouth of the Thames, and wafts the merchandise of the world to the quays of the port of London. In the course of this rapid journey the reader will have noticed how the lines [on the map] in some parts are crowded together closely on each other, while in others they are wide asunder. This indicates that the tide wave is travelling with varying velocity. Across the southern ocean it seems to travel nearly 1000 miles an hour, and through the Atlantic scarcely less; but near some of the shores, as on the coast of India, as on the east of Cape Horn, as round the shores of Great Britain, it travels very slowly ; so that it takes more time to go from Aberdeen to London than over the arc of 120°, which reaches from 60° of southern latitude and 60° north of the Equator. These differences have still to be accounted for ; and the high velocities are invariably found to exist where the water is deep, while the low velocities occur in shallow water. We must therefore look to the conformation of the shores and bottom of the sea, as an important element in the phenomena of the tides." [1]

---

[1] Johnston, *Phys. Atlas.*

The effects of tides on rivers are often very striking; especially is this the case with the Avon at Bristol; when the tide is at its ebb, the river is little better than a shallow ditch, but when the waters have risen to the maximum height, an insignificant stream is converted into a broad and deep channel, navigable by the largest Indiamen.

The instinct of animals in respect of the tides is often very remarkable. A Scotch writer observes: " The accuracy with which cattle calculate the times of ebb and flow, and follow the diurnal variations, is such, that they are seldom mistaken, even when they have many miles to walk to the beach. In the same way they always secure their retreat from these insulated spots in such a manner that they are never surprised and drowned." [1]

In their passage up rivers, tides are gradually extinguished, as will be seen from the following table relating to the Thames [2] :—

	Height.	Distance from Mouth.
London (Docks) . . .	8 ft. 10 in.	60
Putney	10 2	$67\frac{1}{2}$
Kew	7 1	72
Richmond . . .	3 10	75
Teddington	1 $14\frac{1}{2}$	79

At certain places on the coast of Hampshire and Dorsetshire, the waters of the ocean ebb and flow *twice* in 12 hours instead of only once, as is usual elsewhere. Christchurch, Poole, and Weymouth may be mentioned as places where this singular phenomenon has been observed. [3]

The evident connection between the periods of the tides and those of the phases of the Moon, led to these phenomena being attributed to her action long before their true theory was understood. Pytheas of Marseilles is said to have been the first to point this out. [4] Julius Cæsar adverts to the connection existing between the Moon and spring tides. [5]

[1] Mac Culloch, *Highlands and Western Isles of Scotland.*
[2] *Phil. Trans.*, vol. cxxiii. p. 204. 1833.
[3] *Phil. Trans.*, vol. cxxiii. p. 226. 1833.
[4] Plutarch, *De Placitis*, lib. iii. cap. 17.
[5] *De Bello Gallico*, lib. iv. cap. 29.

Another well-known ancient philosopher says : " Æstus maris accedere et reciprocare maxime est, pluribus quidem modis accidit *verum causâ in sole lunâque*." [1] Kepler clearly indicated the principle of gravitation being concerned [2], an opinion which Galileo strongly dissented from.[3] Wallis, in 1666, also published a tidal theory.[4] Before Sir Isaac Newton turned his attention to this subject, the explanations given were at best but vague surmises; " to him was reserved the glory of discovering the true theory of these most remarkable phenomena, and of tracing, in all its details, the operation of the cause which produces them."

---

[1] Pliny, *Hist. Nat.*, lib. ii. cap. 99.
[2] *Epist. Ast.*, p. 555.   [3] *Dialoghi.*
[4] *Phil. Trans.*, vol. i. p. 263.   1666.

BOOK IV.

MISCELLANEOUS ASTRONOMICAL PHENOMENA.

—◆—

CHAPTER I.

*Variation in the Obliquity of the Ecliptic.—Precession.—Its Value.—
Its Physical Cause.—Correction for Precession.—History of its
Discovery.—Nutation.—Herschel's Definition of it.—Connection
between Precession and Nutation.*

Variation in the Obliquity of the Ecliptic. — Although it is
sufficiently near for all general purposes to consider the inclination of the plane of the ecliptic as invariable; yet this is
not strictly the case, inasmuch as it is subject to a small but appreciable change of about 48″ per century. This phenomenon
has long been known to astronomers, on account of the increase
it gives rise to, in the latitude of all stars in some situations,
and corresponding decrease in the opposite regions. Its
effect at the present time is to diminish the inclination of the
two planes of the equator and the ecliptic to each other; but
this diminution will not go on beyond certain very moderate
limits, after which it will again increase, and thus oscillate
backwards and forwards through an arc of 1° 21′: the time
occupied in one oscillation being about 10,000 years. One
effect of this variation of the plane of the ecliptic—that which
causes its nodes on a fixed plane to change—is associated with
the phenomena of the precession of the equinoxes, and undis-
tinguishable from it, except in theory.[1]

---

[1] The inclination of the ecliptic for the epoch of January 1, 1860, is
23° 27′ 27·36″.

Precession.—The precession of the equinoxes is a slow but continual shifting of the equinoctial points from East to West.[1] Celestial longitudes and right ascensions are reckoned from the Vernal Equinox, and if this was a fixed point, the longitude of a star would never vary, but would remain the same from age to age. Such, however, is not the case, as it has been found that *apparently* all the stars have changed their places since the first observations were made by the astronomers of antiquity. Two explanations only can be given to account for this phenomenon : we must either suppose that the whole firmament has advanced forwards, or that the equinoctial points have gone backwards. And as these points depend on the Earth's motion, it is far more reasonable to suppose that the phenomenon is owing to some perturbation of our globe rather than that the starry heavens should have a real motion relative to these points. The latter explanation is accordingly adopted, namely, that the equinoxes have a periodical retrograde motion from *East* to *West*, thereby causing the Sun to arrive at them sooner than it otherwise would had these points remained stationary. The annual amount of this motion is, however, exceeding small, being only equal to $50 \cdot 1''$ [2]; and since the circle of the ecliptic is divided into 360°, it follows that the time occupied by the equinoctial points in making a complete revolution of the heavens is 25,868 years. It is owing to precession that the pole-star varies from age to age, and also that whilst the sidereal year, or *actual* revolution of the Earth

---

[1] It may be well to mention, that the equinoxes are the two points where the ecliptic cuts the equator; and so called, because when the Sun, in its annual course, arrives at either of them, day and night are equal throughout the world. The point where the Sun crosses the equator, going north, is known as the *vernal equinox ;* and the opposite point, through which the Sun passes going south, as the *autumnal equinox.* These intersecting points are also termed Nodes, and an imaginary line joining the two, *the line of nodes.* The ascending node (☊) answers to the Vernal equinox; and the descending (☋), to the Autumnal.

[2] Bessel, by a careful discussion of the most reliable observations, fixed the value of general precession for the epoch of 1750 at $50 \cdot 21129''$, and the value of luni-solar precession at $50 \cdot 37572''$.

round the Sun, is 365d. 6h. 9m. 9·6s., the equinoctial, solar, or tropical year is only 365d. 5h. 48m. 49·7s. The successive returns of the Sun to the same equinoctial points must therefore precede its return to the same point on the ecliptic by 20m. 20s. of time, or 50·1″ of arc. It is also on account of the precession of the equinoxes that the signs of the ecliptic do not now correspond with the constellations of the same name, but lie about 30° westward of them. Thus, that division of the ecliptic known as the *sign* of Taurus lies in the *constellation* Aries, the sign of Aries having passed into Pisces. It should be remarked, however, that the signs and constellations coincided with one another about 370 B.C. In recent times, the attempts that have been made to establish the motion of the solar system through space have rendered an accurate knowledge of precession indispensable ; and the elaborate researches of the Russian astronomers, MM. Peters and Struve, have led to a slight modification in the value of the constants of precession adopted by Bessel [1], which may lead to important results.

" The cause of precession is to be found in the combined action of the Sun and Moon [2] upon the protuberant mass of matter accumulated at the Earth's equator, the attraction of the planets being scarcely sensible.[3] The attracting force of the Sun and Moon upon this shell of matter is of a two-fold character ; one parallel to the equator, and the other perpendicular to it. The tendency of the latter force is to diminish the angle which the plane of the equator makes with the ecliptic ; and were it not for the rotatory motion of the Earth, the planes would soon coincide ; but, by this motion, the planes remain nearly constant to each other. The effect produced by the action of the force in question is, however, that the plane of the equator is constantly, though slowly, shifting its place in the manner we have endeavoured to describe."

---

[1] *Tabulæ Regiomontanæ.*

[2] Called hence, *luni-solar* precession.

[3] When the value of the constant of precession, given at any time, includes the variation caused by the planets, it is called the constant of *general* precession.

In the reduction of astronomical observations the correction for precession in right ascension to be applied, is almost always additive; increasing in the regions round the poles of the heavens, but dwindling to zero near the poles of the ecliptic. It is in the space included between these poles in each hemisphere that the correction becomes subtractive; in the northern hemisphere, this small space comprehends the constellations lying near the XVIIIth hour of R.A., that being the R.A. of the north ecliptic pole; and in the southern hemisphere, the constellation lying near the VIth hour, that being the R.A. of the south ecliptic pole. The remarks we have just made apply only to those stars whose declination north or south exceeds 67°. The annual precession in declination, however, depends on the star's right ascension, both as to amount and direction. At VI and XVIII hours it is at zero; at XII hours it reaches the northern maximum of 20″; and at XXIV it reaches a similar southern maximum. From XVIII to XXIV hours, and from XXIV to VI hours, the precession is N., consequently additive to stars of north declination, but subtractive from those of south declination: but from VI to XVIII, the precession being S., it is additive to southern, and subtractive from northern stars.

The discovery of precession dates about 125 B.C., when it was detected by Hipparchus, by means of a comparison of his own observations with those of Timocharis and Aristyllus, made about 178 years previously: its existence was afterwards confirmed by Ptolemy.[1] It was Copernicus, however, who first gave the true explanation of the phenomenon, and Newton who discovered its physical cause.

Nutation.[2]—It must be borne in mind that the effect of precession varies according to the time of year, on account of the ever-varying distance of the Earth from the Sun and planets. Twice a year at the equinoxes, the influence of the Sun is at zero; and twice a year also, at the solstices, it is at its maximum. On no two successive days is it of exactly the same value, and consequently the precession of the equinoctial points must be

---

[1] *Almagest*, lib. vii.   [2] *Nutatio*, nodding.

uneven, and the obliquity of the ecliptic subject to a half-yearly variation; since the Sun's force which changes the obliquity is constantly varying, while the rotation of the Earth is continuous. This then gives rise to a small oscillating motion of the Earth's axis, termed the *solar nutation :* of a far more considerable amount, however, is the value of the nutation arising from the agency of the Moon; so much so that it was detected by Bradley before its existence had been inferred from theory.[1]

We cannot better explain the nature of precession than in nearly the words of Sir J. Herschel, who says : — " The nutation of the Earth's axis is a small and slow gyratory movement, by which, if subsiding alone, the pole would describe among the stars, in a period of $18\frac{1}{2}$ years, a minute ellipse having its longer axis equal to $18\cdot5''$, and its shorter to $13\cdot74''$ (the longer being directed towards the pole of the ecliptic, and the shorter of course at right angles to it) ; the semi-axis major is, therefore, equal to $9\cdot25''$, which quantity is called the *coefficient* of nutation. The consequence of this real motion of the pole is an apparent advance and recess of all the stars in the heavens to the pole in the same period. Since, also, the place of the equinox on the ecliptic is determined by the place of the pole in the heavens, the same agency will cause a small alternating motion to and fro of the equinoctial points, by which, in the same periods, both the longitudes and the right ascensions of the stars will be alternately increased and diminished.

" Precession and nutation, although for convenience here considered separately, in reality exist together; they are, in fact, constituent parts of the same general phenomenon : and since, while in virtue of this nutation, the pole is describing its little ellipse of $18\cdot5''$ in diameter, it is carried on by the greater and regularly progressive motion of precession over so much of its circle round the pole of the ecliptic as corresponds to $18\frac{1}{2}$ years — that is to say, over an angle $18\frac{1}{2}$ times $50\cdot1''$ round the centre (which, in a small circle of $23°\ 28'$ in diameter, corresponds to $6'\ 20$, as seen from the centre of the sphere);

[1] *Phil. Trans.,* vol. xlv. p. 1, *et seq.* 1748.

the path which it will pursue in virtue of the joint influence
of the 2 motions will be neither an ellipse nor an exact circle,
but a slightly undulating ring.

" These movements of precession and nutation are common
to all the celestial bodies, both fixed and erratic ; and this
circumstance makes it impossible to attribute them to any
other cause than the real motion of the Earth's axis, as we
have described.   Did they only affect the stars, they might,
with equal plausibility, be considered as arising from a *real*
rotation of the starry heavens as a solid shell round our axis,
passing through the poles of the ecliptic in 25,868 years, and
a real elliptic gyration of *that* axis in rather more than 18
years : but since they also affect the Sun, Moon, and planets,
which, having motions independent of the general body of the
stars, cannot without extravagance be supposed to be *attached
to* the celestial conclave, this idea falls to the ground ; and
there only remains, then, a real motion of the Earth by which
they can be accounted for." [1]

---

[1] *Treatise on Ast.*, p. 172.   1833.

CHAPTER II.

Aberration. — The Constant of Aberration. — Familiar Illustration. — History of the Circumstances which led to its discovery by Bradley. — Parallax. — Explanation of its Nature. — Parallax of the Heavenly Bodies. — Parallax of the Moon. — Importance of a correct determination of the Parallax of an Object.—Leonard Digges on the Distance of the Planets from the Earth.

Aberration. — The aberration of light is another important phenomenon which requires to be taken into consideration in the reduction of rough astronomical observations. Although light travels with the enormous velocity of 192,000 miles per second—a speed so great, that for all practical terrestrial purposes we may consider it to be propagated instantaneously ; yet the astronomer, who has to deal with distances of millions of miles, is obliged to be more particular. A simple illustration will show this : the mean distance of our globe from the Sun is 95,000,000 of miles, and since light travels at the rate of 192,000 miles, we can ascertain by a simple arithmetical process that the time occupied by a ray of light reaching us from the Sun is 8m. 18s., so that in point of fact, in looking at the Sun at a given moment, we do not see it shining as it is, but as it did 8m. 18s. previously. If the Earth were at rest, this would be all very well ; but since the Earth is in motion, when the solar ray enters the eye of a person on its surface, he will be some way removed from the point in space at which he was situated when the ray left the Sun ; he will consequently see that luminary behind the true place it actually occupies when the ray enters his eye. In the course of 8m. 8s. the Earth will have advanced in its orbit 20˙4192″ ; this quantity is called the *constant of aberration*. Aberration may be summed up as a phenomenon resulting from the combined effect of the motion of light and the Earth's motion in its orbit. Suppose a ball let fall from a point P above the horizontal line AB, and that a tube of which A is the lower extremity were placed to receive it; if the tube were fixed the ball would strike it on the lower side, but if the tube were carried forward in the direction AB, with a velocity properly adjusted at every instant to that of the ball, while *preserving its inclination* to the horizon, so that when the ball in its

Fig. 20.

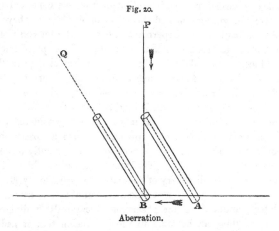

Aberration.

natural descent reached B, the tube would have been carried into the position BQ, it is evident that the ball throughout its whole descent would be found in the tube; and a spectator referring to the tube the motion of the ball, and carried along with the former, unconscious of its motion, would fancy that the ball had been moving in an inclined direction and had come from Q. The following similes are frequently made use of to explain aberration: a shower of rain descending perpendicularly will appear to fall in its true direction to a person at rest, but if he move rapidly through it, it will meet him in a slanting direction : in other words, it will have an apparent as well as a real motion. A cannon-ball fired from a shore-battery at a vessel passing up a river, will not pass through the ship in a line coincident with the direction of the ball, but will emerge on the other side at a point differing more or less from this line; the amount of the variation, however, will depend on the relative velocities of the ball and ship at the time. If we suppose the cannon-ball to represent light, and the movement of the ship the motion of the Earth in its orbit, we have an excellent illustration of the nature of aberration.

This unquestionably grand discovery resulted more immediately from an attempt to detect stellar parallax. Although the facts revealed by the invention of the telescope and the discovery of gravitation had the effect of establishing beyond doubt the truth of the Copernican theory of the universe, still it was much to be desired that some more *direct* proof might be adduced. The absence of any appreciable change in the positions of the fixed stars, when examined from opposite sides of the Earth's orbit, was one of the earliest, and at the same time most serious, arguments brought against the adherents of Copernicus and his system; as it was always considered that the detection of such a change would furnish an irresistible proof that the Earth was not at rest, and consequently was not the centre of the system. The first observation which ultimately led to the discovery of aberration was made by Hook, who selected the star γ Draconis as a suitable one by which to detect parallax.[1] After observing it carefully at different seasons of the year, he came to the conclusion that it had a sensible parallax. It was soon found, however, that the star was subject to a displacement in a direction contrary to what ought to have resulted had the star been affected by parallax only; and it was for the purpose of endeavouring to ascertain the physical cause of this strange phenomenon that induced Bradley to obtain an instrument for himself, that he might more conveniently study the subject. His observations completely confirmed those of Hook, and " at length the happy

---

[1] Hook considered it desirable to observe stars as near the zenith as possible, in order to avoid the effects arising from any uncertainty as to the value of refraction; and γ Draconis happened to be the only bright star passing within a few minutes of the zenith of Gresham College, when his instrument was erected. (*Attempt to prove the Motion of the Earth*, p. 7.)

idea occurred to him, that the phenomenon might be completely accounted for by the gradual propagation of light combined with the motion of the Earth in its orbit."

Parallax " is the apparent change of place which bodies undergo by being viewed from different points." This is the general signification of the word; but with the astronomer it has a conventional meaning, and implies the difference of the apparent positions of any celestial object when viewed from the surface of the Earth and from the centre, the latter being

Fig. 21.

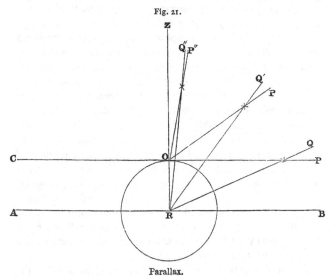

Parallax.

the point to which it is found convenient to refer all astronomical observations. The position of a heavenly body, as seen from the Earth's surface, is called its *apparent place*; and that in which it would be seen, were the observer stationed at the Earth's centre, is known as the *true* or *mean place*. It follows, therefore, that the altitudes of the heavenly bodies are depressed by parallax, which is greatest at the horizon[1], and

---

[1] Such is the case: because imaginary lines, drawn from the object to the observer, and to the centre of the Earth respectively, will then have the greatest possible inclination to each other.

decreases as the altitude of the object increases, until it disappears altogether at the zenith.   In figure 21, Z is the zenith,
C P, the visible horizon, A B, the rational horizon, O, the position of an observer, and R, the centre of the Earth.   From O,
the observer will see these stars projected on the sky at P, P',
and P'' (apparent places); but referred to the centre of the
Earth, the points of projection will be Q, Q', and Q'' (mean
places).   The general nature of parallax may be readily understood by supposing 2 persons placed individually at the end of a
straight line, to look at a carriage standing in front of a house
at the distance (say) of 50 yards from each other.   It is evident that the carriage will appear to each spectator projected
upon a different part of the house.   The angle which this
difference of position gives rise to is the angle of parallax.
Let us suppose the 2 observers (still at the same distance
from each other) to recede from the carriage; the angle of
parallax will become more and more acute, until at length
it will become insensible.   The example we have here adduced holds good also for the heavenly bodies.   The Sun,
Moon, and planets, though separated from us by millions of
miles, are affected by parallax to a small but nevertheless
appreciable amount.   With but a few exceptions, however,
this is not the case with the fixed stars : in only 9 instances
has parallax been detected, and we are enabled to ascertain
that the star *nearest* to us is α Centauri, whose parallax is equal
to only 0·913'', which is equivalent to 21,508,000,000,000
miles, — but we are anticipating.

Of all the heavenly bodies, the horizontal parallax of the
Moon is the most considerable, that luminary being the nearest
to the Earth.   It is found in the following way : — Suppose
that 2 astronomers take their stations in the same meridian,
but one south of the equator, as at the Cape of Good Hope, and
another north of the equator, as at Berlin in Prussia, which
2 places lie nearly in the same meridian : the observers
would severally refer the Moon to different points on the face
of the sky — the southern observer carrying it further to the
north, and the northern observer further to the south — than its
true place as seen from the centre of the Earth.   The observa-

tions thus made at the 2 places furnish the materials for calculating, by means of trigonometry, the value of the horizontal parallax of the Moon, from which we can deduce both its distance and real magnitude. The parallax thus obtained is called the *diurnal*, or geocentric, in contradistinction to the *annual*, or heliocentric; by which, in general, is understood the displacement of a celestial object according as it is viewed from the Earth or the Sun: in particular, however, it denotes the angle formed by 2 imaginary lines drawn from each extremity of the diameter of the Earth's orbit to a fixed star. This angle is generally too small to be appreciable, as we have before mentioned. It was this fact of the non-detection of annual parallax which for a long period of time prior to the invention of the telescope formed a great obstacle to the progress of the Pythagorean and Copernican opinions relative to the system of the universe.

We may obtain some idea of the importance attaching to a correct determination of the parallax of an object, by an inspection of the following table:—

If the Sun's horizontal parallax were $11''$, the mean distance of the following planets from the Sun would be:

The Earth.	Mars.	Jupiter.	Saturn.
75,000,000	114,276,750	390,034,500	715,504,500

If the Sun's parallax were $10''$, the above distance would become:

82,000,000	124,942,580	426,478,720	782,284,920

Errors arising from a mistake of only $1''$:

7,000,000	10,665,830	36,444,220	66,780,420 [1]

The Sun's true parallax being $8\cdot5776$ [2], the real distances are:

95,298,260	144,750,000	470,450,000	906,190,000 [3]

It is only comparatively within the last few years that the efforts of astronomers to detect stellar parallax have been

---

[1] Ferguson's *Astronomy*, p. 76, 2nd Edition, London, 1757

[2] According to Encke. *Der Venusdurchgang von* 1769, p. 108. Gotha: 1824.

[3] We may consider that the mean distance of the Sun from the Earth is now known to within $\frac{1}{300}$ of the whole amount, or about 300,000 miles.

attended with any amount of success. Planetary parallax is of course of an older date. Pliny considered such investigations to be but little better than madness, and Riccioli remarks, " Parallaxis et distantia fixarum, non potest certa et evidenti observatione humanitas comprehendi." Leonard Digges, an old English writer, however, seems to have found no difficulty in the matter; he gives the following table of distances, which, however, unfortunately for his reputation, has turned out to be slightly incorrect. He adds, " Here demonstration might be made of the distance of these orbs, but that passeth the capacity of the cŏmŏ sort." Here are his results[1] :—

				Myles.
From the Earthe to the Moone	.	.	.	15,750
From the Moone to Mercurie	.	.	.	12,812
From Mercurie to Venus	.	.	.	12,812
From Venus to the Sunne	.	.	.	23,437$\frac{1}{2}$
From the Sunne to Mars	.	.	.	15,725
From Mars to Jupiter	.	.	.	78,721
From Jupiter to Saturn	.	.	.	78,721
From Saturn to the Firmament	.	.	.	120,485

Whence it follows, according to Digges, that the distance of the city of London from Sirius is exactly 358,463$\frac{1}{2}$ miles!

---

[1] *Prognostication Everlasting,* 1556.

CHAPTER III.

Refraction. — Its Nature. — Importance of a Correct Knowledge of its Amount. — Table of the Correction for Refraction. — Effect of Refraction on the Position of Objects in the Horizon. — History of its Discovery. — Twilight. — How Caused. — Its Duration. — The Zodiacal Light. — Its Appearance and Extent. — Sir J. Herschel's Opinion. — Historical Notices.

Refraction. — Besides the change of place, to which the heavenly bodies are subjected in consequence of the effects of parallax, atmospheric refraction gives rise to a considerable displacement; and it is this power which the air, in common with all transparent media, possesses, which renders a knowledge of the constitution of the atmosphere very important to the astronomer. "In order to understand the nature of refraction, we must consider that an object always appears in the direction in which the *last* ray of light comes to the eye. If the light which comes from a star were bent into 50 directions before it reached the eye, the star would nevertheless appear in a line described by the ray nearest the eye. The operation of this principle is seen when an oar, or any stick, is thrust into the water. As the rays of light by which the oar is seen have their direction changed as they pass out of water into air, the apparent direction in which the body is seen is changed in the same degree, giving it a bent appearance — the part below the water having apparently a different direction from the part above." [1]

The direction of this refraction is determined by a general law in optics, from which we learn that, when a ray of light passes out of a rarer into a denser medium — *e. g.*, out of air

---

[1] Olmsted, *Mechanism of the Heavens*, p. 94. Edin. Ed.

into water, or out of space into the Earth's atmosphere—it is
turned *towards* a perpendicular to the surface of the medium;
but when it passes out of a denser into a rarer medium, it is
turned *from the perpendicular*. It accordingly follows from
this, that the effect of refraction is, that the apparent altitude
of a heavenly body is greater than the true altitude; so that
any object situated actually *in* the horizon will appear *above*
it. Indeed, some that are actually below the horizon, and
which would be otherwise invisible were it not for the existence of refraction, are thus brought into sight. It was in
consequence of this that, on April 20, 1837, the Moon rose
eclipsed before the Sun had set.

Fig. 22.

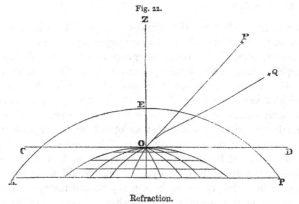

Refraction.

In fig. 22 Z is the zenith, C D, the visible horizon, A B,
the rational horizon, A E B, the boundary of the Earth's
atmosphere. Then the light of the star Q, will, to the observer at O, seem to come from the point P.

A correct determination of the exact amount of atmospheric
refraction, or the angular displacement of a celestial object at
any altitude, is of considerable practical importance, but a very
difficult subject of inquiry, owing to the fact that the density
of any stratum of air (on which its refractive power depends)
is dependent on the operation of certain meteorological phenomena with which we are at present but very imperfectly acquainted. Thus the quantity of refraction at any given alti-

tude depends, not only on the density, but also on the thermo-
metric and hygrometric condition of the air through which
the visual ray passes. And although we know the general
fact that the barometric pressure[1] and temperature[2] constantly
diminish as we recede from the Earth's surface, yet the law of
this diminution is not fully ascertained. In consequence of
our ignorance on these points, a certain degree of uncertainty
is introduced into the determination of the amount of refraction, whence some embarrassment arises in cases when the
quantities to be detected in astronomical observations are extremely minute. Nevertheless it must be remembered that,
inasmuch as the total amount of refraction is never considerable, and in most cases very small, it can be so nearly
estimated as to offer no serious impediment to the astronomer.

Tables are now in general use, constructed partly from
observation and partly from theory, by means of which we
can ascertain approximately the refraction for any given altitude; additional rules being given by which this average
refraction may be corrected according to the physical state of
the air at the time of observation. At the zenith, or at an
altitude of 90°, there is no refraction whatever, objects being
seen in the same position they would have were the Earth
surrounded by no atmosphere at all. In descending from the
horizon towards the zenith, the refraction constantly increases,
objects near the horizon being displaced in a greater degree
than those at high altitudes. Thus the refraction, which at
an altitude of 45° is only equal to 57″, at the horizon, is in-

---

[1] Since the barometer rises with an increase in the weight and density
of the air, its rise causes an augmentation, and its fall a decrease, of
refraction. It will be tolerably near the truth if we assume that the
refraction at any given altitude is increased or diminished by $\frac{1}{300}$ of
its mean amount, for every 10th of an inch by which the barometer
exceeds or falls short of 30 inches.

[2] Also as an increase of temperature causes a decrease of density, it
follows that the elevation of the thermometer diminishes the effect of
refraction, the barometer remaining stationary. We may assume that
the refraction at any given altitude is increased or diminished by $\frac{1}{420}$
of its mean amount, for each degree by which the thermometer exceeds
or falls short of the mean temperature of 55°.

creased to no less than 33'. The rate of the increase at high
altitudes is nearly in proportion to the tangent of the apparent
angular distance of the object from the zenith; but in the
vicinity of the horizon this rule ceases to hold good, and the
law becomes much more complicated in its expression. The
mean diameter both of the Sun and Moon being rather less
than 33', it follows that, when we see the lower edge of either
of these luminaries apparently just *touching* the horizon, in
reality its outer disc is completely *below* it, and would be
altogether hidden by the convexity of the Earth were it not
for the refraction.

It is also under these circumstances that one of the most
curious effects resulting from atmospheric refraction may
often be noticed. We refer to the oval outline presented by
the Sun and Moon when near the horizon. This arises from
the unequal refraction of the upper and lower limbs. The
latter being nearer the horizon, is more affected by refraction,
and consequently is raised in a greater degree than the former,
" the effect being to bring the two limbs apparently closer
together by the difference of the two refractions. The form
of the disc is therefore affected as if it were pressed between
two forces, one acting above and the other below, tending to
compress its vertical diameter, and to give it the form of an
ellipse, the lesser axis of which is vertical and the greater
horizontal."

The dim and hazy appearance of objects seen in the horizon,
is not only occasioned by the rays of light having to traverse
a larger space in the atmosphere, but also from their having
to pass through the lower and denser part. " It is estimated
that the solar light is diminished 1300 times in passing
through these lower strata, and we are thereby enabled to
gaze upon the Sun, when setting, without being dazzled by
his beams."

The famous astronomer Claudius Ptolemy was the first who
remarked that a ray of light proceeding from a star to the
Earth underwent a change of direction in proceeding through
the atmosphere.[1] He moreover stated that the displacement was

---

[1] *Almag.*, lib. viii. cap. 6.

greatest at the horizon, would be diminished as the altitude increased, and finally vanish altogether at the zenith—an assertion which we have already seen to be perfectly correct. In the 16th century, Tycho Brahe also investigated the subject; and his results, though by no means so accurate as Ptolemy's, are interesting from the fact that they were the first which were reduced to the form of a table.[1] Since this period many eminent astronomers have devoted their attention to the matter, and the tables now in most general use are those of the late illustrious Bessel.

MEAN REFRACTION OF CELESTIAL OBJECTS FOR TEMPERATURE 50°, AND PRESSURE 29·6 INCHES.

Alt.	Refr.	Alt.	Refr.	Alt.	Refr.	Alt.	Refr.	A lt.	Refr.
° '	' "	° '	' "	° '	' "	° '	' "	° '	' "
0 0	33 0	20	17 04	40	10 29	7 0	7 20	20	5 6
5	32 10	25	16 44	45	10 20	5	7 15	25	5 34
10	31 22	30	16 24	50	10 11	10	7 11	30	5 31
15	30 36	35	16 04	55	10 02	15	7 06	35	5 28
20	29 50	40	15 45	5 0	9 54	20	7 02	40	5 25
25	29 06	45	15 27	5	9 46	25	6 57	45	5 23
30	28 23	50	15 09	10	9 38	30	6 53	50	5 20
35	27 41	55	14 52	15	9 30	35	6 49	55	5 18
40	27 00	3 0	14 36	20	9 23	40	6 45	10 0	5 15
45	26 20	5	14 20	25	9 15	45	6 41	10	5 10
50	25 42	10	14 04	30	9 08	50	6 37	20	5 05
55	25 05	15	13 39	35	9 01	55	6 33	30	5 00
1 0	24 29	20	13 34	40	8 54	8 0	6 29	40	4 56
5	23 54	25	13 20	45	8 47	5	6 25	50	4 51
10	23 20	30	13 06	50	8 41	10	6 22	11 0	4 47
15	22 47	35	12 53	55	8 34	15	6 18	10	4 43
20	22 15	40	12 40	6 0	8 28	20	6 15	20	4 39
25	21 44	45	12 27	5	8 21	25	6 11	30	4 34
30	21 15	50	12 15	10	8 15	30	6 08	40	4 31
35	20 46	55	12 03	15	8 09	35	6 05	50	4 27
40	20 18	4 0	11 51	20	8 03	40	6 01	12 0	4 23
45	19 51	5	11 40	25	7 57	45	5 58	10	4 20
50	19 25	10	11 29	30	7 51	50	5 55	20	4 16
55	19 00	15	11 18	35	7 45	55	5 52	30	4 13
2 0	18 35	20	11 08	40	7 40	9 0	5 48	40	4 09
5	18 11	25	10 58	45	7 35	5	5 45	50	4 06
10	17 48	30	10 48	50	7 30	10	5 42	13 0	4 03
15	17 26	35	10 39	55	7 25	15	5 39	10	4 00

[1] *Progymnasmata*, Part i. p. 39.

Alt.	Refr.	Alt.	Refr.	Alt.	Refr.	Alt.	Refr.	Alt.	Refr.
° '	' "	° '	' "	° '	' "	° '	' "	° '	' "
20	3 57	30	2 40	40	1 58	36 0	1 18	55 0	0 40
30	3 54	40	2 38	50	1 57	30	1 17	56 0	0 38
40	3 51	50	2 37	26 0	1 56	37 0	1 16	57 0	0 37
50	3 48	20 0	2 35	10	1 55	30	1 14	58 0	0 35
14 0	3 45	10	2 34	20	1 55	38 0	1 13	59 0	0 34
10	3 43	20	2 32	30	1 54	30	1 11	60 0	0 33
20	3 40	30	2 31	40	1 53	39 0	1 10	61 0	0 32
30	3 38	40	2 29	50	1 52	30	1 09	62 0	0 30
40	3 35	50	2 28	27 0	1 51	40 0	1 08	63 0	0 29
50	3 33	21 0	2 27	15	1 50	30	1 07	64 0	0 28
15 0	3 30	10	2 26	30	1 49	41 0	1 05	65 0	0 26
10	3 28	20	2 25	45	1 48	30	1 04	66 0	0 25
20	3 26	30	2 24	28 0	1 47	42 0	1 03	67 0	0 24
30	3 24	40	2 23	15	1 46	30	1 02	68 0	0 23
40	3 21	50	2 21	30	1 45	43 0	1 01	69 0	0 22
50	3 19	22 0	2 20	45	1 44	30	1 00	70 0	0 21
16 0	3 17	10	2 19	29 0	1 42	44 0	0 59	71 0	0 19
10	3 15	20	2 18	20	1 41	30	0 58	72 0	0 18
20	3 12	30	2 17	40	1 40	45 0	0 57	73 0	0 17
30	3 10	40	2 16	30 0	1 38	30	0 56	74 0	0 16
40	3 08	50	2 15	20	1 37	46 0	0 55	75 0	0 15
50	3 06	23 0	2 14	40	1 36	30	0 54	76 0	0 14
17 0	3 04	10	2 13	31 0	1 35	47 0	0 53	77 0	0 13
10	3 03	20	2 12	20	1 33	30	0 52	78 0	0 12
20	3 01	30	2 11	40	1 32	48 0	0 51	79 0	0 11
30	2 59	40	2 10	32 0	1 31	30	0 50	80 0	0 10
40	2 57	50	2 09	20	1 30	49 0	0 49	81 0	0 9
50	2 55	24 0	2 08	40	1 29	30	0 49	82 0	0 8
18 0	2 54	10	2 07	33 0	1 28	50 0	0 48	83 0	0 7
10	2 52	20	2 06	20	1 26	30	0 47	84 0	0 6
20	2 51	30	2 05	40	1 25	51 0	0 46	85 0	0 5
30	2 49	40	2 04	34 0	1 24	30	0 45	86 0	0 4
40	2 47	50	2 03	20	1 23	52 0	0 44	87 0	0 3
50	2 46	25 0	2 02	40	1 22	30	0 44	88 0	0 2
19 0	2 44	10	2 01	35 0	1 21	53 0	0 43	89 0	0 1
10	2 43	20	2 00	20	1 20	30	0 42	90 0	0 0
20	2 41	30	1 59	40	1 19	54 0	0 41		

CORRECTION OF MEAN REFRACTION.

Ap. Alt.	Height of the Thermometer.															
	20°	24°	28°	32°	36°	40°	44°	48°	52°	56°	60°	64°	68°	72°	76°	80°
∘ ′′	′+′′	′+′′	′+′′	′+′′	′+′′	′+′′	′+′′	′+′′	−′′	′−′′	′−′′	′−′′	′−′′	′−′′	′−′′	′−′′
0′ 0	2·40	2·18	1·55	1·33	1·11	51	31	10	1 0	29	48	1· 7	1·25	1·43	2· 1	2·19
0·10	2·32	2·12	1·49	1·28	1· 8	48	29	10	9	27	45	1· 4	1·21	1·38	1·54	2·12
0·20	2·25	2·05	1·44	1·24	1· 4	46	28	9	9	26	44	1· 1	1·17	1·33	1·49	2·05
0·30	2·18	1·59	1·39	1·20	1· 1	44	26	9	8	25	41	58	1·13	1·28	1·43	1·59
0·40	2·11	1·53	1·34	1·16	58	42	25	8	8	24	39	55	1·10	1·24	1·38	1·53
0·50	2·05	1·48	1·29	1·12	55	40	24	8	8	23	37	52	1· 6	1·20	1·34	1·48
1· 0	1·59	1·43	1·25	1· 9	53	38	23	8	7	21	36	50	1. 3	1·17	1·30	1·43
1·10	1·53	1·38	1·21	1· 6	50	36	22	7	7	20	34	48	1· 0	1·13	1·26	1·38
1·20	1·48	1·33	1·17	1· 3	48	34	21	7	6	19	32	45	57	1· 9	1·21	1·33
1·30	1·43	1·29	1·14	1· 0	46	32	20	7	6	18	31	43	54	1· 6	1·18	1·29
1·40	1·39	1·25	1·11	57	44	31	18	6	6	18	30	41	52	1· 4	1·15	1·25
1·50	1·35	1·21	1· 8	55	42	30	17	6	6	17	28	39	50	1· 1	1·11	1·21
2·00	1·31	1·18	1· 5	53	39	29	17	6	5	16	27	37	48	58	1· 8	1·18
2·20	1·23	1·11	1· 0	48	37	26	16	5	5	15	25	35	44	54	1· 3	1·11
2·40	1·17	1·06	55	44	34	24	14	5	5	14	23	32	41	50	58	1· 6
3· 0	1·11	1· 1	51	41	32	22	13	4	4	13	21	30	38	46	54	1· 1
3·20	1·06	57	47	38	29	21	13	4	4	12	20	28	35	43	50	0·57
3·40	1· 2	53	44	36	28	20	12	4	4	11	18	26	33	40	47	53
4· 0	58	49	41	33	26	18	11	4	4	10	17	24	31	37	44	50
4·30	53	45	38	31	24	17	10	3	3	9	16	22	28	34	40	45
5· 0	48	41	35	28	22	16	9	3	3	9	14	20	26	31	36	40
5·30	45	38	32	26	20	14	9	3	3	8	13	19	24	29	34	38
6· 0	41	35	30	24	19	13	8	3	2	7	12	17	22	26	31	35
6·30	38	33	28	22	17	12	7	2	2	7	11	15	20	24	29	33
7· 0	36	31	26	21	16	12	7	2	2	6	10	14	19	23	27	31
8	32	27	23	19	15	10	6	2	2	5	9	13	16	20	24	27
9	28	24	20	16	13	9	5	2	2	5	8	11	14	18	21	24
10	26	22	18	15	12	8	5	2	1	4	7	10	13	16	19	22
11	23	20	17	14	11	8	5	2	1	4	7	9	12	15	18	20
12	21	18	15	13	10	7	4	1	1	4	6	9	11	13	16	18
13	20	17	14	12	9	7	4	1	1	3	6	8	10	12	15	17
14	18	16	13	11	8	6	4	1	1	3	5	7	9	11	14	16
15	17	15	12	10	8	6	3	1	1	3	5	7	9	11	13	15
16	16	14	12	9	7	5	3	1	1	3	5	6	8	10	12	14
17	15	13	11	9	7	5	3	1	1	3	4	6	8	9	11	13
18	14	12	10	8	6	5	3	1	1	2	4	6	7	9	10	12
19	13	11	9	8	6	4	3	1	1	2	4	5	7	8	10	11
20	13	11	9	7	6	4	2	1	1	2	4	5	6	8	9	11
25	10	8	7	6	5	3	2	1	1	2	3	4	5	6	7	8
30	8	7	6	5	4	3	2	1	0	1	2	3	4	5	6	7
35	7	6	5	4	3	2	1	0	0	1	2	3	3	4	5	6

Ap. Alt.	Height of the Thermometer.															
	20°	24°	28°	32°	36°	40°	44°	48°	52°	56°	60°	64°	68°	72°	76°	80°
° ''	'+''	'+''	'+''	'+''	'+''	'+''	'+''	'+''	'−''	'−''	'−''	'−''	'−''	'−''	'−''	'−''
40	6	5	4	3	3	2	1	0	0	1	2	2	3	3	4	5
45	5	4	3	3	2	2	1	0	0	1	1	2	2	3	3	4
50	4	3	3	2	2	1	1	0	0	1	1	2	2	2	3	3
55	3	3	2	2	2	1	1	0	0	1	1	1	2	2	2	3
60	3	2	2	2	1	1	1	0	0	0	1	1	1	2	2	2
65	2	2	2	1	1	1	0	0	0	0	1	1	1	1	2	2
70	2	1	1	1	1	1	0	0	0	0	0	1	1	1	1	1
80	1	1	1	0	0	0	0	0	0	0	0	0	0	1	1	1
90	0	0	0	0	0	0	0	0	0	0	0	0	0	0	0	0
Height of Barometer.	−	−	−	−	−	−	+	+	+	+	+					
	28·26	28·56	28·85	29·15	29·45	29·75	30·05	30·35	30·64	30·93						

Twilight.—This is another phenomenon depending on the agency of the atmosphere with which the Earth is surrounded. It is due partly to refraction and partly to reflection, but chiefly to the latter cause. After sunset the Sun still continues to illuminate the clouds and upper strata of the air, in the same manner that it may be seen shining on the summits of lofty mountains long after it has disappeared from the view of the inhabitants of the adjacent plains. The air and clouds thus illuminated reflect back part of that light to the surface beneath them, and thus produce, after sunset and before sunrise, that light, more or less feeble according as the Sun is more or less depressed, which we call "twilight." Immediately after the Sun has disappeared below the horizon, all the clouds in the vicinity are so highly illuminated as to be able to reflect an amount of light but little inferior to the direct light of the Sun. As the Sun, however, sinks lower and lower, less and less of the visible atmosphere receives its light, and consequently less and less of it is reflected to the Earth's surface surrounding the position where the observer is stationed, until at length, though by slow degrees, all reflection is at an end, and night ensues. The same thing occurs before sunrise; the darkness of night gradually giving place to the faint light of dawn, until the Sun appears above the horizon, producing the full light of day.

The duration of the light is usually reckoned to last until the Sun's depression below the horizon amounts to 18°; this, however, varies; in the tropics a depression of 16° or 17° is sufficient to put an end to the phenomenon, but in England a depression of 17° to 21° is required. The duration of twilight differs in length in different latitudes, it varies also in the same latitude at different seasons of the year, and depends in some measure on the meteorological condition of the atmosphere. Strictly speaking, in the latitude of Greenwich, there is no true night from May 22 to July 21, but constant twilight from sunset to sunrise. It reaches its minimum 3 weeks before the Vernal equinox, and 3 weeks after the Autumnal equinox, when its duration is 1h. 50m. At midwinter it is longer by about 17m., but the augmentation is frequently not perceptible, owing to the greater prevalence of clouds, &c. at that season of the year, which intercept the light and hinders it from reaching the Earth. The duration is least at the equator (1h. 12m.), and increases as we approach the poles, since at the former there are 2 twilights every 24 hours, but at the latter only 2 in a year. At the North Pole the Sun is below the horizon for 6 months; but from January 29 to the Vernal equinox, and from the Autumnal equinox to Nov. 12, the Sun is less than 18° below the horizon, so that there is twilight during the whole of these intervals, and thus the length of the actual night is reduced to 2½ months. The length of the day in these regions is about 6 months, during the whole of which time the Sun is constantly above the horizon. The general rule is, *that to the inhabitants of an oblique sphere the twilight is longer in proportion as the place is nearer the elevated pole.*

The Zodiacal Light.—This is a peculiar nebulous light of a conical or lenticular form[1], which may very frequently be noticed in the evening soon after sunset, during the 2 or 3 months preceding and following the Vernal equinox. It extends upwards from the western horizon nearly in a line with the ecliptic, or, more exactly, in the plane of the Sun's

---

[1] *Lens*, a lentil.

equator. The apparent angular distance of its vertex from
the Sun's plane varies, according to circumstances, from 50°
to 70°, or even more; and the breadth of its base, at right
angles to the major axis, from 8° to 30°. It usually reaches
to a point in the heavens situated not far from the Pleiades in
Taurus. It is always so extremely ill defined at the edges,
that great difficulty is experienced in satisfactorily determining
its limits. Besides being seen in the evening, it is more
rarely to be noticed early in the morning, previous to sunrise,
in the months of September to December. In these northern
latitudes the Zodiacal Light is generally, though not always,
inferior in brilliancy to the Milky Way; but in the tropics it
is seen to far greater advantage. Humboldt says it is of
almost constant occurrence in those regions, and that he has
himself seen it sufficiently luminous to cause a sensible-glow
on the opposite quarter of the heavens.[1] In the winter of
1842-43 it was remarkably well seen in this country, the apex
of the cone attaining a length of no less than 105° from the
Sun.[2] Mr. Lassell also mentions having seen the light very
prominent at Malta in January, 1850.[3]

No satisfactory explanation has yet been given of this phenomenon; it is, however, very generally considered to be a
kind of envelope surrounding the Sun, and extending perhaps
nearly or quite as far as the Earth's orbit. Sir J. Herschel's
opinion is, that it may be conjectured to be no other than the
denser parts of that medium which we have some reason to
believe resists the motions of comets; loaded, perhaps, with
the actual materials of the tails of millions of these bodies, of
which they have been stripped in their successive perihelion
passages [! !]. An *atmosphere* of the Sun, in any proper sense
of the word, it cannot be; since the existence of a gaseous
envelope propagating pressure from part to part—subject to

---

[1] See *infra*, p. 184.
[2] Detailed particulars will be found in the *Greenwich Observations*,
1842.
[3] For observations by E. J. Lowe, vide *Month. Not.* R.A.S., vol. x.
p. 124, and vol. xi. p. 132.

mutual friction in its strata, and thereby rotating in the same, or nearly the same, time with the central body, and of such dimensions and ellipticity — is utterly incompatible with dynamical laws.[1]

The Zodiacal Light was first treated of by Kepler; afterwards by Descartes, about the year 1630; and then by Childrey, in 1659[2]; it was not, however, till D. Cassini published some remarks on it, that much attention was paid to this phenomenon. He saw it first on March 18, 1683. It has been suggested that the Zodiacal Light is identical with what Pliny calls the "Trabes."[3] In connection with Sir J. Herschel's theory, given above, it may be mentioned that, during the visibility of the great comet of 1843, in March of that year, the Zodiacal Light was unusually brilliant; so much so, that by many persons it was mistaken for the comet.

Within the last few years some curious particulars relating to the Zodiacal Light have transpired. In the year 1855, the Rev. G. Jones, Chaplain of the U.S. steam-frigate "Mississippi," published some remarks on this phenomenon[4], as brought under his notice during a cruise round the world in the 2 preceding years. He states: "I was also fortunate enough to be twice near the latitude of 23° 28′ north, where the Sun was at the opposite solstice, in which position the observer has the ecliptic at midnight at right angles with his horizon, and bearing east and west. Whether this latter circumstance affected the result or not, I cannot say ; but I there had the extraordinary spectacle of the Zodiacal Light, simultaneously at both east and west horizons, from 11 to 1 o'clock, for several nights in succession."

Mr. Jones concludes his very interesting letter as follows: —
"You will excuse my prolixity in stating these varieties of

---

[1] *Outlines of Ast.*, p. 658.
[2] *Natural History of England*, 1659. *Brit. Bacon.*, p. 183. 1661.
[3] *Hist. Nat.*, lib. ii. cap. 26.
[4] Gould's *Astronomical Journal*, No. 84, May 26, 1855. In the *Month. Not.* R.A.S., vol. xvii. p. 204–5, are some remarks on this communication, which the reader should refer to.

observations, for the conclusion from all the data in my possession is a startling one. It seems to me that those data can
be explained only by the supposition of *a nebulous ring with
the Earth for its centre, and lying within the orbit of the
Moon.*"

On the publication of the foregoing, Humboldt transmitted
to the Berlin Academy[1] some unpublished observations made
by him at sea in March 1803, to the effect that, on one or two
occasions he also saw a 2nd light in the east, contemporaneously with the principal beam in the west; he, however,
then thought that the 2nd light was merely due to reflection.
He concludes by saying that, " the variations in the brightness
of the phenomenon cannot, according to my experience, be
accounted for solely by the constitution of our atmosphere.
There remains much still to be observed relative to the
subject."

---

[1] *Monatsbericht der kön. Preuss. Akademie der Wissenschaften*, July,
1855.

BOOK V.

COMETS.[1]

—✦—

CHAPTER I.

GENERAL REMARKS.

*Comets long Objects of Popular Interest — and Alarm. — Usual Phe-
nomena attending the Developement of a Comet. — Telescopic Comets.
— Comets diminish in Brilliancy at each Return. — Period of Revo-
lution. — Density. — Mass. — Lexell's Comet. — General Influence of
Planets on Comets. — Comets move in 1 of 3 Kinds of Orbits. — Ele-
ments of a Comet's Orbit. — For a Parabolic Orbit, 5 in number. —
Direction of Motion. — Eccentricity of an Elliptic Orbit. — Early
Speculations as to the Paths Comets moved in. — Comets visible in the
Daytime. — Breaking up of a Comet into Parts. — Instance of Biela's
Comet. — Comets probably self-luminous. — Existence of Phases,
doubtful. — Comets with Planetary Discs. — Phenomena connected
with the Tails of Comets. — Usually a Prolongation of the Radius
Vector. — Vibration sometimes noticed in Tails. — Olbers' Hypothesis.
— Transits of Comets across the Sun's Disc. — Variation in the Ap-
pearance of Comets exemplified in the case of that of* 1769.

THE class of bodies which will now come under our notice are
among the most interesting with which the astronomer has to
deal. Appearing suddenly in the nocturnal sky, and often
dragging after them tails of immense size and brilliancy, they
were well calculated, in the earlier ages of the world, to attract
the attention of all, and still more to excite the fear of many.
It is the unanimous testimony of history, during a period of
upwards of 2000 years, that comets were always considered
to be peculiarly " ominous of the wrath of heaven, and as

---

[1] See Appendices III. and IV. A portion of this Book appeared some
time since in a London periodical, called *Recreative Science*.

harbingers of wars and famines, of the dethronement of monarchs, and the dissolution of empires." We shall hereafter examine this question at greater length. Suffice it for us, here, to quote the words of the Poet, who speaks of

> " The blazing Star,
> Threat'ning the world with famine, plague, and war;
> To princes, death; to kingdoms, many curses;
> To all estates, inevitable losses;
> To herdsmen, rot; to ploughmen, hapless seasons;
> To sailors, storms; to cities, civil treasons."

However little attention might have been paid by the ancients to the more ordinary phenomena of nature (which, however, were pretty well looked after), yet certain it is, that comets and total eclipses of the Sun were not easily forgotten or lightly passed over; hence the aspect of remarkable comets that have appeared at various times have been handed down to us, often with circumstantial minuteness.

A comet usually consists of 3 parts, developed somewhat in the following manner: — A faint luminous speck is discovered by the aid of a good telescope; the size increases gradually, and after some little time, a *nucleus* appears, that is, a part which is more condensed in its light than the rest, sometimes circular, sometimes oval, more rarely presenting a radiated appearance. Both the size and brilliancy of the object still progressively increase, the *coma* or cloudlike mass around the nucleus becomes less regular, and a tail begins to form, becoming fainter as it recedes from the body of the comet. This tail increases in length so as sometimes to spread across a large portion of the heavens; sometimes there are more tails than one, and frequently the tail seems broken off, or much narrower in parts. The comet approaches the Sun in a curvilinear path, which frequently hardly differs from a right line. It generally crosses that part of the heavens in which the Sun is situated, so near the latter body as to be lost in its rays; but it emerges again on the other side, frequently with increased brilliancy and length of tail. The phenomena of disappearance are then (but in the reverse order) the same as those of its appearance.

Telescopic Comet without a Nucleus.

Telescopic Comet with a Nucleus.

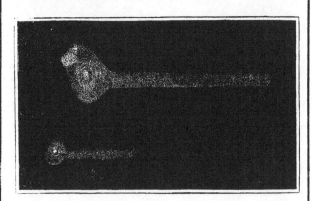

Biela's Comet, 1846.   (*O. Struve.*)

Encke's Comet, 1828, Nov. 30.
(*W. Struve.*)

The 1st Comet of 1847.
(*Hind.*)

COMETS.

In magnitude and brightness comets exhibit great diversity; some are so bright as to be visible in the daytime; others, indeed the majority, are quite invisible, except with powerful optical assistance. Such are usually called *telescopic comets.* The appearance of the same comet at different periods of its return is so varying, that we can never identify a given comet with any other by any mere physical peculiarity of size or shape, until its elements have been calculated and compared. It is now known that " the same comet may, at successive returns to our system, sometimes appear tailed, and sometimes without a tail, according to its position, with respect to the Earth and the Sun; and there is reason to believe that comets in general, from some unknown cause, decrease in splendour in each successive revolution."[1]

The periods of comets in their revolutions vary greatly, as also do the distances to which they recede from the Sun. Whilst the orbit of Encke's comet is contained within that of Jupiter, the orbit of Halley's extends far beyond Neptune. Some comets indeed proceed to a much greater distance than this, whilst others are supposed to pass into curves which do not, like the ellipse, return into themselves. In this case they never come back to the Sun. Such orbits are either parabolic, or hyperbolic. The density or *mass* of comets is exceedingly small, and their tails consist of matter of such extreme tenuity that the smallest stars are seen through them — a fact first recorded by Seneca. That the matter of comets is exceedingly small is abundantly proved by the fact, that they have at times passed very near to some of the planets without disturbing their motions in any appreciable degree. Thus the comet of 1770 (Lexell's) in its advance towards the Sun, got entangled amongst the satellites of Jupiter, and remained near them for 4 months, without in the least affecting them as far as we know. It can therefore be shown that this comet's mass could not have been so much as $\frac{1}{5000}$ that of the Earth. The same comet also came very near the Earth on July 1; its distance at 5h. on that day being about 1,400,000 miles;

---

[1] Smyth, *Cycle*, vol. i. p. 235.

so that had its quantity of matter been equal to that of the
Earth, it would, by its attraction, have caused our globe to
revolve in an orbit so much larger than at present, as to have
increased the length of the year by 2h. 47m., yet no sensible
alteration took place. The comet of 837 remained for a period
of 4 days within 3,700,000 miles without any untoward consequence. A very little argument, therefore, suffices to show
the absurdity of the idea of any danger happening to our planet
from the advent of any of these wandering strangers. Indeed,
instead of comets exercising any influence on the motions of
planets, there is the most conclusive evidence that the contrary
influence prevails—of planets on comets. This fact is strikingly
exemplified in the history of the comet of 1770, just referred
to. At its appearance in that year, this body was found to
have an elliptical orbit, requiring for a complete revolution
only 5½ years; yet although this comet was a large and bright
one, it had never been observed before, and has moreover never
been seen since; the reason being that the influence of the
planet Jupiter, in a short period, completely changed the character of its path.

A comet may move either in an elliptic, parabolic, or
hyperbolic orbit; but for reasons with which our mathematical
readers are acquainted, no comet can be periodical which
does not follow an elliptic path. In consequence, however, of
the comparative facility with which the parabola can be calculated, astronomers are in the habit of applying that curve to
represent the orbit of any newly discovered body. Parabolic
elements having been obtained, a search is then made through
a catalogue of comets, to see whether the new elements bear any
resemblance to those of any object that has been previously
observed; if so, an elliptic orbit is calculated, and a period
deduced. The elements of a parabolic orbit are 5 in number.

1. *The Time of Perihelion Passage*, or the moment when the
comet arrives at its least distance from the Sun[1], denoted by
the symbol PP, or τ.

---

[1] In an elliptic orbit, the corresponding extreme distance from the
Sun is called the *Aphelion*.

2. *The Longitude of the Perihelion*, or the longitude of the comet when it reaches that point. — π.

3. *The Longitude of the Ascending Node* of the comet's orbit, as seen from the Sun. — Ω.

4. *The Perihelion Distance*, or the distance of the comet from the Sun expressed in radii of the Earth's orbit. — q.

5. *The Inclination of the Orbit*, or the angle between the plane of the orbit and the ecliptic. — ι.

It is also necessary to know whether the comet moves in the order of the signs, or in the contrary direction: in the former case, its movement (μ) is said to be *direct* ($+$), in the latter, *retrograde* ($-$). In an elliptic orbit we require to know the eccentricity (ε); this is sometimes expressed by the angle φ, of which the previous quantity ε, is the sine. From this, with the perihelion distance, we can ascertain the length of the major axis, and consequently the comet's periodic time. We should remark that the eccentricity is not the linear distance of the centre of the ellipse from the focus, but the ratio of that quantity to the semi-axis major.

Up to the present time the orbits of nearly 280 comets have been calculated: these will be found in Appendix III.

To the early astronomers the motions of comets gave rise to great embarrassment. Tycho Brahe thought they moved in circular orbits; Kepler, on the other hand, suggested right lines. Hevelius seems to have first remarked that cometary orbits were much curved near the perihelion, the concavity being towards the Sun. He also threw out an idea relative to the parabola, as being the form of a comet's path, though it does not seem to have occurred to him that the Sun was likely to be the focus. Borelli suggested an ellipse or a parabola. Sir William Löwer was probably the first to hint that comets sometimes moved in very eccentric ellipses; this he did in his letter to his "especiall goode friend, Mr. Thomas Harryot," dated Feb. 6, 1610. Dörfel, a native of Upper Saxony, was the first practical man; he showed that the comet of 1681 moved in a parabolic orbit.

History informs us that some comets have shone with such splendour as to have been distinctly seen in the daytime.

The comets of B.C. 45, A.D. 575, 1106, 1402, 1532, 1577, 1618, 1744, 1843 (i), 1847 (i), and 1853 (iii), are the principal ones which have been thus observed.

There are several well-established instances of the separation of a comet into 2 or more distinct portions. Seneca mentions an instance.[1] Such was the case with Biela's comet in 1845. When first detected on Nov. 28, it presented the appearance of a faint nebulosity, almost circular, with a slight condensation towards the centre; on Dec. 19, it appeared somewhat elongated, and by the end of the month the comet had actually separated into 2 distinct nebulosities which travelled together for more than 3 months: the maximum distance (157,240 miles) was attained on March 3, 1846, after which it began to diminish until the comet was lost sight of in the middle of April.[2] At its return in 1852, the separation was still maintained, but the interval had increased to 1,250,000 miles. At its return in 1859, the comet does not appear to have been detected, owing to its unfavourable position.

The question whether or not comets are self-luminous, has never been satisfactorily settled. The high magnifying power that may sometimes be brought to bear on them tends to show that they shine by their own light. Sir W. Herschel was of this opinion from his observations of the comets of 1807 and 1811 (i).[3] It is manifest, however, that if the existence of phases could be certainly known, this would furnish an irrefragable proof that comets shine by reflected light. It has been asserted, from time to time, that such phases have been seen, but the statements made are unsupported. Delambre mentions that the Registers of the Royal Observatory at Paris exhibit undoubted evidence of the existence of phases in the comet of 1682; but neither Halley nor any other astronomer who observed this comet has given the slightest intimation of their having seen any phase phenomena. James Cassini mentions the existence of phases

[1] *Quæst. Nat.*, lib. vii. cap. 16.
[2] Comet i, of 1860, discovered by Liais in Brazil, had a double nebulosity, resembling, it would seem, that of Biela's comet.
[3] *Phil. Trans.*, vol. cii. p. 115, *et seq.* 1812.

in the comet of 1744[1] ; on the other hand, Heinsius and Chésaux, who paid particular attention to this comet, positively deny having seen anything of the kind. More recently Cacciatore, of Palermo, expressed a decided conviction that he had seen a crescent in the comet of 1819.

Arago sums up by saying that the observations of M. Cacciatore prove only that the nuclei of comets are sometimes very irregular.[2]   Sir W. Herschel states that he could see no signs of any phases in the comet of 1807, although he fully ascertained that a portion of its disc was not illuminated by the Sun at the time of observation.[3]   The general opinion now is against the existence of phases, and thus we must consider that comets shine by their own inherent light; but the observations of Airy and others, on Donati's comet in 1858, point to exactly the opposite conclusion, at least as regards the *tail* of that comet.[4]

Some comets have been observed with round and well-defined planetary discs. Seneca relates that one appeared after the death of Demetrius, king of Syria, but little inferior to the Sun [in size ?] being a circle of red fire, sparkling with such a light as to surmount the obscurity of night. The comet of 1652, seen by Hevelius, was almost as large as the Moon, though not nearly so bright.

The following diagrams will convey to the eye some idea of the lengths of the tails of comets.

Fig. 23

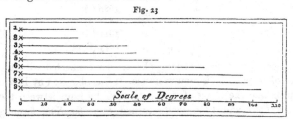

Relative APPARENT Lengths of the Tails of certain Large Comets.

---

[1] *Mém. Acad. des Sciences*, 1744, p. 303.
[2] *Pop. Ast.*, vol. i. p. 627. Eng. Ed.
[3] *Phil. Trans.*, vol. xcviii.. p. 156.   1808.
[4] *Green. Obs.* 1858.

In Fig. 23, the representations are as follows : —

1.	The Comet of	1744	
2.	,,	,,	1811 (i)
3.	,,	,,	1843 (i)
4.	,,	,,	1858 (vi)
5.	,,	,,	1456
6.	,,	,,	837
7.	,,	,,	1769
8.	,,	,,	390
9.	,,	,,	1618 (ii)

Fig. 24.

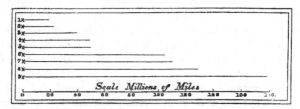

Relative TRUE Lengths of the Tails of certain Large Comets.

In Fig. 24, the representations are as follows : —

1.	The Comet of	1744	
2.	,,	,,	1860 (iii)
3.	,,	,,	1769
4.	,,	,,	1618 (ii)
5.	,,	,,	1858 (vi)
6.	,,	,,	1680 (i)
7.	,,	,,	1811 (i)
8.	,,	,,	1811 (ii)
9.	,,	,,	1843 (i)

There are several curious phenomena connected with the
tails of comets which require notice. It was observed by Pierre
Apian that the trains of 5 comets, seen by him between the
years 1531 and 1539, were turned *from the Sun*, forming
a prolongation of the radius vector, an imaginary line joining
the Sun and the comet ; as a general rule, this has been found
to be the case[1], although exceptions do occur.

---

[1] The researches of M. E. Biot show that this fact was known to the
Chinese long before the time of Apian. *Comp. Rend.*, vol. xvi. p. 751.

Thus the tail of the comet of 1577 deviated 21° from the line of the radius vector. In some few instances, where a comet had more than 1 tail, the 2nd extended more or less *towards* the Sun, as was the case with the comets of 1823 and 1851 (iv). Although comets usually have but 1 tail, yet 2 is by no means an uncommon number, and indeed the great comet of 1825 had 5 tails, and that of 1744 as many as 6, according to Chésaux. The tails of many comets are curved so as frequently to resemble in appearance a sabre; such was the case with the comets of 1843, 1844 (iii), and 1858 (vi), amongst others. The comet of 1769 had a double curved tail, thus ∽.

The trains of some great comets have been seen to vibrate in a manner somewhat similar to the Aurora Borealis. The vibrations commence at the head, and appear to traverse the whole length of the comet in a few seconds. It was long supposed that the cause was connected with the nature of the comet itself, but Olbers has pointed out that such appearances could be only fairly attributed to the effects of our own atmosphere, for this reason : — " The various portions of the tail of a large comet must often be situated at widely different distances from the Earth; so that it will frequently happen, that the light would require several minutes longer to reach us from the extremity of the tail than from the end near the nucleus. Hence, if the coruscations were caused by some electrical emanation from the head of the comet, even if it occupied but 1 second in passing over the whole surface, several minutes must necessarily elapse before *we* could see it reach the tail. This is contrary to observation, the pulsations being almost instantaneous."

Comets have been seen to pass over the Sun's disc. One of the most remarkable instances occurred on the morning of Nov. 18, 1826. The phenomenon was anticipated by Gambart, and said to have been witnessed by him and Flaugergues, though there is some doubt on this point.

The following is an excellent instance of the ever-changing appearance of comets; it relates to the one of 1769. On Aug. 8, Messier, whilst exploring with a 2-foot telescope,

K

perceived a round nebulous body which turned out to be a comet. On the 15th the tail became visible to the naked eye, and appeared to be about 6° in length; the 28th it measured 15°; on Sept. 2, 36°; on the 6th, 49°; and on the 10th, 60°. The comet having now plunged into the Sun's rays ceased to be visible. On Oct. 8, the perihelion passage took place; on the 24th of the some month it reappeared, but with a tail only 2° long; on Nov. 1 it measured 6°; on the 8th it was only $2\frac{1}{2}$°; on the 30th it was $1\frac{1}{2}$°; the comet then disappeared.

[1] *Mém. Acad. des Sciences*, 1775, p. 392.

CHAPTER II.

PERIODIC COMETS.

Periodic Comets conveniently divided into Three Classes. — Comets in Class I.—Encke's Comet. — The Resisting Medium. — Table of Periods of Revolution. — Di Vico's Comet. — Pons' Comet of 1819.— Brorsen's Comet.—Biela's Comet.—D'Arrest's Comet.—Faye's Comet. —Méchain's Comet of 1790. — List of Comets presumed to be of Short Periods but only once observed. — Comets in Class II. — West- phal's Comet.—Pons' Comet of 1812.—Di Vico's Comet of 1846.— Olbers' Comet of 1815.—Brorsen's Comet of 1847.—Halley's Comet. Of special Interest. — Résumé of the Early History of Halley's Labours.—Its Return in 1759.—Its Return in 1835. — Its History prior to 1531 traced by Hind.—Comets in Class III.—Not requiring detailed Notice. — Hyperbolic Comets.

THE comets we propose treating of in the present chapter may be conveniently divided into 3 classes : —

1. Comets of short periods.
2. Comets revolving in about 70 years.
3. Comets of long periods.

The following are the comets belonging to Class I. with which we are best acquainted : —

Name.	Period.	Next Return.
	Year.	
1. Encke	3·296	1862 Feb.
2. Di Vico	5·469	1861 Jan.
3. Winnecke . . .	5·54	1863 Nov.
4. Brorsen	5·581	1862 Oct.
5. Biela	6·617	1866 Jan.
6. D'Arrest	6·64	1864 Feb.
7. Faye	7·44	1866 Feb.
8. Méchain	13·60	1871 Oct.

No. 1 is by far the most interesting comet in the list, and we shall therefore review its history somewhat in detail. On the evening of Nov. 26, 1818, M. Pons, of Marseilles, detected a telescopic comet: it was soon found that its orbit could not be parabolic. Professor Encke was therefore induced to undertake a rigorous investigation of its elements, and he ascertained that the real path of the comet was an ellipse, with a period of about $3\frac{1}{3}$ years. On looking over a catalogue of all the comets then known, he was struck with the similarity which the elements obtained by him bore to those of the comets of 1786 (i), 1795, and 1805; further examination placed beyond doubt the identity of those comets with the one of 1819. Encke then turned his attention to its next return, and he announced that the comet would arrive at perihelion on May 24, 1822, after being retarded about 9 days by the influence of the planet Jupiter.

" So completely were these calculations fulfilled, that astronomers universally attached the name of ' Encke ' to the comet of 1819, not only as an acknowledgment of his diligence and success in the performance of some of the most intricate and laborious computations that occur in practical astronomy, but also to mark the epoch of the first detection of a comet of short period — one of no ordinary importance in this department of the science." Agreeably to Encke's prediction, the comet arrived in 1822; and has been observed at every successive return since that time, viz., 1825, 1828, 1832, 1835, 1838, 1842, 1845, 1848, 1852, 1855, 1858.

The account of this comet would be incomplete were we not to refer to a peculiarity connected with its motion, which attracted Encke's attention as far back as 1838. He found that, notwithstanding every allowance being made for planetary influence, yet the comet always attained its perihelion distance about $2\frac{1}{2}$ hours sooner than his calculations led him to expect. In order to account for this gradual diminution of the period of revolution (in 1789 it was nearly 1213 days, but in 1858 it was scarcely $1210\frac{1}{2}$), Encke conjectured the existence of a thin ethereal medium, sufficiently dense to produce some impression on a body of such extreme tenuity as the comet in question, but incapable of exercising any sensible influence on the movements of the planets. This contraction of the orbit must be continually progressing, if we suppose the existence of such a medium; and we are naturally led to inquire, What will be the final consequence of this resistance? Though the catastrophe may be averted for many ages by the powerful attraction of the larger planets, especially Jupiter, will not the comet be at last precipitated on to the Sun? The question is full of interest, though widely open to conjecture.

The following table, recently published by Professor Encke, will more clearly illustrate the effect of the resisting medium on the comet's periodic time: —

Year of P. P.					Period Days.
1786	.	.	.	.	
1789	.	.	.	.	1212·79
1792	.	.	.	.	1212·67
1795	.	.	.	.	1212·55
1799	.	.	.	.	1212·44
1802	.	.	.	.	1212·33
1805	.	.	.	.	1212·22
1809	.	.	.	.	1212·10
1812	.	.	.	.	1212·00
1815	.	.	.	.	1211·89
1819	.	.	.	.	1211·78
1822	.	.	.	.	1211·66
1825	.	.	.	.	1211·55
1829	.	.	.	.	1211·44
1832	.	.	.	.	1211·32
1835	.	.	.	.	1211·22
1838	.	.	.	.	1211·11
1842	.	.	.	.	1210·98
1845	.	.	.	.	1210·88
1848	.	.	.	.	1210·77
1852	.	.	.	.	1210·65
1855	.	.	.	.	1210·55
1858	.	.	.	.	1210·44

No. 2. On August 22, 1844, M. Di Vico, at Rome, discovered a telescopic comet, whose orbit was afterwards proved to be an ellipse of moderate eccentricity, with a period of about $5\frac{1}{2}$ years. The return of this comet to perihelion was predicted for March, 1850; but owing to its unfavourable

K 3

position it was not seen.   Le Verrier has made some computations which render it probable that this is a return of the
comet of 1678.

No. 3 was discovered by M. Pons on June 12, 1819.   Professor Encke assigned to it a period of 5½ years, which, as the
table will show, was a very close approximation to the truth.
It was not, however, seen from that time till March 8, 1858,
when it was detected by Winnecke at Bonn, who soon
ascertained the identity of the two objects.

No. 4 was detected by M. Brorsen at Kiel, on February 26,
1846.   The observations showed an elliptic orbit, and the
epoch of the ensuing arrival at perihelion was fixed for September 26, 1851.   It escaped observation in that year, however, owing to its proximity to the Sun; but was rediscovered
by D'Arrest in February, 1857.

No. 5.   This is another very remarkable periodic comet,
and second only in interest to Encke's.   It was discovered in
Bohemia by M. Biela, on February 27, 1826; the path it
pursued was observed to be similar to that of the comets of
1772 and 1807, with which it was subsequently identified.   It
was followed by Olbers and M. Gambart until the end of
April.   Soon after its disappearance, Professor Santini of
Padua undertook to investigate its motion, and announced that
its perihelion passage would take place on November 27, 1832.
The first glimpse of the comet on its return was obtained at
Rome on August 23, and in a few weeks it became generally
visible.   It arrived at perihelion only 12 hours before the time
named, a much closer fulfilment of Santini's prediction than
could have been expected under the circumstances.   Considerable sensation was created among the general public on
this occasion, on account of an apprehended collision with the
Earth; but none of course took place, a result which astronomers in vain tried to prove beforehand.   We have already
adverted to a very curious phenomenon which took place at
the apparition of this comet in 1845, shortly after its discovery
in that year.

No. 6.   On June 27, 1851, D'Arrest discovered a faint
comet in the constellation Pisces.   It was soon remarked by

this astronomer that the path followed by the comet appeared
to be an elliptic one—a surmise which subsequent observations
fully confirmed.   Calculation showed that it would probably
reappear about the end of 1857, or beginning of 1858; and
although not seen in Europe, it was discovered by Sir T.
Maclear at the Cape of Good Hope, near the time fixed.

No. 7 was discovered by M. Faye, on November 22, 1843,
at the Paris Observatory.   It exhibited a bright nucleus, with
a short tail, but was never sufficiently brilliant to be seen by
the unaided eye.   It was soon found by several observers
to be moving in an elliptic orbit, with a period of about $7\frac{1}{2}$
years.   M. Le Verrier predicted the comet's return to perihelion for April 3, 1851 ; and, true enough, it was detected
in the great Northumberland telescope at Cambridge, by Professor Challis, on November 28, 1850, and passed through the
perihelion near the time named some years before.

No. 8 was detected by Méchain, January 9, 1790; but the
elliptic nature of its orbit does not appear to have been then
suspected.   It was not reobserved until its return, at the commencement of 1858.

Short periods have also been assigned to the following
comets ; but, since much uncertainty prevails about them,
they have not been included in the above list :—

Clausen (1743, i).	Pigott (1783).
Burckhardt (1766, ii).	Blainpain (1819, v).
Lexell (1770, i).	Peters (1846, vi).

In Class II. we have the following comets :—

Name.	Period.	Probable next Return.
	Years.	
1. Westphal (1852, iv)	67·77	1920
2. Pons (1812)	70·68	1883
3. Di Vico (1846, iv)	73·25	1919
4. Olbers (1815)	74·05	1889
5. Brorsen (1847, v)	74·97	1922
6. Halley	76·78	1912

It has been thought by some astronomers that 4 of the above may have originally constituted a single comet.

No. 6. The comet whose history is the most interesting is undoubtedly that which bears the name of our illustrious countryman, Halley, which has a period of about 75 years; and as it will, moreover, serve to exemplify what we have already said on the nature and appearance of comets, we cannot do better than give a summary of its history, from the time of its last appearance, in 1835, back to the earliest ages. Four years after the advent of the celebrated comet of 1680, Sir Isaac Newton published his *Principia*, in which he applied to that body the general principles of physical investigation first promulgated in that work. He explained the method of determining, by geometrical construction, the visible portion of the path of a body of this kind, and invited astronomers to apply these principles to the various recorded comets. Such was the effect of the force of analogy upon the mind of the great philosopher, that, without awaiting the discovery of a periodic comet, he boldly assumed these bodies to be analogous to planets in their revolutions round the Sun. Startling as this theory might have been when first propounded, yet it was not long before it was fully substantiated. Halley, who was then a young man, undertook the labour of examining the circumstances attending all the comets previously recorded, with a view to ascertain whether any, and which of them, appeared to follow the same path. Careful investigation soon pointed out the similarity of the orbits of the comets of 1531 and 1607, and that they were, in fact, the same as that followed by the comet of 1682, seen by himself; he suspected therefore (and rightly too, as the sequel showed), that the appearances at these 3 epochs were produced by the 3 successive returns of one and the same body, and that, consequently, its period was somewhere about 75½ years. There were, nevertheless, 2 circumstances which might be supposed to offer some difficulty, inasmuch as it appeared that the intervals between the successive returns were not precisely equal; and, 2ndly, the inclination of the orbit was not exactly the same in each case. Halley, however, "with a degree of sagacity which, considering the state of knowledge at the time, cannot fail to excite unqualified admiration, observed that it was natural to suppose that the same causes which disturbed the planetary motions would likewise act on comets ; " in other words, the attraction of the planets would exercise some other influence on comets and their motions.   The truth of this idea we have already seen exemplified in the case of the comet of 1770.   In fine, Halley found that in the interval between 1607 and 1682, the comet passed so near Jupiter that its velocity must have been considerably increased, and its period consequently shortened; he was, therefore, induced to *predict* its return about the end of 1758 or the beginning of 1759. Although Halley did not survive to see his prediction fulfilled, yet, as the time drew near, great interest was manifested in the result, more especially as Clairaut had named April 13, 1759, as the day on which the perihelion passage would take place. It was not destined, however, that a professional astronomer should be the first to detect the comet on its anticipated return ; that honour was reserved for a farmer, near Dresden, named Palitzch, who saw it on the night of Christmas-day, 1758.   But few observations were made before the perihelion passage (on March 12), owing to the comet's proximity to the Sun; during the months of April and May, however, it was seen throughout Europe, although to the best advantage only to the southern hemisphere.   On May 5 it had a tail 47° long.

Previous to the last return of this comet, in 1835, numerous preparations were made to receive it.   Early in that year Professor Rosenberger, of Halle, published a memoir, in which he announced that the perihelion passage would take place on November 11, though Damoiseau and Pontécoulant both fixed upon a somewhat earlier period.

Let us now see how far these expectations were realised. The comet was seen at Rome on August 5 ; as it approached the Sun it gradually increased both in magnitude and brightness, but did not become visible to the naked eye till September 20.   On October 19 the tail had attained a length of fully 30°.   The comet soon after this was lost in the rays of the

Sun, and passed through its perihelion on November 15, or within 4 days of the time named by M. Rosenberger. It reappeared early in January, 1836; and was observed in the south of Europe and at the Cape till the middle of May, when it was finally lost to view, not to be seen again till the year 1912.

We have seen above that Halley traced his comet back to the year 1531; we must now, therefore, briefly review its probable history prior to that date, as made known by the labours of modern astronomers. Halley surmised that the great comet of 1456 was identical with the one observed by him in 1682, and M. Pingré has converted Halley's suspicion into a certainty. The preceding return took place, as M. Laugier has shown, in 1378, when it was observed both in Europe and China; but it does not appear to have been so bright as in 1456. In September, 1301, a great comet is mentioned by nearly all the historians of the period. It was seen as far north as Iceland. It exhibited a bright and extensive tail, stretched across a considerable part of the heavens. This was most likely Halley's. The previous apparition is not so well ascertained, but most likely occurred in July, 1223, when it is recorded in an ancient chronicle that a wonderful sign appeared in the heavens shortly before the death of Philip Augustus of France, of which event it was generally considered to be the precursor. It was only seen for 8 days. Although but little information is possessed about it, and that of a very vague character, yet it seems probable that this was Halley's comet. In April, 1145, a great comet is mentioned by European historians, which is one of the most certain of our series of returns. There is considerable probability in favour of the appearance of the comet in the year of the Norman Conquest, or in April, 1066. The famous body which astonished Europe in that year is minutely, though not very clearly, described in the Chinese annals; and its path there assigned is found to agree with elements which have great resemblance to those of Halley's comet. In England it was considered the forerunner of the victory of William of Normandy, and was looked upon with universal dread. It was equal to the Full Moon in size, and its train, at first small, increased to a wonderful length. Almost every historian and writer of the 11th century bears witness to the splendour of the comet of 1066, and there can be but little doubt but that it was Halley's. Previous to this year the comet appeared in 989, 912, 837, 760, 684, 608, 530, 451, 373, 295, 218, 141, 66 A.D., and 11 B.C., all of which apparitions have been identified by Mr. Hind.[1]

Concerning the comets belonging to Class III. it is not necessary to notice them further here; they can be inspected in the catalogue.

Though not belonging to this chapter, as they are so few in number, we may just mention that the following are the only hyperbolic comets yet known:—

1729	. . .	1818 (iii).
1771	. . .	1840 (i).
1774	. . .	1843 (ii).

---

[1] *Month Not.* R.A.S., vol. x. p. 51, *et seq.*

CHAPTER III.

REMARKABLE COMETS.

Donati's Comet of 1858. — *Its Insignificance when first discovered.* —
Résumé of the Circumstances attending the Increase of its Brilliancy.
— *Length of Time Visible.* — *Probably Periodic.* — *Measurements of*
G. P. Bond. — *The Great Comet of* 1843. — *Thought to be iden-*
tical with several Comets previously observed. — *The Great Comet of*
1811. — *Carefully observed by Sir W. Herschel.* — *The Comet of* 1860
(iii).

THE comets which might be included under the above head
are so numerous as to make it impossible for all to receive
proper attention; we must therefore limit ourselves to some of
the *most* interesting, commencing with—

The Comet of 1858 (vi). On June 2 in that year Dr. G. B.
Donati, at Florence, descried a faint nebulosity slowly advancing towards the north, and near the star λ Leonis. Owing
to its immense distance from the Earth (240,000,000 miles),
great difficulty was experienced in laying down its orbit. By
the middle of August, however, its future course, and great
increase of brightness in September and October, had been
ascertained with entire certainty. Up to this time (middle of
August) it had remained a faint object, not discernible by the
unaided eye. It was distinguished from ordinary telescopic
comets only by the extreme slowness of its motion (in singular
contrast to its subsequent career), and by the vivid light of its
nucleus: " the latter peculiarity was of itself prophetic of a
splendid destiny." Traces of a tail were noticed on August 20,
and on August 29 it was faintly perceptible to the naked eye;
for a few weeks it occupied a northern position in the heavens,
and was thus seen both in the morning and evening sky. On
September 6 a slight curvature of the tail was noticed, which

Fig. 93.   *Plate* XXII.

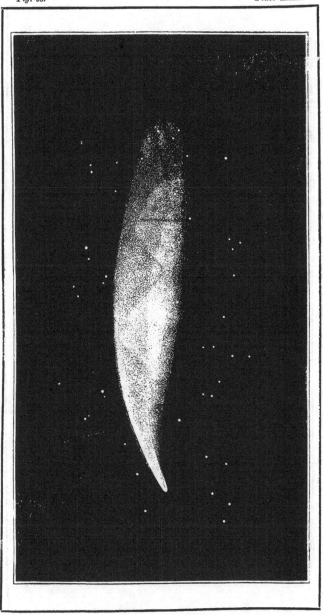

DONATI'S COMET: OCTOBER 5, 1858.
DRAWN BY PAPE.

Fig. 94.   Plate XXIII.

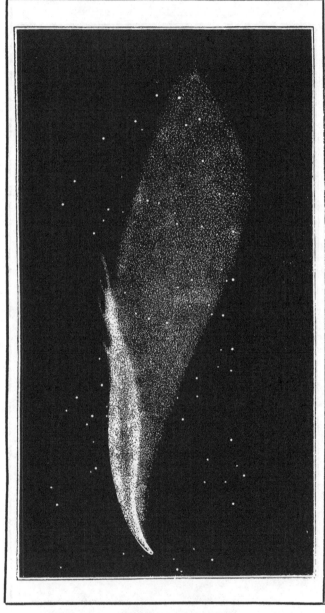

DONATI'S COMET: OCTOBER 9, 1858.
DRAWN BY PAPE.

September 22.

September 29.

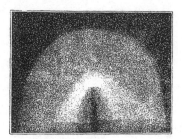

October 6.

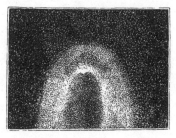

October 12.

DONATI'S COMET, 1858: VIEWS of the COMA.
DRAWN BY PAPE.

subsequently became one of its most interesting features. On September 17 the head equalled in brightness a star of the 2nd magnitude, the length of the tail being 4°. The comet passed through perihelion on September 29, and was at its least distance from the Earth on October 10. Its rapid passage to the southern hemisphere rendered it invisible in Europe after the end of October, but it was followed at Santiago de Chili, and the Cape of Good Hope Observatories for some months afterwards, and was last seen by Sir T. Maclear at the latter place in March 4, 1859. When these southern observations come to be fully discussed, it is probable that we shall get a much more certain determination of the comet's periodic time than we at present possess.

"Its early discovery enabled astronomers, while it was yet scarcely distinguishable in the telescope, to predict, some months in advance, the more prominent particulars of its approaching apparition, which was thus observed with all the advantage of previous preparation and anticipation. The perihelion passage occurred at the most favourable moment for presenting the comet to good advantage. When nearest the Earth, the direction of the tail was nearly perpendicular to the line of vision, so that its proportions were seen without foreshortening. Its situation in the latter part of its course afforded also a fair sight of the curvature of the train, which seems to have been exhibited with unusual distinctness, contributing greatly to the impressive effect of a full-length view."

This comet, though surpassed by many others in size, has not often been equalled in the intense brilliancy of the nucleus, which the absence of the Moon, in the early part of October, permitted to be seen to the very best advantage. There is no doubt but that the comet of Donati revolves in an elliptic orbit; the period, however, is at present somewhat uncertain, but is probably about 2500 years.

The following is a table of the dimensions of the comet's nucleus and tail, at the under-mentioned dates[1] :—

----

[1] G. P. Bond. *Math. Month. Mag.*, Boston, U. S., Nov. and Dec. 1858.

Date.	Diameter of Nucleus.		Length of Tail.	
1858.	*"*	Miles.	°	Miles.
July 19 . .	5 =	5600	...	...
Aug. 30 . .	6 =	4660	2 =	14,000,000
Sept. 8 . .	3 =	1980	4 =	16,000,000
,, 12 . .	...		6 =	19,000,000
,, 23 . .	3 =	1280	5 =	12,000,000
,, 25 . .	...		11 =	17,000,000
,, 27 . .	...		13 =	18,000,000
,, 28 . .	...		19 =	26,000,000
,, 30 . .	...		22 =	26,000,000
Oct. 2 . .	...		25 =	27,000,000
,, 5 . .	1·5 =	400	33 =	33,000,000
,, 6 . .	3·0 =	800	50 =	45,000,000
,, 8 . .	4·4 =	1120	50 =	43,000,000
,, 10 . .	2·5 =	630	60 =	51,000,000
,, 12 . .	...		45 =	39,000,000

The Comet of 1843 (i) was one of the finest that have appeared during the present century. It was first seen in the southern hemisphere, towards the end of the month of February, and during the first fortnight in March it shone with great brilliancy. It was not visible in England until after the 15th, when its splendour was much diminished; but the suddenness with which it made its appearance added not a little to the interest it excited. The general length of the tail during March was about 40°, and its breadth about 1°. The orbit of this comet is remarkable for its small perihelion distance, which did not exceed, according to the most reliable calculations, 538,000 miles; and the immense velocity of the comet in its orbit, when near the perihelion, occasioned some extraordinary peculiarities. Thus between February 27 and 28 it described upon its orbit an arc of 292°. Supposing it to revolve in an ellipse, this would leave only 68° to be described during the time which elapses before its next return to perihelion.

It has been thought by some that this comet was identical with those of 1668 and 1689, but so little is known *for certain*

Fig. 99. *Plate* XXV.

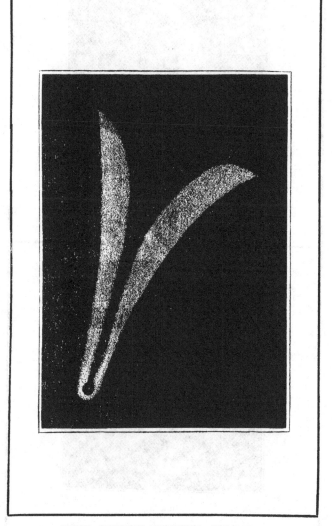

THE GREAT COMET OF 1811.

June 26.

June 28.

COMET III. 1860.
DRAWN BY CAPPELLETTI AND ROSA.

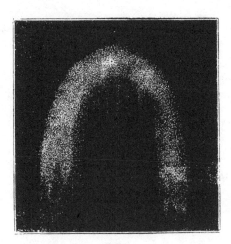

June 30.

July 1.

COMET III. 1860.
DRAWN BY CAPPELLETTI AND ROSA.

July 6.

July 8.

COMET III. 1860.
DRAWN BY CAPPELLETTI AND ROSA.

about the latter, that we are not yet in a position to admit or deny the identity of the two bodies. In the work referred to in the note, the question is entered into with great ability.[1]

The Comet of 1811(i) is one of the most celebrated of modern times. It was discovered by M. Flaugergues, at Viviers, on March 26, 1811, and was last seen by Wisniewski at Neu-Tscherkask, on August 17, 1812. In the autumnal months of 1811, it shone very conspicuously, and its great northern declination caused it to remain visible throughout the night for many weeks. The extreme length of the tail, about the beginning of October, was about 25°, and its breadth about 6°. Sir W. Herschel paid particular attention to it, and the observations which he made are very valuable. He states that it had a well-defined nucleus, the diameter of which, he found by careful measurement, was 428 miles, and of a ruddy hue, though the surrounding nebulosity was of a bluish green tinge.[2] This comet undoubtedly is a periodical one. Argelander's investigation of the orbit, the most complete that has been carried out, assigns to it a period of 3065 years, subject to an uncertainty of only 43 years.[3] The aphelion distance is 14 times that of Neptune, or, more exactly, 40,121,000,000 miles!

The Comet of 1860 (iii). In the latter end of June 1860 a comet of considerable brilliancy suddenly made its appearance in the northern circumpolar regions. Bad weather prevented its being generally observed in England, but in the south of Europe it was well seen; copies of some drawings made at Rome are annexed.

---

[1] E. J. Cooper, *Cometic Orbits*, pp. 159–169.
[2] *Phil. Trans.*, vol. cii. pp. 118, 119, 121.
[3] Berlin, *Ast. Jahrbüch.* 1825, p. 250.

CHAPTER IV.

COMETARY STATISTICS.

Dimensions of the Nuclei of Comets. — Of the Comæ. — Comets contract and expand on approaching to, and receding from, the Sun. — Exemplified by Encke's in 1838. — Lengths of the Tails of Comets. — Dimensions of Cometary Orbits. — Periods of Comets. — Number of Comets Recorded. — Duration of Visibility of Comets.

ALTHOUGH we have hitherto refrained as much as possible from embarrassing the reader with any tedious display of figures, yet we must now say something about the real dimensions of comets ; of the orbits they describe ; also of their number and duration of visibility.

The following are the real diameters in English miles of the nuclei of some of the comets which have been satisfactorily measured within the last century : —

Examples of a Large Nucleus.		Examples of a Small Nucleus.	
	Miles.		Miles.
The Comet of 1845 (iii) .	8000	The Comet of 1798 (i) .	28
The Comet of 1815 . .	5300	The Comet of 1806 . .	30
The Comet of 1858 (vi) .	5600	The Comet of 1798 (ii) .	125
The Comet of 1825 (iv) .	5100	The Comet of 1811 (i) .	428

The dimensions of the *comæ* or heads of comets also vary much, thus : —

Examples of a Large Coma.		Examples of a Small Coma.	
	Miles.		Miles.
The Comet of 1811 (i)	1,125,000	The Comet of 1847 (v) .	18,000
The Comet of 1835 (iii,		The Comet of 1847 (i) .	25,500
Halley)	357,000	The Comet of 1849 . .	51,000
The Comet of 1828			
(Encke)	312,000		
The Comet of 1780 (i)	269,000		

It should be remarked that the real dimensions of comets are found to vary greatly at different periods of the same apparition, for there is no doubt that many of these bodies *contract* as they approach the Sun, and expand again as they recede from it — a fact first noticed by Kepler.

The following measurements of Encke's comet in 1838, when approaching the Sun, will illustrate this : —

Date.		Diameter.	Distance.
1838.		Miles.	From ☉
Oct. 9	. . . .	281,000	1·42
„ 25	. . . .	120,500	1·19
Nov. 6	. . . .	79,000	1·00
„ 13	. . . .	74,000	0·88
„ 16	. . . .	63,000	0·83
„ 20	. . . .	55,500	0·76
„ 23	. . . .	38,500	0·71
„ 24	. . . .	30,000	0·69
Dec. 12	. . . .	6,600	0·39
„ 14	. . . .	5,400	0·36
„ 16	. . . .	4,250	0·35
" 17	. . . .	3,000	0·34

The tails of comets, more especially of those visible to the naked eye, are often of stupendous length, as the following table will show : —

	Greatest Length.		Miles.
The Comet of 1744 . . .	24°	=	19,000,000
The Comet of 1860 (iii) . .	15	=	22,000,000
The Comet of 1769 . . .	97	=	40,000,000
The Comet of 1858 (vi) . .	50	=	42,000,000
The Great Comet of 1618 . .	104	=	50,000,000
The Comet of 1680 . . .	60	=	100,000,000
The Comet of 1811 (i) . . .	25	=	100,000,000
The Comet of 1811 (ii) . .			130,000,000
The Comet of 1843 (i) . . .	65	=	200,000,000

Cometary orbits are usually of immense extent. Thus : —

i. *As to Perihelion Distance.*

Greatest Known.	Miles.	Least Known.	Miles.
The Comet of 1729 .	383,800,000	The Comet of 1843 (i)	538,000

ii. *As to Aphelion Distance.*

Greatest Known.	Miles.
The Comet of 1844 (ii)	406,130,000,000

Least Known	Miles.
The Comet of Encke	388,550,000

We have already seen that the period of the shortest comet yet known is but little more than 3 years; a striking contrast to the periods exhibited in the following table : —

	Years.
The Comet of 1744	122,683
The Comet of 1844 (ii) . . .	102,050
The Comet of 1780 (i)	75,314
The Comet of 1680	15,864
The Comet of 1847 (iii) . . .	13,918
The Comet of 1840 (ii) . . .	13,864

From the earliest period up to the present time, the number of which there is any trustworthy record, is nearly 800; but as it is only within the last 100 years that optical assistance has been made available in searching for them, the real number of comets that have appeared, is probably not less than 4000 or 5000, especially when we consider that many doubtless have only been visible in the southern hemisphere.

TABLE OF NUMBER OF COMETS RECORDED.

Period.	Comets Observed.	Orbits Calculated.	Comets Identified.
Before A.D. . . .	68	4	1
Century 0—100 . .	21	1	1
101—200 . .	24	2	1
201—300 . .	40	3	2
301—400 . .	25	0	1
401—500 . .	18	1	1
501—600 . .	25	4	1
601—700 . .	31	0	2
701—800 . .	15	2	1
801—900 . .	35	1	1
901—1000 . .	24	2	3
1001—1100 . .	31	3	2
1101—1200 . .	26	0	1
1201—1300 . .	27	3	3
1301—1400 . .	31	8	3
1401—1500 . .	35	6	1
1501—1600 . .	31	13	5
1601—1700 . .	25	20	5
1701—1800 . .	69	64	8
1801—1860 . .	141	134	37
	742	271	80

Comets remain visible for periods varying from a few days to more than a year, but the most usual time is 2 or 3 months. Much depends on the apparent position of the comet with respect to the Earth, and especially on its own intrinsic lustre. There are some few comets which have only been seen on one occasion, unfavourable weather preventing further observations: such was the case with one seen in August 1856, by Mr. E. J. Lowe, of Highfield House. Among the comets which have remained longest in sight, we may mention the following : —

				Months.
The Comet of 1811 (i)	.	.	.	17
The Comet of 1825 (iv)	.	.	.	12
The Comet of 1835 (iii, Halley's)		.	.	$9\frac{1}{2}$
The Comet of 1847 (iv)	.	.	.	$9\frac{1}{2}$
The Comet of 1858 (vi)	.	.	.	9
The Comet of 1844 (ii)	.	.	.	8
The Comet of 1847 (ii)	.	.	.	8

213

CHAPTER V.

HISTORICAL NOTICES.

Opinions of the Ancients on the Nature of Comets.—Superstitious Notions associated with them.—Extracts from Ancient Chronicles.— Pope Calixtus III. and the Comet of 1456.—Extracts from the Writings of English Authors of the 16th and 17th Centuries.—Napoleon and the Comet of 1769.—Supposed Allusions in the Bible to Comets. —Conclusion.

In this chapter we shall briefly advert to a few historical remarks, relating to cometary astronomy.

Going back to the early ages of the world, we find that the Chaldæans considered comets to be permanent bodies analogous to planets, but revolving round the Sun in orbits so much more extensive, that they were therefore only visible when near the Earth. This opinion, which, by the by, is the earliest information we have of there being periodical comets, was also held by the Pythagorean school of philosophers. Yet Aristotle, who records this, insists that comets are merely mundane exhalations, carried up into the atmosphere, and there ignited.

Anaxagoras, Apollonius, Democritus, and Zeno considered that these bodies were formed by the clustering of many smaller planets.

It is a somewhat remarkable fact, that Ptolemy, so celebrated for his varied astronomical attainments, should nowhere have made any mention of comets; his omission is, however, made up for, by Pliny, who seems to have paid much attention to them. He enumerates 12 kinds, each class deriving its name from some physical peculiarity of the objects belonging to it.

Seneca considered that comets must be above [beyond]

the Moon, and that by their rising and setting, they had something in common with the stars.

Paracelsus gravely insisted that comets were celestial messen-
gers, sent to foretell good or bad events—an idea which, even
in the present day, has not altogether died out.   The ancient
Romans did not trouble themselves much about astral phenomena; they, nevertheless, looked upon the comet of B.C. 43, as
a celestial chariot carrying away the soul of Julius Cæsar,
who had been assassinated shortly before it made its appearance.

In an ancient Norman Chronicle there occurs a curious exposition of the divine right of William I. to invade England:—
" How a star with 3 long tails appeared in the sky; how the
learned declared that stars only appeared when a kingdom
wanted a king, and how the said star was called a comette."
Another old Chronicler, speaking of the year 1065, says:—
" Soon after [the death of Henry, king of France, by poison], a
comet — a star denoting, as they say, change in kingdoms—
appeared, trailing its extended and fiery train along the sky.
Wherefore, a certain monk of our monastery, by name Elmer,
bowing down with terror at the sight of the brilliant star, wisely
exclaimed, ' Thou art come ! a matter of lamentation to many
a mocker art thou come.   I have seen thee long since; but I
now behold thee much more terrible, threatening to hurl
destruction on this country.' " [1]

The superstitious dread in which comets were held during
the middle ages, is well exemplified in the case of the comet
of 1456 (Halley's). We find that the then Pope, Calixtus III.,
ordered the church bells to be rung daily at noon, and extra
Ave Marias to be repeated by everybody. By the way,
whilst the comet was still visible, Hunniades, the Papal general,
gained an advantage over Mahomet, and compelled him to
raise the siege of Belgrade, the remembrance of which, the
silly Pope immortalised by ordering the Festival of the Trans-
figuration to be scrupulously observed throughout the Christian
world. " Thus was established the custom which still exists

[1] Will. Malmes. *Hist. Nov.*, iii. cap. 13.

in Romish countries, of ringing the bells at noon; and perhaps
it is from this circumstance that the well-known cakes made
of sliced nuts and honey, sold at the church-doors in Italy, on
Saints' days, are called *comete*." [1]

Leonard Digges says that "cometes signifie corruptions of
the ayre. They are signs of earthquakes, of wars, of changy-
ing kyngdoms, great dearthe of corne, a common death of
man, and beast." [2]

Old John Gadbury says : — "Experience is an eminent
evidence, that a comet like a sword, pretendeth war ; and a
hairy comet, with a beard, denoteth the death of kings." Mr.
G. gives us a chronological register of cometary announcements
for upwards of 600 years, and in large Roman capitals, "as if
God and nature intended by comets to ring the knells of
princes, esteeming bells in changes upon the Earth not sacred
enough for such illustrious and eminent performances."

A great English poet says : —

> "Satan stood
> Unterrified, and like a comet burned;
> That fires the length of Ophiuchus huge
> In th' Arctic sky, and from its horrid hair
> Shakes pestilence and war." [3]

The last comet employed in an astrological character was
that of 1769, which Napoleon I. looked upon as his protecting
genie. Indeed as late as 1808, Messier published a work on
it, named below.[4]

It would be quite impossible for us to allude to a tithe of
the cometary theories which have been advanced since the
introduction of telescopic observation. Some of the more important will be found in Grant's *History*, to which the reader
is referred.

During the visibility of the great comet of 1858, in the
autumn of that year, the question was mooted whether the
Bible contained any reference to these objects: the following
passages were adduced in support of the idea : —

---

[1] Smyth, *Cycle*, vol. i. p. 2.
[2] *Prognostications* for 1556.   [3] Milton, *Paradise Lost*.
[4] *La Grande Comète qui a paru à la Naissance de Napoléon le Grand.*

1. In Leviticus xvii. 7 it is said, " They shall no more offer there sacrifices unto Seirim " or Shoirim, which is rendered in the authorised version "devils," and in other versions "goats." We are informed by Maimonides, that the Zabian astrologers worshipped these *seirim*, which seems to confirm the idea of their being astral bodies.

2. In Isaiah xiv. 12 we find, "How art thou fallen from heaven, O Hillel, Sun of the morning! how art thou cut down to the ground, which didst weaken all nations! For thou hast said in thy heart, I will ascend into heaven, I will exalt my throne above the stars of God." In this passage a certain Hillel is said to have fallen from heaven ; but it is quite un- known who or what this Hillel was. Some interpreters derive the word from Hebrew verbs signifying to glorify, boast, agitate, howl, &c. Hillel *may* therefore signify a comet, for it answers to the ideas of brightness, swift motion, and calamity. Comets may be said, in a peculiar sense, to fall from heaven on account of the great rapidity of their motions ; they may be called "sons of the morning" on account of their great brightness ; they may be said, to some extent anyhow, to weaken nations, because of the terror they inspire ; they also ascend and traverse the heavens with great rapidity, and moreover, often by their lustre, overpower the "stars of God."

3. In the Catholic Epistle of S. Jude, certain impious im- postors are compared to " wandering stars, to whom is reserved the blackness of darkness for an æon [age]." In all probability the passage may be taken to refer to comets.

4. The last quotation we shall make is from the Apocalypse of S. John the Divine, xii. 3 : — " There appeared another wonder in heaven ; and behold a great red dragon . . . . and his tail drew the third part of the stars of heaven." Satan is here likened to a comet, because a comet resembles a dragon (or serpent) in form, and its tail frequently does compass or take hold of the stars.

These ideas are given for what they are worth, which is probably not much.

BOOK VI.

CHRONOLOGICAL ASTRONOMY.

—•—

CHAPTER I.

What Time is. — The Sidereal Day. — Its Length. — Difference between the Sidereal Day and the Mean Solar Day. — The Equation of Time. — The Anomalistic Year. — Use of the Gnomon. — Length of the Solar Year according to different Observers. — The Julian Calendar. — The Gregorian Calendar. — Old Style versus New Style. — Romish Miracles. — Table of Differences of the Styles.

TIME is, strictly speaking, of infinite duration; we are, therefore, obliged to choose some arbitrary unit by means of which a measurement of time may be effected. For short intervals, the diurnal rotation on its axis of the globe we inhabit; for longer intervals, the annual revolution of the Earth around the Sun, are the standards of measurement we employ; but any event which takes place at equal intervals of time may serve the purpose of a chronometrical register. Thus, the ages of certain trees may be ascertained by counting the number of concentric rings in the trunk, so many being formed annually; the ages of certain cattle, by the number of rings on the horns; the ages of horses may in like manner be ascertained by the successive disappearance of marks from their teeth; so also the pulsations of the heart, the flowing of a certain quantity of water from one vessel to another, the oscillations of a pendulum, may all be used as measurers of time: but, in

L

practice, the length of a day is a natural interval of time, which the domestic habits of man force upon him; and accordingly we find that amongst all nations this unit of measurement is, under some form or other, the one adopted.

The interval in which the Earth rotates on its axis is known to us as the *Sidereal Day*; it is determined by 2 consecutive passages of a star across the meridian of the place of observation, and is subdivided into 24 equal portions, called *sidereal hours*, which in turn are made up of 60 *sidereal minutes*, &c. The sidereal day may be otherwise defined to be the time occupied by the celestial sphere in making one complete revolution. The duration of this interval can be shown by theory to be invariable, and the actual comparisons of observations made on numerous stars, in widely different ages of the world, most completely corroborate this conclusion. Here we have, then, a chronometric unit far surpassing in accuracy anything that can be artificially contrived. Laplace has ascertained, from a careful comparison of modern with ancient observations, that the length of the sidereal day cannot have altered so much as $\frac{1}{100}$th of a second in upwards of 2000 years; it may, therefore, be regarded as possessing that indispensable qualification for a standard unit, *invariability*.

The *solar* day is reckoned by the interval elapsing between 2 successive meridian passages of the Sun.

The orbit of the Earth not being exactly circular, and its axis being considerably inclined, it follows that the daily velocity of our globe round the Sun, or, what for our purpose is the same, the daily motion of the Sun through the Zodiac, is not uniform[1], and the length of the solar day, therefore, varies at different seasons of the year; this variable interval is called the *apparent solar day*, and time so reckoned is *apparent time*. In order to obviate the inconvenience which would attend such a method of reckoning time, astronomers have agreed to suppose the existence of an imaginary sun moving in the

---

[1] The average daily motion is $0°\ 59'\ 8\cdot2''$, but with the Sun in perigee, in January, it amounts to $1°\ 1'\ 9\cdot9''$; with the Sun in apogee, in July, it is only $0°\ 57'\ 11\cdot5''$.

equator, with a velocity equal to the Sun's average motion in the ecliptic. When this fictitious or *mean sun* comes to the meridian, it is said to be *mean noon;* and when the true meridian passage of the Sun takes place, it is *apparent noon.* If the Sun were, like a star, stationary in the heavens, then it is clear, from what we have said above, that the solar and sidereal days would be equal; but since the Sun passes from west to east, through a whole circle — that is to say, through 360° in 365¼ days — it moves eastward about 59′ 8·2″ daily. While, therefore, the Earth is revolving on its axis, the Sun is moving in the same direction ; so that, when we have come round to the same meridian as we started from, we do not find the Sun there, but nearly 1° to the eastward ; and the Earth must perform a part of her 2nd revolution before we can come under the Sun again. Thus it is that the mean solar day is longer than the sidereal day in the ratio of 1·00273791 to 1, the former being taken at exactly 24h. 0m. 0s. Clocks regulated to keep sidereal time are in general use in astronomical observations — 1 revolution of the hands of the clock through 360°, or 24h., thus representing 1 complete revolution of the heavens ; but it is obvious that a sidereal hour is shorter than a solar hour, the difference being 9·83s. In consequence of the sidereal day being 3m. 55·91s. shorter than the mean solar day, the stars all pass the meridian 3m. 55·91s. earlier every day. This gaining of the stars upon the Sun is called the *acceleration of sidereal upon mean time;* an obvious consequence of this acceleration is, that the aspect of the heavens varies at different times of the year, those stars which at one time are seen on the meridian at midnight, passing it at 9 h. in the evening, after about 6 weeks.

The clocks we have in common use are all regulated to mean time, and will, therefore, show 12 o'clock sometimes before, and sometimes after the true Sun has reached the meridian; this difference between Mean time and Apparent time is called the *Equation of time,* and tables are constructed for the purpose of reducing the one to the other. Four times a year, the correction is zero, and the true and imaginary Suns coincide. Twice in the same period the clock is before the Sun, and twice after it. The equation reaches its maximum about February 10, when a correction of about 14m. 32s. is required to reduce apparent to mean solar time; or, in other words, the mean Sun is on

APPROXIMATE EQUATION-OF-TIME TABLE.

Days and Months	Minutes.		Days and Months.	Minutes		Days and Months.	Minutes.		Days and Months.	Minutes.	
Jan. 1	4		April 1	4	Clock faster.	Aug. 9	5		Oct. 27	16	
3	5		4	3		15	4	Clock faster.	Nov. 15	15	
5	6		7	2		20	3		20	14	
7	7		11	1		24	2		24	13	
9	8		15	0		28	1		27	12	Clock slower than the Dial.
12	9		*	—		31	0		30	11	
15	10	Clock faster than the Dial.	19	1		*	—		Dec. 2	10	
18	11		24	2		Sept. 3	1		5	9	
21	12		30	3	Clock slower.	6	2		7	8	
25	13		May 13	4		9	3		9	7	
31	14		29	3		12	4	Clock slower than the Dial.	11	6	
Feb. 10	15		June 5	2		15	5		13	5	
21	14		10	1		18	6		16	4	
27	13		15	0		21	7		18	3	
Mar. 4	12		*	—		24	8		20	2	
8	11		20	1		27	9		22	1	
12	10		25	2	Clock faster.	30	10		24	0	
15	9		29	3		Oct. 3	11		*	—	
19	8		July 5	4		6	12		26	1	Cl. faster.
22	7		11	5		10	13		28	2	
25	6		28	6		14	14		30	3	
28	5					19	15				

the meridian 14m. 32s. before the true Sun. On April 15, there is no equation, the real and fictitious Suns being on the meridian at the same moment. Towards the middle of May the equation again reaches a maximum of 3m. 54s., but becomes reduced to zero by June 15. Another maximum occurs about July 27, when a correction of 6m. 11s. is required to be added to the apparent solar time. On Sept. 1, the equation is again at zero, but increases from that time until the beginning of November, when the correction amounts to 16m. 18s., subtractive from apparent time. Another zero takes place about December 25, completing the round.

In France, until 1816, apparent time was used, and the confusion arising from this practice may be readily imagined. Arago relates that he was once told by Delambre, that he had frequently heard the different public clocks striking the same hour at an interval of half an hour. At the time of the change, the Prefect of the Seine refused to sign the necessary order, fearing an insurrection amongst the lower classes ; none however took place, and the worthy magistrate's fears were groundless. Especially were the watch-makers thankful for the change ; under the old system, all in vain was it that they tried to explain to their enraged customers, when they came to complain of the watches they had bought, that it was the Sun and not the watches which were in fault.

The interval of time elapsing from the moment when the Sun leaves a fixed star until it returns to it again, constitutes the *sidereal year*, and consists of 365d. 6h. 9m. 9·6s., which is somewhat longer than the mean solar year. The latter is the interval of time which elapses between 2 successive passages of the Sun through the same equinox. If the equinoxes were fixed points, then this period would be identical with the sidereal revolution of the Earth ; but since these points are possessed of a retrograde motion from east to west of 50·1″ annually, it follows that the Sun returns to the equinox sooner every year by a period equal to the time it takes to traverse 50″ of arc, or by 20m. 19·9s. of time. The mean solar year is therefore 20m. 19·9s. shorter than the sidereal year, or its length is 365d. 5h. 48m. 49·7s. In consequence of the motion of the equinoctial points not being uniform (on account of planetary perturbation), a variation takes place in the length of the mean solar year, which is now being diminished in length at the rate of 0·6s. per century. This, however, will not always be the case.

The line of apsides of the Earth's orbit is subject to an annual motion of 11·8″. If the Earth, then, be supposed to start from her perihelion, she will require a longer interval of time than her sidereal period to reach it again, and the excess will be equal to the time necessary for the Earth to describe 11·8″ of her orbit; this she would do in 4m. 32·7s., which quantity must be added to the sidereal period before we can ascertain the interval between 2 successive returns to perihelion. The result then is a period of 365d. 6h. 13m. 49·3s. which is called the *anomalistic year*.

The manner in which the ancients ascertained the length of the year, was by means of the *gnomon* or *stylus*, a vertical rod standing on a smooth plane, with a meridian line described on it. The time when the shadow was shortest would indicate the day of the summer solstice, and the number of days which elapsed until the shadow returned to the same length, would be the number of days in the year. This interval having been found to be 365 days, 365 days was the period adopted for the length of the common year, or nearly 6 hours less than the true length. Such a difference would, after the lapse of some time, throw everything into confusion; for supposing that in any one year the summer solstice fell on June 21, after the lapse of 4 years it would fall on the 22nd, in another period of 4 years it would happen on the 23rd and so on. The inhabitants of Thebes in Egypt are said to have been the first to have noticed the necessity of an addition of 6 hours to the 365 days in order to make the year coincide with the annual course of the Sun. In the time of Democritus, 450 B. C., 365¼ days was supposed to be the length of the year; Eudoxus made it somewhat longer, and, according to Diodorus Siculus, Œnopides of Chios fixed it at 365d. 8h. 48m. Hipparchus, by means of his own observations, found the length then in use (365¼ days) to be too long by 4m. 48s. Ptolemy also examined the subject, but came to no definite determination. Towards the close of the 9th century, the Arabian prince, Albategnius, from observations of his own, considered that the length given by Hipparchus was still too great by some 8¾m.; he accordingly assigned a new determination in his work, *De Scientiâ Stellarum*. The following table exhibits some of the principal determinations which have been arrived at, both in ancient and modern times : —

	D.	H.	M.	S
Ancient Egyptian . . .	365	0	0	0
Euctemon and Meton . .	365	6	18	57
Calippus, &c. . . .	365	6	0	0
Hipparchus	365	5	55	12
Hindu	365	5	50	30
Albategnius	365	5	46	24
Alphonsine Tables, 1252 .	365	5	49	16
Walther	365	5	48	50
Copernicus, 1543 . . .	365	5	49	6
Tycho Brahe, 1602 . .	365	5	48	45·3
Kepler	365	5	48	57·6
J. Cassini, 1743 . . .	365	5	48	52·4
Flamsteed	365	5	48	57·5
Halley	365	5	48	54·8
La Caille	365	5	48	49
Delambre	365	5	48	51·6
Laplace	365	5	48	49·7
Bessel	365	5	48	47·8

We have seen that the mean solar year does not contain a whole number of days, but that a fractional quantity is appended ; that is to say, its length is thus expressed, 365·24225 days. 100 years of 365 days each would contain 36,500 days, or would fall short of 365 revolutions of the Sun by about 24 days. In order to remedy this state of things, Julius Cæsar, who was as distinguished for the varied nature of his mental attainments as for his skill in military affairs, called to his aid the Egyptian astronomer Sosigenes, and they both set to work to reform the calendar. They introduced an additional day every 4th year into the month of February, thereby making 25 additional days in the century. This 4th year was termed *bissextile*, because the 6th day before the calends of March was reckoned twice, and in this year, therefore,

February was made to consist of 29 days. This almost perfect arrangement, which was called from its author the Julian
style, prevailed generally throughout the Christian world till
the close of the 16th century.[1] The calendar of Julius Cæsar
was defective in this particular, — that the solar year, consisting of 365d. 5h. 48m., and not of 365d. 6h., as was supposed
in his time, there was a difference of 11m. between the apparent and true years. At the time of Gregory XIII. this difference had so accumulated as to amount to more than 10
days, the Vernal Equinox falling, in 1582, on March 11 instead of 21, at which it was in the year A. D. 325, when the
Council of Nicæa was held. At this council it was decreed that
Easter should be kept upon the first Sunday after the first
full Moon next following the Vernal Equinox; and "as certain other festivals of the Romish Church were appointed at
particular seasons of the year, confusion would result from
such want of constancy between any fixed date and a particular season of the year. Suppose, for example, a festival,
accompanied by numerous religious ceremonies, was decreed
by the Church to be held at the time when the Sun crossed
the equator in the spring (an event hailed with great joy as
the harbinger of the return of summer), and that in the year
325, March 21 was designated as the time for holding the
festival, since at that period it was on March 21 when the Sun
reached the Equinox; the next year the Sun would reach the
Equinox a little sooner than the 21st, only 11m. indeed, but
still amounting to 10 days in 1200 years; that is, in 1582 the
Sun reached the Equinox on March 11. If, therefore, they
continued to observe the 21st as a religious festival in honour of
this event, they would commit the absurdity of celebrating it
10 days after it had passed by." This anomaly Gregory XIII.
undertook to correct, which he did with perfect success; he
ordained that 10 days should be left out of the current year
in order to bring back the Equinox to March 21, and in order
to keep it on that day, he prescribed the following rule : —

---

[1] The Julian error amounted to + 0·00778 day annually, or 1 whole
day in 129 years.

Every year divisible by 4, to be a bissextile or leap year, containing 366 days : every year not so divisible to consist of only 365 days: every secular year (1800, 1900, &c.,) divisible by 400, to be a bissextile or leap year, containimg 366 days : every secular year not so divisible to consist of 365 days. If, every 4th year were to consist of 366 days, a century would be too long by $\frac{1}{4}$ of a day; that is to say, we should have allowed 1 whole day in 100 years instead of only $\frac{3}{4}$: in 400 years this would have amounted to 1 day ; this will explain why a day is to be dropped every 4th century. We will now perform some short calculations to see how near this rule is to the truth, by comparing 10,000 Gregorian with 10,000 Tropical (or solar) years. In 10,000 the numbers not divisible by 4 will be $\frac{3}{4}$ of 10,000 or 7500; those divisible by 100, but not by 400, will in like manner be $\frac{3}{4}$ of 100, or 75 ; so that in the 10;000 years in question, 7575 consists of 365 days, and the remaining 2425 of 366, producing in all 3,652,425 days. Dividing this number by 10,000 we get 365·2425 as the mean Gregorian length of the solar year. The actual value of the latter being 365·2422, the error in 10,000 years amounts to 2·6, or 2d. 14h. 24m., or 1 day in 3846 years, or 22s. annually, a quantity which we can, without inconvenience, disregard. But even this error, trifling as it is, may be still further eliminated by declaring those years divisible only by 4000, to be common and not bissextile years ; this would make the error but 1 day in 100,000 Gregorian years.

The Julian calendar was introduced in the year 44 B. C., which Cæsar ordered should commence on January 1st, being the day of the New Moon immediately following the winter solstice the preceding year, which year was made to consist of 445 days, and was known as "the year of confusion." Cæsar did not live to carry out in person the reform he had decreed, and the consequence was that great confusion ensued. We are not acquainted with the terms of the edict which he promulgated, but we are led to infer, that it was not so explicit as it ought to have been, and that it contained some expressions relating to " every 4th year " which were not clearly enunciated. The consequence was, that Cæsar's successors counted the leap year just elapsed as No. 1 of the quadrennial period, and intercalated every *third* instead of every 4th year. " This erroneous practice continued during 36 years, in which, therefore, 12 instead of 9 days were intercalated, and an error of 3 days produced; to rectify which Augustus ordered the suspension of all intercalation during three complete *quad-rennia*, — thus restoring, as may be presumed his intention to have been, the Julian dates for the future, and re-establishing the Julian system for the future, which was never after vitiated by any error till the epoch when its own inherent defects gave occasion to the Gregorian reformation. And starting from *this* (the period of the Augustan reform) as a certain fact, (for the statements of the transaction by classical authors are not so precise as to leave *absolutely no doubt* as to the previous intermediate year), astronomers and chronologists have agreed to reckon backwards in unbroken succession on this principle, and thus to carry the Julian chronology into past time, *as if* it had never suffered such interruption, and *as if* it were certain (which it is not, though we conceive the balance of probabilities to incline that way) that Cæsar, by way of securing the intercalation as a matter of precedent, made his initial year, 44 B. C., a leap year. Whenever, therefore, in the relation of any event, either in ancient history or in modern, previous to the change of style, the time is specified in our modern nomenclature, it is always to be understood as having been identified with the assigned date by threading the mazes (often very tangled and obscure ones) of special and national chronology, and referring the day of its occurrence to its place in the Julian system *so interpreted*."[1] The reformed Gregorian calendar was published to the world in 1582, the Pope at the same time issuing a decree commanding its observance throughout Christendom. His mandate met with great opposition from the Catholic powers of Europe, who did not recognise the Papal supremacy; but in Romish countries it was soon adopted. It was not established in Great Britain till 1752,

---

[1] Herschel, *Outlines of Ast.*, p. 675.

when an Act of Parliament was passed for the purpose of enjoining its use. As 170 years had elapsed since the new style had been first brought into use, it became necessary to suppress 11 days, in order to set right the Equinox; this was effected by calling the day after September 3 the 14th. By the same Act it was decreed that January 1 should be the commencement of the year, instead of March 25, as it had been heretofore. We shall now see the practical effect of these changes. In order to make any date given in the o. s. comparable with our present mode of reckoning, we must add thereto 11 days : thus, April 24 o. s. is equivalent to May 5 N. S. If any event happened between January 1 and March 25, the date of the year will be advanced 1; thus, March 21, 1864, o. s. will be April 2, 1865, N. s.; bearing in mind that the difference between the two styles, which in 1752 was 11 days, is now 12. Russia and the Greek Church generally still adhere to the old style, consequently their dates are thus expressed :—

$$1860 \ \frac{\text{May 23.}}{\text{June 4.}}$$

Such a sweeping change as this measure of Pope Gregory was, as might have been anticipated, received with great dissatisfaction by the English nation at large, but more especially by the lower orders, who considered they had been robbed of 11 clear days. The inconveniences everybody was subjected to, on account of the disturbances of every kind of festival and anniversary, were, moreover, by no means agreeable to the feelings of most people. Professor De Morgan, on the authority of a scientific friend since deceased, relates the following : — "A worthy couple in a country town, scandalised by the change of style in 1752, continued for many years to attempt the observance of Good Friday on the old day. To this end they walked seriously and in full dress to the church door, at which the gentleman rapped with his stick; on finding no admittance, they walked as seriously back again, and read the service at home. But on the new and spurious Good Friday they took

pains to make such a festival at their house as would con-
vince the neighbours that their Lent was either ended or in
abeyance." But there must have been some days of comfort,
for between 1752 and 1800 there were 18 years in which the
old and new Easter Day coincided. This may happen occasionally, and will do so, though less and less frequently, till
A.D. 2698, when it will occur for the last time.

"Previous to the change of style there existed a wide-
spread superstition in England that, at the moment when
Christmas Day began, the cattle always fell on their knees in
their stables; it was averred, however, that the animal creation
refused to obey the Papal Bull, and still continued their
prostrations on the Christmas eve, according to the old style.
In Romish countries, however, inanimate things even agreed
to change their habits; for Riccioli positively assures us that
the blood of S. Januarius, which under the old order of things
always liquefied punctually on September 19 (?), immediately
changed its day of liquefaction to Sept. 9, O.S., in order to con-
form to Sept. 19, N.S., thereby putting itself back 10 days,
that it might obey the Pope's mandate! But this was not
all; for Riccioli goes on to add that a certain twig, which
always budded on Christmas Day O.S., thence budded on
January 6, N.S., for the same reason!"[1]

In England the members of the Calendar Reforming Government were pursued and mobbed in the streets of London, the
populace demanding the restoration of the 11 days of which
they supposed they had been illegally deprived. The illness
and subsequent death of Bradley, the well-known astronomer,
who had assisted the Government with his advice, was, as a
matter of course, looked upon as a judgment from Heaven.

The following is a table of the differences between the old
and new styles, for the under-mentioned periods : —

---

[1] *Companion to the Almanac*, 1845, p. 19.

Date.	Difference.	Date.	Difference.
1500—1700	10 days.	2900—3000	20 days.
1700—1800	11	3000—3100	21
1800—1900	12	3100—3300	22
1900—2100	13	3300—3400	23
2100—2200	14	3400—3500	24
2200—2300	15	3500—3700	25
2300—2500	16	3700—3800	26
2500—2600	17	3800—3900	27
2600—2700	18	3900—4100	28
2700—2900	19	&c.	&c.

CHAPTER II.

Hours. — Commencement of the Days. — Usage of different Nations. — Days. — Weeks. — Origin of the English Names for the Days of the Week. — The Egyptian 7-day Period. — The Roman Week. — Months. — Memoranda on the Months. — Years. — The Egyptian Year. — The Jewish Year. — The Greek Year. — The Roman Year. — The Roman Calendar and the Reforms it underwent. — The French Revolutionary Calendar. — The Year. — Its Sub-divisions into Quarters. — Quarter Days.

WE have now to consider the different divisions of time which are in use, beginning with —

Hours. — We have already mentioned that a day is divided into 24 equal portions, called hours; each of these contains 60 minutes, and each minute 60 seconds.[1] It is now quite impossible to assign any date to the origin of this custom, so completely is it lost in the obscurity of antiquity. Although the duodecimal division of the day is so universal, yet different usages have prevailed in different countries relative to the enumeration of those hours. Some nations have counted the hours consecutively from 1 to 24; others have divided the hours into two series of 12 each; whilst in France, during the revolutionary period following the year 1793, a decimal system was introduced, the day being divided into 10 hours, each hour into 100 minutes, and each minute into 100 seconds. When, however, after the lapse of a few years, the French nation recovered its sanity, this, together with many other equally absurd innovations, was given up. The 24 hours into which

---

[1] The old sub-division of thirds and fourths have fallen into disuse, every odd part of a second being expressed decimally. Thus: 13h. 17m. 24s. 18t. 13f. would now be expressed as 13h. 17m. 24·303s.

the day is divided were usually intended to be equal, each comprising $\frac{1}{24}$th part of the whole; but there were exceptions to this. For instance, at one period in the history of Greece, the interval between sunrise and sunset was divided into 12 equal portions, termed "hours of the day;" the other interval between sunset and sunrise being also similarly divided into the "hours of the night." It is clear that at the Equinoxes only would the former be equal to the latter, that from the Vernal to the Autumnal Equinox the diurnal hours would be the longest, and that from the Autumnal to the Vernal Equinox the nocturnal hours would be the longest. A variation in the length of each also took place from day to day. Such a system, inconvenient as it doubtless was for the ordinary affairs of life, was positively useless for all scientific purposes; hence we find that Ptolemy was in the habit of transforming these common hours into equinoctial hours — so called, probably, on account of their being at the Equinoxes equal in duration to the vulgar hour. Even with this improvement, we find that the above-named distinguished astronomer was unable to indicate the time of any celestial phenomenon within a quarter of an hour of its true time. This conclusively shows us how imperfect were the chronometric appliances then in use: we are now able to obtain observations within $\frac{1}{10}$th of a second of the absolute truth.[1]

Having determined on the unit we are going to employ as a chronometric register, it is also necessary to determine

---

[1] An interesting instance of the surprising accuracy now attainable in astronomical observations was afforded by the occultation of the planet Saturn by the Moon on May 8, 1859. The phenomenon was watched at the Greenwich Observatory by 5 persons, with different telescopes, and the following are the times of disappearance recorded by them :—

	H.	M.	S.
Rev. R. Main	8	19	42·4
Mr. Glaisher	8	19	42·5
Mr. Dunkin	8	19	42·6
Mr. Ellis	8	19	42·9
Mr. Criswick	8	19	42·2

conventionally some particular moment when each successive unit shall commence and end. The Jews, the Chinese, the ancient Athenians, and the Oriental nations, as well as the inhabitants of the Italian peninsula, fixed upon sunset as the termination of one day, and the commencement of the following, counting the hours from o to 24. As the hours of sunrise and sunset vary from day to day, it is manifest that 4 o'clock one day will not be the same as 4 o'clock the day previously; so that for a clock to indicate such time it must be set from day to day, or from week to week, since the hour of sunset will be constantly later during one half year, and earlier during the other. This system of reckoning has been defended upon the ground of the convenience it affords to travellers and others, in telling them, without trouble, how much time they have left at their disposal before nightfall. This is no doubt perfectly correct, as far as it goes; but then, on the other side of the question, there is the constant necessity of resetting the clocks every day to be considered, to say nothing of the other " obvious inconveniences attending such a system, such as the constant variation of the hours of meals, of going to bed and rising, of all description of regular labour, the hours of opening and closing all public offices, of commencing and terminating all public business," &c. Notwithstanding all this, however, the force of habit is so strong, that this system is still in use; though in many places it is customary to set up 2 clocks side by side, one indicating Italian and the other common time. This system, with the unimportant modification of starting from sun*rise*, was also used by the Babylonians, Syrians, and Persians, and is at the present time adopted by the modern Greeks and the inhabitants of the Balearic Islands.

Hipparchus (about 150 B.C.) adopted the plan of commencing the day at midnight, and dividing it into 2 equal series of 12 hours each: this system was followed by Copernicus, and is now in general use throughout all civilised parts of the globe. According to this plan of reckoning, whenever an hour is named, it is requisite to state in what position it stands as regards noon. The hours previous to noon are indicated by the letters A.M., and those after noon by P.M.; the former being the initial letters of the Latin words *ante meridiem* ("before midday"), and the latter of *post meridiem* ("after midday"). The ancient Egyptians commenced their day with the Sun's passage over the meridian; in this they were followed by Ptolemy and by astronomers in modern times, who divide the day into 24 hours. We must, therefore, carefully distinguish between civil and astronomical time, the former being 12 hours ahead of the latter.

Days.—A day is the standard unit of measurement now universally adopted, all shorter intervals of time being reckoned by some of its fractional sub-divisions, and all longer intervals by some or other multiple of it; it is, therefore, of primary importance that its absolute length should be certainly and precisely ascertained. Being a standard measure, it is indispensably necessary that its length should be invariable, and that facilities should be possessed for verification from time to time. Such operations have been conducted by eminent mathematicians, and we believe we can say with truth that, if it is not immutable, it is almost the only thing in the universe whose mutability never has been, and is never likely to be, detected.

"Tempora mutantur, et nos mutamur in illis."

Astronomy and geology bear witness to the truth of this rule, though it would seem that we are acquainted with one exception. Precession of the Equinoxes, the Secular Variation in the Elements of the planets and comets of our system, the Proper Motions of the stars, are examples of astronomical change. The Deinotherium, the Mastodon, and the huge Saurian Reptiles, now only to be seen stored up in museums, testify to geological change.

Weeks.—The historical origin of the well-known period of 7 days is quite lost in antiquity, and much difference of opinion still exists as to the date and prevalence of it, as a mode of reckoning. It has commonly been regarded as a memorial of the creation of the world, reference being made to it in the account of that event handed down to us in the Book of Genesis. It is also an obvious and convenient subdivision of the lunar month, besides being so

nearly an exact aliquot part of a solar year of 365 days
(7 × 52 = 364) — two good reasons for its adoption.

The English names of the week are derived as follows: —

1. Dies Solis (L.) Sun's day	whence	Sunday.
2. Dies Lunæ (L.) Moon's day	,,	Monday.
3. Tiues daeg (Sax.) Tiu's day	,,	Tuesday.
4. Wodnes daeg (Sax.) Woden's day	,,	Wednesday.
5. Thunres daeg (Sax.) Thor's day	,,	Thursday.
6. Frige daeg (Sax.) Friga's day	,,	Friday.
7. Dies Saturni (L.) Saturn's day	,,	Saturday.

In all parliamentary and judicial documents the Latin
names are still retained ; the Quakers, however, do not use
either, but call Sunday the 1st day of the week, Monday the
2nd, and so on. The reason why the 1st day of the week is
kept as the Christian Sabbath is that the resurrection of our
LORD took place on that day, and it was accordingly determined that the day for the observance of the Sabbath should
be changed, to symbolise the displacement of the Mosaic by
the Christian dispensation.

By Dion Cassius the use of a 7-day period is ascribed to
the Egyptians, and from them in after times copied by the
Greeks and other nations. He makes the following remarks
on the manner in which the Egyptians derived the names of
the days of the week and their order from the 7 members of
the solar system known to the ancients.[1] The series of hours,
without reference to days, was resolved into periods of 7, each
dedicated to a planet. Thus the 1st hour was dedicated to
Saturn, the next to Jupiter, the 3rd to Mars, &c.[2] The day
being divided into 24 hours, which is not a multiple of 7, it

---

[1] *Hist. Roman.*, lib. xxxvii. 19, *et seq.*

[2] It may be well to remark, that the ancient Egyptians included the
Sun and Moon under the general name of planets, the order of their
distance from the Earth being, as they supposed, as follows :—

1. Saturn.	5. Venus.
2. Jupiter.	6. Mercury.
3. Mars.	7. The Moon.
4. The Sun.	

followed that necessarily each successive day would begin with an hour dedicated to a different planet. The day which begins with the hour of Saturn would evidently end with the hour of Mars, for the 24 hours would consist of 3 complete periods of 7, and the 24th hour would be the 3rd of the 4th period, and would consequently be the hour of Mars. The 1st hour of the next day would be that of the Sun. In like manner, the day beginning with the hour of the Sun, and consisting of 3 hours more than 3 complete periods, would end with the hour of Mercury, and the next would begin with the hour of the Moon. The succeeding day would in like manner commence with the 3rd in order from the Moon, that is Mars; the next with the 3rd in order from Mars, that is Mercury; the next with the 3rd in order from Mercury, that is Jupiter; the next with the 3rd in order from Jupiter, that is Venus; and after Venus the series would recommence with the hour dedicated to Saturn. Thus the 1st hour of each successive day, in each successive period of 7 days, would be dedicated to the planets in the following order : —

1. Saturn.	5. Mercury.
2. The Sun.	6. Jupiter.
3. The Moon.	7. Venus.
4. Mars.	

This order was retained by the Romans, and we find that the following were the names of the Roman days of the week : —

1. Dies Saturni	.	.	Saturn's Day.
2. Dies Solis	.	.	Sun's Day.
3. Dies Lunæ	.	.	Moon's Day.
4. Dies Martis	.	.	Mars's Day.
5. Dies Mercurii	.	.	Mercury's Day.
6. Dies Jovis	.	.	Jupiter's Day.
7. Dies Veneris	.	.	Venus's Day.

From the above have been derived the modern names used in the different countries of Europe, either by a literal translation as in the Italian, French, Spanish, and other languages of Latin origin, or, as in the Teutonic languages, by a

substitution for the name of the classical, of that of the corresponding Teutonic deity.

Months.—The relation of this division of time to the
Moon is singularly apparent in many languages, notwithstanding that the Moon's period of revolution is unsuitable as a
measure of time, both on account of its not being marked by
any easily observed phenomena, and also from its not being a
multiple of either a day or year. The lunar month, however,
has even within the last few years, been used by the inhabitants of many of the South Sea Islands.

A few memoranda on the months as they now stand may be
useful. Every one is familiar with the following lines. We
are now quoting from a version published in 1596 : —

> " Thirtie daies hath September,
> Aprill, June, and November,
> Februarie hath eight and twentie alone,
> All the rest thirtie and one.
> Except in Leap year, at which time
> Februarie's daies are twentie and nine."

To find the day of the month without an almanac, it may
be useful to know that the following are all of the same
name : —

> 1st of January, October.
> 2nd of April, July.
> 3rd of September, December.
> 4th of June.
> 5th of February, March, November.
> 6th of August.
> 7th of May.

The common year begins and ends on the same day of the
week ; leap year ends on the following day. Many persons
who call the year 52 weeks are not aware that it is always 52
weeks 1 day, and in leap year 52 weeks 2 days.

A nation possessed of two such well-defined chronometric
units as the solar day and year, would doubtless soon find the
inconvenience of not having some period intermediate between
them. Let us now see what intermediate subdivisions there
are which would answer the purpose. A year does not con-

tain any exact whole number of days, it contains $365\frac{1}{4}$ (nearly); but since the ancients reckoned it at 365 exactly, we will do the same. Now it is clear that the only factors of 365, are 5 and 73 ; and we must, therefore, either divide the year into 5 equal periods of 73 days each, or 73 equal periods of 5 days each. That neither of these subdivisions will meet the requirements of mankind, we have conclusive evidence, in the fact that during more than 5000 years neither of them has ever been adopted. No other equal subdivision of the year being possible, some different plan must be devised; we may either divide the year into a certain number of equal parts, with a remainder, and then consider the remainder as a supplemental part; or we may resolve the year into some convenient number of unequal parts.

The Egyptian year was arranged according to the 1st of these plans, by dividing it into 12 months, each of 30 days, with 5 days added at the end of the 12th month.

The Jewish year is regulated according to the 2nd expedient. The following are the months: —

	Days.			Days.
Tisri	30	Nisan . . .		30
Marchesvan	29 or 30	Ijar . . .		29
Kislev	29 or 30	Sivan . . .		30
Tebeth	29	Thamuz . .		29
Schebat	30	Ab . . .		30
Adar	29	Elul . . .		29 or 30
Veadar (intercalary)	29			

The division of the year into months by the Greeks was not only very unmethodical, but it would seem that no 2 states agreed either in the number, names, or lengths of their months. Some months were designated by particular names, while others were known only by the numerical order in which they followed one another. These numbers, however, did not correspond in different states on account of the year beginning at different times. Thus the 5th Delphic, Lacedæmonian, Bœotian, and Attic months corresponded severally with January, February, May and November, according to our mode of reckoning. The confusion arising from such a state of things in a small country, inhabited by one nation speaking one language, may be readily imagined.

Athens being the capital of Greece, we subjoin a table of the Attic months : —

	Days.		Days.
Ἑκατομβαιὼν	30	Γαμελιὼν	30
Μεταγειτνιὼν	29	Ἀνθεστηριὼν	29
Βοηδρομιὼν	30	Ἐλαφηβολιὼν	30
Μαιμακτηριὼν	29	Μουνυχιὼν	29
Πυανεψιὼν	30	Θαργηλιὼν	30
Ποσειδεὼν	29	Σκιροφοριὼν	29

The intercalary month was a second Ποσειδεὼν of 30 days. It is said that formerly the months were all of 30 days each, but that Solon introduced the alternation of 29 and 30 days. Those months which contained only 29 days were termed κοῖλοι (hollow), the others were πλήρεις (full). The year in which an intercalary month was inserted was called ἐμβόλιμος, whence our word embolismic. The months were subdivided into 3 decades or periods of 10 days each, the last decade containing 10 or 9 days as the case might be.

Notwithstanding the advanced civilisation of the Romans, it was not till 700 years after the foundation of their city that they become possessed of a properly arranged system of reckoning. The year, as established by Romulus, contained 10 months, 4 of 31, and 6 of 30 days, making a total of 304, as follows : —

	Days.		Days.
1. Martius	. 31	6. Sextilis	. 30
2. Aprilis	. 30	7. September	. 30
3. Maius	. 31	8. October	. 31
4. Junius	. 30	9. November	. 30
5. Quintilius	. 31	10. December	. 30

It was soon found that a year of 304 days was utterly irreconcilable with the nature of things, and accordingly we find that in the following reign, that of Numa Pompilius, 2 new months were added, Februarius and Januarius. The latter was placed at the beginning, and the former at the end of the year; this arrangement was however afterwards altered, and Januarius made to precede Februarius, leaving Martius the first month of the year. This will account for the circumstance of September and the 3 following months bearing names which do not correspond to the places they now occupy. In order to make the year correspond with tolerable accuracy to the true solar year, Numa resolved on increasing its length by 51 days. This being too long for 1 month, and not long enough for 2, it was decided that 1 day should be taken from each of the months having 30; there being 6 of these, 57 days (51 + 6) were made available for the formation of the 2 new months, which were then arranged as follows: —

	Days.		Days.
Januarius	29	Quintilius (aft. Julius)	31
Februarius	28	Sextilis (aft. Augustus)	29
Martius	31	September	29
Aprilis	29	October	31
Maius	31	November	29
Junius	29	December	29
			355

Januarius derived its name from Janus, the deity who presided over everything. Februarius from Febrius, an ancient Italian deity whose rites were celebrated this month. Martius from Mars, the god of War, and father of Romulus. Aprilis, probably from *aperire*, to open, in allusion to the budding of vegetation in spring. Maius from Maia, the mother of Mercury. Junius from Juno, Queen of Heaven. The names of the months Quintilius and Sextilis were afterwards changed in compliment to the emperors Julius and Augustus Cæsar.

In case it should be asked, why Numa did not take away the odd days from those months which had 31 days, which would be the obvious expedient, we may remark that even numbers (which would have thus resulted) were looked upon as unlucky and inauspicious! Since, however, it was impossible to avoid having 1 month of an even number of days, this unlucky number was assigned to Februarius, over which the Genius of Death presided.

Notwithstanding the modifications introduced by Numa, the year was still 10 days too short; to correct this he decreed that a 13th, or intercalary month, should be introduced every other year, consisting of 22 or 23 days. This was reasonable enough, but the singular thing was the plan adopted for effecting this. It was done by inserting the new month, called Mercedonius, probably derived from *merces*, wages, which perhaps were usually paid at this time of the year, between the 23rd and 24th of February; making the 1st of Mercedonius to follow the 23rd, and the last day of Mercedonius to be followed by the 24th of Februarius.

The days of the Roman month were reckoned in the following way: the 1st day of each month was called the *kalends* [1]; the 7th day of each of the 4 great months (those of 31 days), and the 5th of each of the lesser months (those of 29 days), were called the *nones*; and the 15th of all the great months, and the 13th of all the lesser months, were called the *ides*. The difference in the positions of the two kinds of nones and ides, is owing to the Roman custom of reckoning time backwards.

As frequent allusion is made in classical and other writers to the Roman mode of computation, it may be useful to subjoin the following table, showing the correspondence of the ancient Roman with the modern months : —

---

[1] This word, not being used by the Greeks, gave rise to the saying, that such and such a thing was postponed to the Greek kalends, or *sine die* as we should say.

THE ROMAN CALENDAR.

Dies Mensis.	April. Jun. Sept. Nov.	Jan. Aug. Decemb.	Mar. Mai. Jul. Octob.	Feb.
1	KALENDÆ.	KALENDÆ.	KALENDÆ.	KALENDÆ.
2	IV.	IV.	VI.	IV.
3	III.	III.	V.	III.
4	Prid. Non.	Prid. Non.	IV.	Prid. Non.
5	NONÆ.	NONÆ.	III.	NONÆ.
6	VIII.	VIII.	Prid. Non.	VIII.
7	VII.	VII.	NONÆ.	VII.
8	VI.	VI.	VIII.	VI.
9	V.	V.	VII.	V.
10	IV.	IV.	VI.	IV.
11	III.	III.	V.	III.
12	Prid. Id.	Prid. Id.	IV.	Prid. Id.
13	IDUS.	IDUS.	III.	IDUS.
14	XVIII.	XIX.	Prid. Id.	XVI.
15	XVII.	XVIII.	IDUS.	XV.
16	XVI.	XVII.	XVII.	XIV.
17	XV.	XVI.	XVI.	XIII.
18	XIV.	XV.	XV.	XII.
19	XIII.	XIV.	XIV.	XI.
20	XII.	XIII.	XIII.	X.
21	XI.	XII.	XII.	IX.
22	X.	XI.	XI.	VIII.
23	IX.	X.	X.	VII.
24	VIII.	IX.	IX.	VI.
25	VII.	VIII.	VIII.	V.
26	VI.	VII.	VII.	IV.
27	V.	VI.	VI.	III.
28	IV.	V.	V.	Prid. Kal. ⎱ Martii. ⎰
29	III.	IV.	IV.	
30	Prid. Kal. ⎱ mensis seq. ⎰	III.	III.	
31		Prid. Kal. ⎱ mensis seq. ⎰	Prid. Kal. ⎱ mensis seq. ⎰	

M

The Mahometan year is a lunar one, consisting of 12 months, each of which begins with the first appearance of the new moon, without any intercalation to bring the year to the same season. It is obvious, therefore, that every year will begin earlier than the preceding one by about 11 days. The inconveniences which this gives rise to have already been pointed out. It moreover happens, that as the commencement of each month depends on the first visibility of the new moon, a few cloudy days will produce serious confusion, as it will lead to differences of sometimes a day or two in the reckoning, in parts of the country widely separated from each other.

We have one other system to refer to.—In 1792, the French nation, in its excessive desire to sweep away every vestige of monarchy and of the existing institutions of their country, determined on adopting a new calendar, founded on radical reform principles; but finding that it was unable to produce any plan more accurate or convenient than the one in previous use, it only made some minor alterations. The first year of the new Republican era commenced on September 22, 1792 (N.S.), the day of the Autumnal Equinox. The year consisted of 12 months of 30 days each, with 5 supplementary ones kept as festivals. Every 4th year was a leap year, but called by the demagogues a Franciad or Olympic year. The months were as follows [1] : —

Vindemaire	.	.	Vintage Month commences Sep. 22.		
Brumaire .	.	.	Foggy	,,	,,	Oct. 22.
Frimaire .	.	.	Sleety	,,	,,	Nov. 21.
Nivôse	.	.	Snowy	,,	,,	Dec. 21.
Pluviôse .	.	.	Rainy	,,	,,	Jan. 20.
Ventôse	.	.	Windy	,,	,,	Feb. 19.
Germinal .	.	.	Budding	,,	,,	Mar. 21.

[1] An English wag, given to "chaff," composed the following paraphrase:—

Autumn. — Wheezy, sneezy, breezy;
Winter. — Slippy, drippy, nippy;
Spring. — Showery, flowery, bowery;
Summer. — Hoppy, croppy, poppy.

Brady, Clavis Calendaria, vol. i. p. 38, 2nd ed. 1812.

Floréal	.	.	.	Flowering Month commences	April 20.
Prairial	.	.	.	Pasture „ „	May 20.
Messidor	.	.	.	Harvest „ „	June 19.
Thermidor	.	.		Hot „ „	July 19.
Fructidor	.	.	.	Fruit „ „	Aug. 18.

The Festivals, or *Jours Complementaires*, were dedicated to : —

Virtue	.	.	.	Sept. 17.
Genius	.	.	.	Sept. 18.
Labour	.	.	.	Sept. 19.
Opinion	.	.		Sept. 20.
Reward	.	.	.	Sept. 21.

The bissextile day, every fourth year, was called *La Jour de la Révolution*, and was set apart for the renewal of the oath to live free or die ! Not content to rest here, these misguided men abolished the week, and divided the month into 3 decades, the days of which were named : Primidi, Duodi, Tridi, Quartidi, &c. ; the 10th or Decadi being observed as a sort of Sabbath, though not exactly in a Christian sense of the word ! This state of things, as may be supposed, did not last long ; " for on January 1, 1806, the Gregorian calendar was resumed, and the Republic, which had legislated for the 4000th year of its existence by name, wore its own livery just one day and a quarter for every one of those years." Fabre D'Eglantine was the author of this calendar.

The whole of the decadery days were kept, or ordered to be kept, as secular festivals. The following is a list of the dedications : —

1st. Nature and to the Supreme Being.
2nd. The Human Race.
3rd. The French People.
4th. The Benefactors of Humanity.
5th. The Martyrs of Liberty.
6th. Liberty and Equality.
7th. The Republic.
8th. The Liberty of the World.
9th. The Love of our Country.
10th. The Hatred of Tyrants.
11th. Truth.
12th. Justice.
13th. Chastity.
14th. Glory and Immortality.
15th. Friendship.
16th. Frugality.
17th. Courage.
18th. Good Faith.

19th. Heroism.

20th. Disinterestedness.

21st. Stoicism.

22nd. Love.

23rd. Conjugal Faith.

24th. Parental Love.

25th. Maternal Tenderness.

26th. Filial Piety.

27th. Infancy.

28th. Youth.

29th. Manhood.

30th. Old Age.

31st. Misfortune.

32nd. Agriculture.

33rd. Industry.

34th. Our Ancestors.

35th. Posterity.

36th. Prosperity.

"That one day in the calendar should have been appropriated to the ' *Supreme Being* ' *in conjunction with* ' *Nature* ' was a low conceit of Robespierre, who meant to identify Nature with the Supreme Being as one and the same source; and yet even this slight remembrance of the Almighty power appears to have afforded some consolation to a great majority of the people who had not lost every sense of religion; and to delude them with a belief of his sincerity, that arch-hypocrite himself joined in apparent devotion to *that* Almighty power whose attributes it was his real object to deride. He had even the audacious craft to decree a *fête* for the express purpose of paying *adoration* to the Deity, when for *one* day the fatal guillotine was veiled from public view; and the better to conceal his depravity, a hideous and frightful figure, prepared for public exhibition at the festival, as the type of atheism, was previously destroyed. Part of the community, after these regulations, distinguished *Sunday* in the ancient style of festivity, whereby *to mark* the recurrence of that holy day, though no one had the temerity publicly to oppose the current of error by a more suitable observance. Many, indeed, wholly conformed to the innovation, and hence one part of the people shut up their shops on Sundays, while the *sans culotte* adherents of Robespierre rigidly observed the decades." [1]   Robespierre artfully overcame the difficulty by decreeing that both the Sundays and the decades should be observed as festivals, thus granting to the people 88 days of recreation in the year instead of 36 or 52.

---

[1] Brady, *Clavis Calend.*, vol. i. p. 35.

The year is the largest astronomical chronometric unit, and is used to express all long periods of time. It is, as we have already shown, in some form or other a derivative either of the Moon's revolution round the Earth, or the Earth's revolution round the Sun. Various times have been used to fix the commencement of the year; we call January 1, New Year's Day now, though previous to the introduction of the new style of reckoning in 1752, the 25th of March was usually considered the 1st day of the new year. On referring to the Prayer-book used in our National Church, it will be seen that the ecclesiastical year begins on Advent Sunday, whilst the academical year used at the Universities begins in October. December 25, March 1, Easter, September 22, &c., have all been used at different times for the same purpose.

The year is also subdivided into 4 quarters, which point out the days on which the Sun attains its greatest declination, north or south (called the Solstices, Summer and Winter), and on which it is on the Equator going north or south (called the Equinoxes, Vernal or Autumnal); owing to physical causes, into which we cannot here enter, these events do not now take place at equal intervals, and will not for several thousand years. The following are the dates of the commencement, and the lengths of the seasons, in 1860:—

		d.	h.	m.	d.	h.	m.	
Spring commences	Mar. 19	21	5...92	20	38	length of Spring.
Summer	„	June 20	17	43...93	14	9	length of Summer.
Autumn	„	Sept. 22	7	52...89	17	59	length of Autumn.
Winter	„	Dec. 21	1	51...89	1	2	length of Winter.
				365	5	48	

The following days of the year are used as quarter-days, for leases, &c., in England and Scotland:—

ENGLAND.	SCOTLAND.
Mar. 25, or Lady-day.	Feb. 2, or Candlemas-day.
June 24, or Midsummer-day.	May 15, or Whitsun-day.
Sept. 29, or Michaelmas-day.	Aug. 1, or Lammas-day.
Dec. 25, or Christmas-day.	Nov. 11, or Martimas-day.

CHAPTER III.

Means of Measuring Time. — The Almanac. — Epitome of its Contents. — Times of Sunrise and Sunset. — Positions of the Sun, Moon, and Planets. — The Phases of the Moon. — The Ecclesiastical Calendar. — The Festival of Easter. — Method of Calculating it.

NATURE, though she has supplied us with visible phenomena to measure the larger units of time, such as days, months, and years, has not furnished us with any means whereby we may measure the lesser units of hours, minutes and seconds; artificial contrivances must therefore be sought for. Rough approximations to the true time were at first obtained by setting up *gnomons*, or upright staves[1]; which, in conjunction with a knowledge of the north point of heavens, would afford a tolerably correct indication of noon, or the moment of the Sun's passage over the meridian. An instrument constructed with a gnomon pointing towards the North Pole of the heavens, constitutes a *sun-dial*, and affords a still better mode of ascertaining the hour of the day. According to Herodotus, sun-dials were first introduced into Greece from Chaldæa; the hemisphere of Berosus, who lived 540 B.C., is the oldest recorded in history.[2] The earliest attempt to form a strictly artificial time-keeper, is due to Ctesibius, of Alexandria[3], who invented *Clepsydræ*, or water-clocks, which were contrivances for allowing a continuous stream of water to trickle through a small aperture in the pipe of a funnel, the time being

---

[1] Ptolemy describes one erected at Alexandria. (*Almag.* lib. iii. cap. 2.)

[2] Described by Vitruvius, *De Architecturâ*, lib. ix. cap. 9. Stuart mentions one found on the south side of the Acropolis at Athens, which is probably of the same kind as the above. (*Antiquities of Athens*, vol. ii.)   [3] Vitruvius, in loc. cit.

measured by the quantity of fluid discharged.   A species of
Clepsydra, in which mercury is employed, has been introduced
with great success by the late Captain Kater, for the measurement of minute intervals of time, by persons engaged in delicate philosophical experiments.   *Hour glasses,* still used for
boiling eggs, and in the Houses of Parliament to fix the termination of the time allotted to members to prepare for a
division, are merely small clepsydræ in which sand is used
instead of water.   All the above contrivances have now, more
or less, fallen into disuse, being supplanted by the pendulum
clock and the watch, into a description of which it would be
foreign to our present purpose to enter.

Of all the books to be found in every library, however
humble, there is none, the Holy Scriptures excepted, of such
indispensable value as the Almanac.   We accordingly find an
immense variety in the character, size, and price of those now
in circulation, more especially since the abolition of the 15-
penny stamp, about 25 years ago.   One might imagine that a
book so generally used by everybody would be fully understood by all; but this is by no means the case, and there are
probably few things about which so much ignorance prevails,
notwithstanding the frequency with which the almanac is consulted.

The word "Almanac" is derived from the Arabic word,
"Manah," to reckon, and is applied to publications which
describe the astronomical, civil, and ecclesiastical phenomena
of the year.[1]   The word "Calendar," applied, in a more
limited sense, to the events of the several months, comes from
the Greek, καλέω, to proclaim.

The astronomical phenomena which usually find a place in
good almanacs are the following:—(1) The times of sunrise
and sunset; (2) the right ascension and declination of the Sun,
Moon, and principal planets; (3) the equation of time; (4) the
time of high and low water; (5) the time of the rising, meridian
passage, and setting of the Moon; (6) the Moon's phases and

---

[1] Arago gives as the derivation, *man,* the Moon.   (*Pop. Ast.* vol. ii.
p. 721, Eng. Ed.)

age ; concerning each of which we shall proceed to give a short explanation.

(1.) If the Earth's axis were perpendicular to the plane of the ecliptic, the Sun would always (at the Equator) rise in the east and set in the west; but as it is inclined at an angle of 66° 32′ 24″, this is not the case ; it is this inclination which gives rise to the alternation of the seasons and the ever-varying length of each successive day. It also follows from this that the hour at which the heavenly bodies rise, culminate, and set, differs on the same day at places having different latitudes or different longitudes. Celestial objects which will be visible from one place will be invisible from another. Thus the constellation of the Great Bear, which we see every night, is invisible at the Cape of Good Hope ; and, conversely, the Southern Cross, which glitters in the other hemisphere, is never seen so far north as England : so that spectators in northern latitudes look down at objects which spectators in southern latitudes look up at, and which spectators in the tropics see directly over their heads. These circumstances must be borne in mind when we are going to ascertain beforehand the aspect of the heavens at any particular place, on any given night. From observation and theory, we learn that the aspect of the heavens changes from hour to hour; and, knowing the period occupied by the Earth in her diurnal rotation and annual revolution, it becomes an easy matter to foretell, by calculation, what celestial objects will be visible at any proposed day and hour. The presence of an atmosphere around the Earth gives rise to the phenomenon of refraction[1], by which the heavenly bodies, when in and near the horizon, suffer a considerable displacement, equal to about 33′, by which quantity the apparent exceeds the true altitude. Now, since the diameter of the Sun is only 32′, it follows that the Sun is elevated through a space equal to more than its own diameter ; so that when we see the disc of the Sun apparently just above the horizon, the Sun itself is, in reality, just below it ; and would, therefore, were it not for the atmosphere, be

---

[1] See above, page 173.

invisible.   The time of *apparent* sunrise and sunset is what is now given in our almanacs.   At the Equinoxes only is it that the interval between sunrise and sunset is equal to the interval between sunset and sunrise, or that the length of the day is equal to the length of the night [1] : this occurred on March 18 and September 25 in the year 1860.

(2.)   The Right Ascension (R. A.) of a celestial object is its distance, reckoned on the Equator, from the Vernal Equinox, or first point of Aries, either in angular measure, of degrees, minutes, and seconds (° ′ ″) ; or in time, of hour, minutes, and seconds (h. m. s.) — 24 hours making a circumference of a circle ; each hour, therefore, being equal to 15° of arc.   The Declination (δ) is the angular distance of a celestial object from the Equator reckoned North (+), or South (−), towards the poles.   Sometimes the position of an object is indicated by its angular distance being reckoned from the North down to the South Pole : when this is the case, it is indicated by the abbreviation, N. P. D.—North Polar Distance.

(3.)   The Equation of Time has already been explained.

(4.)   The time of High and Low Water is usually given for the port of London ; an additional constant, called the *Establishment*, being supplied, by means of which the tidal phenomena at all the other ports in the list given, can be readily ascertained.

(5.)   When the Moon, in the course of her revolution round the Earth, has the same Right Ascension as, and passes the meridian with, the Sun, it is said to be in *Conjunction* (☌), or " New " (●).   We know that the Moon completes a revolution in 29d. 12h. 44m. 2·873.   It therefore moves round the heavens, from west to east, at the rate of 13° 10′ 55″ daily ; while the Sun moves in the same direction, with a mean daily motion of 59′ 8·2″ : the Moon, therefore, departs eastward from the Sun 12° 10′ 46″ every 24 hours.   On the 8th day, or about a week after conjunction, it will be 90° from the Sun. This phase is the *maximum Elongation East*, or *Eastern Quadrature* (E. □), and is popularly known as the " First quarter "

---

[1] This is not strictly true, but we use the term in a popular sense.

($ \mathrm{)} $). The Moon, previously a crescent, is then halved, with its illuminated limb turned towards the Sun. Proceeding onwards, it becomes gibbous, and about 14 days from conjunction attains an angular distance of 180° from the Sun. It is then in *opposition* to the Sun ($ \mathcal{S} $), and the Moon is said to be "Full" (○). At the end of another period of 7 days, or 21 from conjunction, the Moon, after becoming gibbous, is again halved; but it is the opposite limb which is now illuminated. This phase is the *maximum Elongation West*, or *Western Quadrature* (w. □), and is popularly known as the "Last quarter." ($ \mathrm{(} $) Finally, after the lapse of another 7 days, the Moon again comes into conjunction, and becomes "New."

In nearly every almanac there is a column assigned to the "Moon's age." This is simply the interval in days and parts of a day which have elapsed since the Moon's last conjunction with the Sun. Thus, on Monday, June 4, 1860, we find that at noon the Moon's age was 14·7 : this means that 14·7 days had elapsed since the last previous conjunction on May 20. An examination of the dates of several successive New Moons will show that the lengths of different lunations differ considerably, owing to the varying velocity of the Moon's orbital motion : this is due to numerous and complicated physical causes, which we cannot further advert to here. In associating any lunation with any particular month, it may be well to mention that the Moon does not take its name from the month in which it passes the principal part of its time, but from the one in which the lunation terminates. Thus, in the year 1859, the "June Moon" is the one which commences on the 31st of May, and terminates on the 30th June. All writers on chronology agree in this arrangement, which is sometimes attended with rather absurd consequences. There were, in fact, in 1859, two "May Moons," the whole of the first, however, with the exception of 2d. 10h., belonging to the month of April. Since the month of February in a common year only contains 28 days, and a lunar month always exceeds 29 days, it will sometimes happen that there will be no "February Moon" at all.

It is not necessary to advert to the civil portion of the calendar further than to mention that Quarter-days, Law and

University Term-days, and anniversaries of important by-gone events, &c., all find a place in every well-appointed almanac.

The Ecclesiastical calendar has for its object the regulation of the different Sundays and Festivals ordained by the Church to be kept holy. Some religious Festivals — such as the Feast of S. Andrew, the Nativity of our Lord, the Annunciation of the Blessed Virgin, &c., are observed on the same day of the month every year; others, such as Easter, return on different days in different years, whence they are termed *Moveable Festivals*. Easter is the most important of all, for upon this depend nearly all the rest.

The Jewish Feast of the Passover was observed in accordance with the following commands : — "In the 1st month, on the 14th day of the month at even, ye shall eat unleavened bread, until the 1 and 20th day of the month at even." (Exodus xii. 18.) Again : "In the 14th day of the 1st month, at even, is the Lord's Passover." (Leviticus xxiii. 5.) And since our Saviour was crucified at the time of the Jewish Passover, our festival of Easter has ever been a moveable one. The word is probably of Saxon origin, for the ancient Saxons sacrificed in the month of April to a goddess whom they called Eoster (in Greek Astarte, and in Hebrew Ashtaroth[1]); whose name was given to the month in question. It has been suggested, that the word *East* in Saxon refers to "rising," and that the point of the compass now known by that name, derived it from the rising of the Sun, and the festival from the rising of our Saviour. Another derivation is the Saxon *yst*, a storm, on account of the tempestuous weather which frequently prevailed at this season of the year. That the observance of Easter as a Christian institution is as ancient as the times of the Holy Apostles, there can be no doubt; but in the 2nd century a controversy arose as to the exact time at which it ought to be celebrated. The Eastern Churches elected to keep it on the 14th day of the first Jewish month; and the Western, on the night which preceded the anniversary of our Saviour's resurrection. The objection to

[1] *Vide* Milton, *Paradise Lost*, b. i. l. 438, where it is referred to as a Phœnician deity.

the former plan was, that the festival was commonly held on some other day than the 1st day of the week, which was undoubtedly the proper one. The disputing churches each had their own way until the year 325 A.D., when the General Council of Nicæa ordered *that it should be kept on the Sunday which falls next after the first full Moon following the 21st of March the Vernal Equinox. If a Full Moon fall on the 21st of March, then the next Full Moon is the Paschal Moon; and if the Paschal Moon fall on a Sunday, then the next Sunday is Easter-Day.*

By common consent, it is not the Apparent or real Sun and Moon which is employed in finding Easter, but the Mean or fictitious Sun and Moon of astronomers. We must, therefore, not be surprised at finding sometimes the Easter of any year not agreeing with the above definition. Such was the case in 1845 and in 1818, when a violent controversy took place about it. Suppose, for instance, that the real Opposition of the Sun and Moon took place at 11h. 59m. P.M., March 21, and the Mean Opposition 2m. afterwards. It is clear that, counting by the real bodies, the Full Moon in question would not be the Paschal Moon, while that of the Mean bodies would be so.[1] However, the following rules will determine the Easter Day of chronologists for any year of the Christian era, and this is all that is necessary : —

    I. Add 1 to the given year.

    II. Take the quotient of the given year, divided by 4, neglecting the remainder.

    III. Take 16 from the centurial figure of the given year if it can be done.

    IV. Take the quotient of III. divided by 4, neglecting the remainder.

    V. From the sum of I. II. and IV. subtract III.

    VI. Find the remainder of V. divided by 7.

    VII. Subtract VI. from 7; this is the number of the DOMINICAL LETTER.

A	B	C	D	E	F	G
1	2	3	4	5	6	7

---

[1] The investigation of this question is too long and complicated to interest the general reader. Those who wish for it will find it in a valuable memoir, by Prof. De Morgan, in the *Companion to the Almanac* for 1845, p. 1, *et seq.*

VIII. Divide I. by 19, the remainder (or 19 if there is no remainder) is the GOLDEN NUMBER.

IX. From the centurial figures of the year subtract 17, divide by 25, and keep the quotient.

X. Subtract IX. and 15 from the centurial figures, divide by 3, and keep the quotient.

XI. To VIII. add 10 times the next less number, divide by 30, and keep the remainder.

XII. To XI. add X. and IV. and take away III. throwing out the thirties, if any. If this gives 24, change it into 25. If 25, change it into 26, whenever the Golden number exceed II. If 0, change it into 30. Thus we get the EPACT.

When the Epact is 23 or less.	*When the Epact is greater than 23.*
XIII. Subtract XII. from 45.	XIII. Subtract XII. from 75.
XIV. Subtract the Epact from 27, divide by 7, and keep the remainder.	XIV. Subtract the Epact from 57, divide by 7, and keep the remainder.

XV. To XIII. add VII. (and 7 besides, if XIV. be greater than VII.) and subtract XIV., the result is the day of March, or if more than 31 subtract 31, and the result is the day of April on which Easter Day falls.

The following exemplifies the above rule : —

To find when Easter falls in 1865.

I.　$1865 + 1 = 1866.$

II.　$\dfrac{1865}{4} = 466 - 1 \text{ rem.}$

III.　$18 - 16 = 2.$

IV.　$\dfrac{2}{4} = 0 - 2 \text{ rem.}$

V.　$1866 + 466 + 0 - 2 = 2330.$

VI.　$\dfrac{2330}{7} = 332 - 6.$

VII.　$7 - 6 = 1$, whence A is the DOMINICAL LETTER.

VIII.　$\dfrac{1866}{19} = 98 - 4 \text{ rem.} \;\therefore\; 4$ is the GOLDEN NUMBER.

IX.　$\dfrac{18 - 17}{25} = 0 - 1 \text{ rem.}$

X.　$\dfrac{18 - (0 + 15)}{3} = 1.$

XI.　$\dfrac{4 + 30}{30} = 1 - 4 \text{ rem.}$

XII.　$4 + 1 + 0 - 2 = 3$, which is the EPACT.

XIII. $45 - 3 = 42$.

XIV. $\dfrac{27 - 3}{7} = 3 - 3$ rem.

XV. $42 + 1 + 7 - 3 = 47$; subtract 31, and we get April 16 as Easter Day.

Easter Day being known, any of the other days depending on it can readily be found.

Septuagesima Sunday is 9 weeks
Sexagesima Sunday is 8 weeks
Shrove or *Quinquagesima Sunday* is 7 weeks
Shrove Tuesday and *Ash Wednesday* follow *Quinquagesima Sunday*
Quadragesima Sunday is 6 weeks
Palm Sunday is 1 week
Good Friday is 2 days

} Before Easter.

Low Sunday is 1 week
Rogation Sunday is 5 weeks
Ascension Day or *Holy Thursday*, follows *Rogation Sunday*
Whitsun-Day is 7 weeks
Trinity Sunday is 8 weeks

} After Easter.

CHAPTER IV.

The Dominical or Sunday Letter.—Method of finding it.—Its Use.—
The Lunar or Metonic Cycle.—The Golden Number.—The Epact.—
The Solar Cycle.—The Indiction.—The Dionysian Period.—The
Julian Period.

THE *Dominical Letter*, called also the Sunday Letter, is an
expedient by means of which we can readily find out the day
of the week on which any day of the year falls, knowing the
day of the week on which New Year's Day falls. To the first
seven days of January are affixed the first seven letters of the
alphabet—A, B, C, D, E, F, G; and the one of these which
denotes Sunday is the Dominical letter. Thus, if Sunday is
New Year's Day, then A is the Dominical letter; if Monday,
that letter is G; and so on. If there were 364 days, or 52
weeks exactly in the year, then the Dominical letter would
always be the same; but as the year contains about 365¼
days, or 1¼ more than 364, this excess has to be taken into
account every year, and the ¼ makes a day in every 4 years;
so that the Dominical letter falls backward *one letter* every
common year, and two letters every Bissextile or Leap year.
Knowing the Dominical letter, we can ascertain all the
Sundays, all the Mondays, &c., in the year. The reason why
Leap years have two letters may be thus explained:—Take,
for example, the year 1860. The year begins on a Sunday,
so A is the Sunday letter; but the intercalary day, February
29, throws back the 1st of March a day later than it would
otherwise have been, and therefore the Sunday letter for the
following 10 months is thrown back 1—that is to say, to G;
so that the Dominical letters for 1860 are A and G. The
following examples, worked according to the 1st part of the

rule already given to find Easter, illustrate the practical use of a knowledge of the Dominical letter :—

1. What day of the week is June 4, 4779?

I. $4779 + 1 = 4780$.

II. $\dfrac{4779}{4} = 1194 - 3$ rem.

III. $47 - 16 = 31$.

IV. $\dfrac{31}{4} = 7 - 3$ rem.

V. $4780 + 1194 + 7 - 31 = 5950$.

VI. $\dfrac{5950}{7} = 850 - 0$ rem.

VII. $7 - 0 = 7 \therefore$ G is the Dom. letter.

Now June 3 has G affixed to it, and is Monday; whence June 4, 4779, will fall on a Monday.

2. What is the first Sunday in June, 1865?

I. $1865 + 1$.

II. $\dfrac{1865}{4} = 466 - 1$ rem.

III. $18 - 16 = 2$.

IV. $\dfrac{2}{4} = 0 - 2$ rem.

V. $1866 + 466 + 0 - 2 = 2330$.

VI. $\dfrac{2330}{7} = 332 - 6$ rem.

VII. $7 - 6 = 1 \therefore$ A is the Dom. letter.

Now June 4 has A affixed to it, consequently that day is the first Sunday in 1865.

Since the Solar year consists of 365d. 5h. 48m. 48s., 19 Solar years will consist of about—

6939d. 14h. 27m. 12s.;

and since the Mean duration of a Lunar month is 29d. 12h. 44m. 3s., 235 Mean Lunar months will consist of about—

6939d. 16h. 31m. 45s.

Thus we see that 19 Solar years fall short of 235 Lunar months by only 2h. 4m. 33s.; if, therefore, any given length of time be resolved into periods of 19 years each, the same phases of the Moon which are presented in any year of 1 cycle will be reproduced in the corresponding year of the following cycle, but 2h. 4m. 33s. later. If we reckoned time by Solar years, and if the, length of each lunation were always 29d. 12h. 44m. 3s., the days of the Moon's changes in any 1 cycle being known, the days of the Moon's changes in every succeeding and every preceding cycle could be easily ascertained. But since the Solar year of 365d. 5h. 48m. 48s. is not employed, and the duration of different lunations varies, the reproduction of the Lunar phases on corresponding days does not take place.

"Unlike the Astronomical year, the Civil year is not constantly of the same length. It consists, as has been already explained, sometimes of 365 and sometimes of 366 days. Neither is a cycle of 19 successive Civil years always of the same length. Such a cycle contains sometimes 5 and sometimes only 4 Leap years, and consists, therefore, sometimes of 6940 and sometimes of 6939 days. It, therefore, sometimes exceeds a cycle of 19 Astronomical years by nearly a quarter of a day, and sometimes falls short of such a cycle by more than three-quarters of a day. If 4 successive cycles of 19 Civil years be taken, 3 of them will exceed 1 Astronomical year by something less than a quarter of a day, and the 4th will fall short of an Astronomical year by something more than three-quarters of a day. The total length of the 4 successive cycles of 19 Civil years will be as nearly as possible equal to 4 cycles of 19 Astronomical years.[1]

"Thus it is evident that the Civil year, though variable in length, oscillates alternately on one side and the other of the Astronomical year; and, in like manner, the cycle of 19 Civil years, which is also variable by 1 day, oscillates at each side of the cycle of 19 Astronomical years. The Civil year and the Civil cycle are alternately overtaking and overtaken by the Astronomical year and cycle, and their average lengths are

---

[1] This cycle of 76 years (19 × 4) is known as the *Calippic period*, from Calippus its discoverer.

respectively equal in the long run to the average length of the latter. In like manner, the Lunar month is subject to a certain limited variation; so that the phases of the real Moon are alternately overtaking and overtaken by those of the average Moon.

" Now, let us imagine a fictitious Moon to move round the heavens in the path of the real Moon, but with such a motion that its periodical phases shall take place in exact accordance with the Civil years, and with the cycles of 19 Civil years, in the same manner as the phases of the real Moon recur in the succession of Astronomical years, and in the cycles of 19 Astronomical years. Such a fictitious Moon is then the Ecclesiastical Moon, and is the Moon whose phases are predicted in the Calendar. It will be evident from all that has been explained, that this Ecclesiastical Moon will alternately pursue, overtake, and outstrip the real Moon, and be pursued, overtaken, and outstripped by it; that they will thus make together their successive revolutions of the heavens, and that they will never part company, nor either outstrip nor fall behind the other beyond a certain distance, which is limited by the extent of the departure of the Civil from the Astronomical year, and by that of the real from the average Lunar Month."

The course of time is now considered as made up of so many cycles each of 19 Civil years; and it has been agreed that each cycle shall commence with a year, the first day of which shall be the last of the preceding lunation, or the day on which the age of the following Moon is 0. The number which marks the place of any given year in the cycle is termed the *Golden Number*, and the period the *Lunar* or *Metonic Cycle*, from its discoverer, Meton. When the discovery of this cycle was first published, so great was the popular favour which was bestowed upon it, that it was ordered to be written up in letters of gold.[1] The age of the Ecclesiastical Moon on the 1st day of the 1st year of the cycle being known, its age upon the 1st day of any succeeding

---

[1] The metonic cycle was adopted on July 16, 433, B.C. and the New Moon, which happened at 7h. 43 P.M., was the actual commencement of it.

year of the cycle may be determined. The number which expresses the age of the Moon upon the 1st day of any year of the cycle is called the *Epact*. The series of Epacts corresponding to the years of a Lunar cycle are given in the following table : —

Year of Cycle	1	2	3	4	5	6	7	8	9	10	11	12	13	14	15	16	17	18	19
Epact	0	11	22	3	14	25	6	17	28	9	20	1	12	23	4	15	26	7	18

The method of finding the Golden number and Epact for any given year has already been shown in the rule for finding Easter.

The *Solar Cycle* is the period of years that elapse before the days of the week correspond to the same days of the month. If there were 364 days in the year, this would happen every year; if 365, it would happen every 7th year; but because the $\frac{1}{4}$ of a day makes an alteration of a whole day in 4 years, the cycle must extend to 7×4, or 28 years. Nine years of this cycle had elapsed before the birth of Christ. Therefore, *to find the cycle of the Sun, add 9 to the given year, and divide by 28; the quotient is the number of cycles since the commencement of the Christian era, and the remainder is the cycle of the Sun.*

Example. To find the Cycle of the Sun for 1865 : —

$$\text{I.}\quad 1865 + 9 = 1874.$$

$$\text{II.}\quad \frac{1874}{28} = 66 - 26 \text{ rem.}$$

whence 26 is the number in the Solar Cycle for the given year.

The cycle of the *Indiction* has no immediate connection with the motions of the Sun and Moon, but it may, however, as well be referred to here. It is a period of 15 years, first established as a conventional division of time in the Roman empire by the Emperor Constantine. It has been conjectured that it was designed to supersede the Pagan method of reckoning time by the olympiads, a period of 4 years, familiar to every classical reader. Unlike the Golden number and the Epact, the Indiction was purely a Civil period. Gregory VII. finally fixed upon the 1st day of the year 313 as the

commencement of the Indiction, whence it follows that the 1st
year of the Christian era was the 4th of the current Indiction.
*To find the Indiction, add 3 to the given year, and divide by
15, and the remainder will be the Indiction.*

Example. To find the Indiction for 1865 : —

$$\text{I.}\quad 1865 + 3 = 1868.$$

$$\text{II.}\quad \frac{1868}{15} = 124 - 8 ;$$

whence 8 is the Indiction of the given year.

The *Dionysian Period* is obtained by a combination of the
Lunar and Solar cycles, forming a period of 532 years
($19 \times 28 = 532$); at the end of which time the changes of the
Moon take place on the same days of the week and month as
before.   This period is valuable for testing the accuracy of
chronological statements.

The *Julian Period*, a useful period in chronology, is obtained by multiplying together the Lunar cycle, the Solar
cycle, and the Indiction, forming a period of 7980 years —

$$(19 \times 28 \times 15 = 7980.)$$

It received its name from Scaliger.   Since this period contains
an exact number of each of the above cycles, its most obvious
commencement would be some year which was at the same
time the 1st of each of the 3.   This, Scaliger found, would
have been the year 4711 B.C., which he accordingly fixed as
the commencement of the Julian Period.

261

CHAPTER V.

Table for the Conversion of Solar into Sidereal Time. — Table for the Conversion of Sidereal into Solar Time. — Table for Reducing Longitude to Time. — Table for Reducing Time to Longitude. — Table of the Mean Motion of the Sun for Periods of Mean Time.

TABLE FOR CONVERTING INTERVALS OF MEAN SOLAR TIME

HOURS.		MINUTES.				SECONDS.			
Hours of Mean Time.	Equivalents in Sidereal Time.	Minutes of Mean Time.	Equivalents in Sidereal Time.	Minutes of Mean Time.	Equivalents in Sidereal Time.	Seconds of Mean Time.	Equivalents in Sidereal Time.	Seconds of Mean Time.	Equivalents in Sidereal Time.
	h. m. s.		m. s.		m. s.		s.		s.
1	1 0 9.8565	1	1 0.1643	31	31 5.0925	1	1.0027	31	31.0849
2	2 0 19.7130	2	2 0.3286	32	32 5.2568	2	2.0055	32	32.0876
3	3 0 29.5694	3	3 0.4928	33	33 5.4211	3	3.0082	33	33.0904
4	4 0 39.4259	4	4 0.6571	34	34 5.5853	4	4.0110	34	34.0931
5	5 0 49.2824	5	5 0.8214	35	35 5.7496	5	5.0137	35	35.0958
6	6 0 59.1388	6	6 0.9857	36	36 5.9139	6	6.0164	36	36.0986
7	7 1 8.9953	7	7 1.1499	37	37 6.0782	7	7.0192	37	37.1013
8	8 1 18.8518	8	8 1.3142	38	38 6.2424	8	8.0219	38	38.1040
9	9 1 28.7083	9	9 1.4785	39	39 6.4067	9	9.0246	39	39.1068
10	10 1 38.5647	10	10 1.6428	40	40 6.5710	10	10.0274	40	40.1095
11	11 1 48.4212	11	11 1.8070	41	41 6.7353	11	11.0301	41	41.1123
12	12 1 58.2777	12	12 1.9713	42	42 6.8995	12	12.0329	42	42.1150
13	13 2 8.1342	13	13 2.1356	43	43 7.0638	13	13.0356	43	43.1177
14	14 2 7.9906	14	14 2.2998	44	44 7.2281	14	14.0383	44	44.1205
15	15 2 27.8471	15	15 2.4641	45	45 7.3924	15	15.0411	45	45.1232
16	16 2 37.7036	16	16 2.6284	46	46 7.5566	16	16.0438	46	46.1259
17	17 2 47.5600	17	17 2.7927	47	47 7.7209	17	17.0465	47	47.1287
18	18 2 57.4165	18	18 2.9569	48	48 7.8852	18	18.0493	48	48.1314
19	19 3 7.2730	19	19 3.1212	49	49 8.0495	19	19.0520	49	49.1342
20	20 3 17.1295	20	20 3.2855	50	50 8.2137	20	20.0548	50	50.1369
21	21 3 26.9859	21	21 3.4498	51	51 8.3780	21	21.0575	51	51.1396
22	22 3 36.8424	22	22 3.6140	52	52 8.5423	22	22.0602	52	52.1424
23	23 3 46.6989	23	23 3.7783	53	53 8.7066	23	23.0630	53	53.1451
24	24 3 56.5554	24	24 3.9426	54	54 8.8708	24	24.0657	54	54.1479
		25	25 4.1069	55	55 9.0351	25	25.0685	55	55.1506
		26	26 4.2711	56	56 9.1994	26	26.0712	56	56.1533
		27	27 4.4354	57	57 9.3637	27	27.0739	57	57.1561
		28	28 4.5997	58	58 9.5279	28	28.0767	58	58.1588
		29	29 4.7640	59	59 9.6922	29	29.0794	59	59.1615
		30	30 4.9282	60	60 9.8565	30	30.0821	60	60.1643

INTO EQUIVALENT INTERVALS OF SIDEREAL TIME.

	FRACTIONS OF A SECOND.				
Seconds of Mean Time.	Equivalents in Sidereal Time.	Seconds of Mean Time.	Equivalents in Sidereal Time.	Seconds of Mean Time.	Equivalents in Sidereal Time.
	s.		s.		s.
0·01	0·01003	0·34	0·34093	0·67	0·67183
0·02	0·02006	0·35	0·35096	0·68	0·68186
0·03	0·03008	0·36	0·36099	0·69	0·69189
0·04	0·04011	0·37	0·37101	0·70	0·70192
0·05	0·05014	0·38	0·38104	0·71	0·71194
0·06	0·06016	0·39	0·39107	0·72	0·72197
0·07	0·07019	0·40	0·40110	0·73	0·73200
0·08	0·08022	0·41	0·41112	0·74	0·74203
0·09	0·09025	0·42	0·42115	0·75	0·75205
0·10	0·10027	0·43	0·43118	0·76	0·76208
0·11	0·11030	0·44	0·44120	0·77	0·77211
0·12	0·12033	0·45	0·45123	0·78	0·78214
0·13	0·13036	0·46	0·46126	0·79	0·79216
0·14	0·14038	0·47	0·47129	0·80	0·80219
0·15	0·15041	0·48	0·48131	0·81	0·81222
0·16	0·16044	0·49	0·49134	0·82	0·82225
0·17	0·17047	0·50	0·50137	0·83	0·83227
0·18	0·18049	0·51	0·51140	0·84	0·84230
0·19	0·19052	0·52	0·52142	0·85	0·85233
0·20	0·20055	0·53	0·53145	0·86	0·86235
0·21	0·21057	0·54	0·54148	0·87	0·87238
0·22	0·22060	0·55	0·55151	0·88	0·88241
0·23	0·23063	0·56	0·56153	0·89	0·89244
0·24	0·24066	0·57	0·57156	0·90	0·90246
0·25	0·25068	0·58	0·58159	0·91	0·91249
0·26	0·26071	0·59	0·59162	0·92	0·92252
0·27	0·27074	0·60	0·60164	0·93	0·93255
0·28	0·28077	0·61	0·61167	0·94	0·94257
0·29	0·29079	0·62	0·62170	0·95	0·95260
0·30	0·30082	0·63	0·63173	0·96	0·96263
0·31	0·31085	0·64	0·64175	0·97	0·97266
0·32	9·32088	0·65	0·65178	0·98	0·98268
0·33	0·33090	0·66	0·66181	0·99	0·99271

This TABLE is useful for the conversion of MEAN SOLAR into SIDEREAL Time.

Sidereal Time *required* = Sidereal Time at the *preceding* Mean Noon + the Equivalent to the *given* Mean Time.

EXAMPLE.—To convert 2h. 22m. 25s·62s. Mean Time at Greenwich, Jan. 7, 1859, into Sidereal Time.

Sidereal Time at the *preceding* Mean Noon, viz. January 7 .
	h.	m.	s.
	19	0	5·24
	2	22	19·713
			3·614
			0·622

The Table gives the Equivalent Sidereal Intervals.

For Mean Intervals.
	2h.	0m.	0s.	25·069
	22	0		
		25		
		0·62		

The Sum is the Sidereal Time required .   21   28   54·26

TABLE FOR CONVERTING INTERVALS OF SIDEREAL TIME

HOURS.		MINUTES.				SECONDS.			
Hours of Sidereal Time.	Equivalents in Mean Time.	Minutes of Sidereal Time.	Equivalents in Mean Time.	Minutes of Sidereal Time.	Equivalents in Mean Time.	Seconds of Sidereal Time.	Equivalents in Mean Time.	Seconds of Sidereal Time.	Equivalents in Mean Time.
	h. m. s.		m. s.		m. s.		s.		s.
1	0 59 50·1704	1	0 59·8362	31	30 54·9214	1	0·9973	31	30·9154
2	1 59 40·3409	2	1 59·6723	32	31 54·7576	2	1·9945	32	31·9126
3	2 59 30·5113	3	2 59·5085	33	32 54·5937	3	2·9918	33	32·9099
4	3 59 20·6818	4	3 59·3447	34	33 54·4299	4	3·9891	34	33·9072
5	4 59 10·8522	5	4 59·1809	35	34 54·2661	5	4·9864	35	34·9045
6	5 59 1·0226	6	5 59·0170	36	35 54·1023	6	5·9836	36	35·9017
7	6 58 51·1931	7	6 58·8532	37	36 53·9384	7	6·9809	37	36·8990
8	7 58 41·3635	8	7 58·6894	38	37 53·7746	8	7·9782	38	37·8963
9	8 58 31·5340	9	8 58·5256	39	38 53·6108	9	8·9754	39	38·8935
10	9 58 21·7044	10	9 58·3617	40	39 53·4470	10	9·9727	40	39·8908
11	10 58 11·8748	11	10 58·1979	41	40 53·2831	11	10·9700	41	40·8881
12	11 58 2·0453	12	11 58·0341	42	41 53·1193	12	11·9672	42	41·8853
13	12 57 52·2157	13	12 57·8703	43	42 52·9555	13	12·9645	43	42·8826
14	13 57 42·3862	14	13 57·7064	44	43 52·7917	14	13·9618	44	43·8799
15	14 57 32·5566	15	14 57·5426	45	44 52·6278	15	14·9591	45	44·8772
16	15 57 22·7270	16	15 57·3788	46	45 52·4640	16	15·9563	46	45·8744
17	16 57 12·8975	17	16 57·2150	47	46 52·3002	17	16·9536	47	46·8717
18	17 57 3·0679	18	17 57·0511	48	47 52·1364	18	17·9509	48	47·8690
19	18 56 53·2384	19	18 56·8873	49	48 51·9725	19	18·9481	49	48·8662
20	19 56 43·4088	20	19 56·7235	50	49 51·8087	20	19·9454	50	49·8635
21	20 56 33·5792	21	20 56·5597	51	50 51·6449	21	20·9427	51	50·8608
22	21 56 23·7497	22	21 56·3958	52	51 51·4810	22	21·9399	52	51·8580
23	22 56 13·9201	23	22 56·2320	53	52 51·3172	23	22·9372	53	52·8553
24	23 56 4·0906	24	23 56·0682	54	53 51·1534	24	23·9345	54	53·8526
		25	24 55·9044	55	54 50·9896	25	24·9318	55	54·8499
		26	25 55·7405	56	55 50·8257	26	25·9290	56	55·8471
		27	26 55·5767	57	56 50·6619	27	26·9263	57	56·8444
		28	27 55·4129	58	57 50·4981	28	27·9236	58	57·8417
		29	28 55·2490	59	58 50·3343	29	28·9208	59	58·8389
		30	29 55·0852	60	59 50·1704	30	29·9181	60	59·8362

INTO EQUIVALENT INTERVALS OF MEAN SOLAR TIME.

	FRACTIONS OF A SECOND.				
Seconds of Sidereal Time.	Equivalents in Mean Time.	Seconds of Sidereal Time.	Equivalents in Mean Time.	Seconds of Sidereal Time.	Equivalents in Mean Time.
	s.		s.		s.
0·01	0·00997	0·34	0·33907	0·67	0·66817
0·02	0·01995	0·35	0·34904	0·68	0·67814
0·03	0·02992	0·36	0·35902	0·69	0·68812
0·04	0·03989	0·37	0·36899	0·70	0·69809
0·05	0·04986	0·38	0·37896	0·71	0·70806
0·06	0·05984	0·39	0·38894	0·72	0·71803
0·07	0·06981	0·40	0·39891	0·73	0·72801
0·08	0·07978	0·41	0·40888	0·74	0·73798
0·09	0·08975	0·42	0·41885	0·75	0·74795
0·10	0·09973	0·43	0·42883	0·76	0·75793
0·11	0·10970	0·44	0·43880	0·77	0·76790
0·12	0·11967	0·45	0·44877	0·78	0·77787
0·13	0·12965	0·46	0·45874	0·79	0·78784
0·14	0·13962	0·47	0·46872	0·80	0·79782
0·15	0·14959	0·48	0·47869	0·81	0·80779
0·16	0·15956	0·49	0·48866	0·82	0·81776
0·17	0·16954	0·50	0·49864	0·83	0·82773
0·18	0·17951	0·51	0·50861	0·84	0·83771
0·19	0·18948	0·52	0·51858	0·85	0·84768
0·20	0·19945	0·53	0·52855	0·86	0·85765
0·21	0·20943	0·54	0·53853	0·87	0·86762
0·22	0·21940	0·55	0·54850	0·88	0·87760
0·23	0·22937	0·56	0·55847	0·89	0·88757
0·24	0·23934	0·57	0·56844	0·90	0·89754
0·25	0·24932	0·58	0·57842	0·91	0·90752
0·26	0·25929	0·59	0·58839	0·92	0·91749
0·27	0·26926	0·60	0·59836	0·93	0·92746
0·28	0·27924	0·61	0·60833	0·94	0·93743
0·29	0·28921	0·62	0·61831	0·95	0·94741
0·30	0·29918	0·63	0·62828	0·96	0·95738
0·31	0·30915	0·64	0·63825	0·97	0·96735
0·32	0·31913	0·65	0·64823	0·98	0·97732
0·33	0·32910	0·66	0·65820	0·99	0·98730

This TABLE is useful for the conversion of SIDEREAL into MEAN SOLAR Time.

Mean Solar Time *required* = Mean Time at the *preceding* Sidereal Noon + the Equivalent to the *given* Sidereal Time.

EXAMPLE. — To convert 21h. 28m. 54s·26 Sidereal Time at Greenwich, Jan. 7, 1859, into Mean Time.

```
                                                               h.  m.   s.
Mean Time at the preceding Sidereal Noon, viz. January 6  .    20  56  33·579

                        ⎧ 21h  0m  0s                       4  57   2·52
For Sidereal            ⎪    28  0                              27  55·413
    Intervals.          ⎨    54        The Table gives the          53·853
                        ⎪     0·26     Equivalent Mean Intervals,    ·259
                        ⎩

The Sum is the Mean Time required, Jan. 7  .    2  21  25·62
```

N

FOR REDUCING LONGITUDE TO TIME.

° ' "	H. M. M. S. S. T.		° ' "	H. M. M. S. S. T.		Degrees.	Hours.	Minutes.
1	0	4	31	2	4	70	4	40
2	0	8	32	2	8	80	5	20
3	0	12	33	2	12	90	6	0
4	0	16	34	2	16	100	6	40
5	0	20	35	2	20	110	7	20
6	0	24	36	2	24	120	8	0
7	0	28	37	2	28	130	8	40
8	0	32	38	2	32	140	9	20
9	0	36	39	2	36	150	10	0
10	0	40	40	2	40	160	10	40
11	0	44	41	2	44	170	11	20
12	0	48	42	2	48	180	12	0
13	0	52	43	2	52	190	12	40
14	0	56	44	2	56	200	13	20
15	1	0	45	3	0	210	14	0
16	1	4	46	3	4	220	14	40
17	1	8	47	3	8	230	15	20
18	1	12	48	3	12	240	16	0
19	1	16	49	3	16	250	16	40
20	1	20	50	3	20	260	17	20
21	1	24	51	3	24	270	18	0
22	1	28	52	3	28	280	18	40
23	1	32	53	3	32	290	19	20
24	1	36	54	3	36	300	20	0
25	1	40	55	3	40	310	20	40
26	1	44	56	3	44	320	21	20
27	1	48	57	3	48	330	22	0
28	1	52	58	3	52	340	22	40
29	1	56	59	3	56	350	23	20
30	2	0	60	4	0	360	24	0

FOR REDUCING TIME TO LONGITUDE.

Hours.	Degrees.	M. S. T.	° ' "	' " '''	M. S. T.	° ' "	' " '''
1	15	1	0	15	31	7	45
2	30	2	0	30	32	8	0
3	45	3	0	45	33	8	15
4	60	4	1	0	34	8	30
5	75	5	1	15	35	8	45
6	90	6	1	30	36	9	0
7	105	7	1	45	37	9	15
8	120	8	2	0	38	9	30
9	135	9	2	15	39	9	45
10	150	10	2	30	40	10	0
11	165	11	2	45	41	10	15
12	180	12	3	0	42	10	30
13	195	13	3	15	43	10	45
14	210	14	3	30	44	11	0
15	225	15	3	45	45	11	15
16	240	16	4	0	46	11	30
17	255	17	4	15	47	11	45
18	270	18	4	30	48	12	0
19	285	19	4	45	49	12	15
20	300	20	5	0	50	12	30
21	315	21	5	15	51	12	45
22	330	22	5	30	52	13	0
23	345	23	5	45	53	13	15
24	360	24	6	0	54	13	30
		25	6	15	55	13	45
		26	6	30	56	14	0
		27	6	45	57	14	15
		28	7	0	58	14	30
		29	7	15	59	14	45
		30	2	30	60	15	0

MEAN MOTION OF SUN FOR PERIODS OF MEAN TIME.			
Mean Time.	Mean Motion.	Mean Time.	Mean Motion.
Hours.	′ ″	Days.	° ′ ″
1	2 27·847	1	0 59 8·32
2	4 · 55·694	2	1 58 16·65
3	7 23·541	3	2 57 24·98
4	9 51·388	4	3 56 33·30
5	12 19·235	5	4 55 41·62
6	14 47·081	6	5 54 49·95
7	17 14·928	7	6 53 58·28
8	19 42·775	8	7 53 6·60
9	22 10·622	9	8 52 14·92
10	24 38·469	10	9 51 23·25
11	27 6·316	11	10 50 31·58
12	29 34·163	12	11 49 39·90
13	32 2·010	13	12 48 48·23
14	34 29·857	14	13 47 56·55
15	36 57·703	15	14 47 4·87
16	39 25·550	16	15 46 13·20
17	41 53·397	17	16 45 21·53
18	44 21·244	18	17 44 29·85
19	46 49·091	19	18 43 38·17
20	49 16·938	20	19 42 46·50
21	51 44·784	21	20 41 54·83
22	54 12·631	22	21 41 3·15
23	56 40·478	23	22 40 11·48
24	59 8·325	24	23 39 19·80
		25	24 38 28·12
		26	25 37 36·45
		27	26 36 44·78
		28	27 35 53·10
		29	28 35 1·42
		30	29 34 9·75
		31	30 33 18·08

MEAN MOTION OF SUN FOR PERIODS OF MEAN TIME.

Mean Time.	Mean Motion.	Mean Time.	Mean Motion.	Mean Time.	Mean Motion.	Mean Time.	Mean Motion.
min.	′ ″	m.	′ ″	sec.	″	s.	″
1	0 2·464	31	1 16·387	1	0·041	31	1·273
2	0 4·928	32	1 18·851	2	0·082	32	1·314
3	0 7·392	33	1 21·315	3	0·123	33	1·355
4	0 9·856	34	1 23·779	4	0·164	34	1·396
5	0 12·321	35	1 26·244	5	0·205	35	1·437
6	0 14·785	36	1 28·708	6	0·246	36	1·478
7	0 17·249	37	1 31·172	7	0·287	37	1·519
8	0 19·713	38	1 33·636	8	0·328	38	1·560
9	0 22·177	39	1 36·100	9	0·369	39	1·601
10	0 24·641	40	1 38·565	10	0·411	40	1·643
11	0 27·105	41	1 41·029	11	0·452	41	1·684
12	0 29·569	42	1 43·493	12	0·493	42	1·725
13	0 32·033	43	1 45·957	13	0·534	43	1·766
14	0 34·497	44	1 48·421	14	0·575	44	1·807
15	0 36·962	45	1 50·885	15	0·616	45	1·848
16	0 39·426	46	1 53·349	16	0·657	46	1·889
17	0 41·890	47	1 55·813	17	0·698	47	1·930
18	0 44·354	48	1 58·277	18	0·739	48	1·971
19	0 46·818	49	2 0·741	19	0·780	49	2·012
20	0 49·282	50	2 3·206	20	0·821	50	2·053
21	0 51·746	51	2 5·670	21	0·862	51	2·094
22	0 54·210	52	2 8·134	22	0·903	52	2·135
23	0 56·674	53	2 10·598	23	0·944	53	2·176
24	0 59·138	54	2 13·062	24	0·985	54	2·217
25	1 1·603	55	2 15·526	25	1·027	55	2·259
26	1 4·067	56	2 17·990	26	1·068	56	2·300
27	1 6·531	57	2 20·454	27	1·109	57	2·341
28	1 8·995	58	2 22·918	28	1·150	58	2·382
29	1 11·459	59	2 25·382	29	1·191	59	2·423
30	1 13·923	60	2 27·847	30	1·232	60	2·464

BOOK VII.

THE STARRY HEAVENS.

———◆———

CHAPTER I.

———

"O ye Stars of Heaven, bless ye the LORD : praise Him and
magnify Him for ever."—*Benedicite.*

———

*The Pole-Star — Not always the same. — Curious Circumstance con-
nected with the Pyramids of Egypt. — Stars classified into different
Magnitudes. — Antiquity of the Custom of forming them into Groups.
— Anomalies of the present System. —Stellar Photometry. — Dis-
tances of the Stars. — How distinguished. — Antiquity of the Custom
of naming Stars. — Invention of the Zodiac. — Letters introduced by
Bayer. — Effects of the increased Care bestowed on Observations of the
Stars. — Ideas of the Ancients on the Stars. — Remarks by Sir J.
Herschel.*

IF, on some clear evening, the reader will take the trouble to
station himself on the summit of any rising ground, and cast
his eye upwards, he will see the sky spangled with countless
multitudes of brilliant specks of light; these are the *fixed stars*
(we shall presently see that this appellation is not strictly correct); an attentive observer will soon notice, also, that the stars
he is contemplating seem to revolve in a body around one of
their number situated in the north, about midway between the
horizon and the zenith ; this is the Pole-star, so called from
its being near the pole of the celestial equator. On account,
however, of the *precession of the equinoxes*, the present Pole-
star (a Ursæ Minoris) will not always be so ; the true pole is

N 3

now about $1\frac{1}{2}°$ from this star; this distance will be gradually diminished until it is reduced to about half a degree; it will then increase again, and after the lapse of a long period of time, the pole will depart from this star, which will then cease to bear the name or serve the purposes of a Pole-star. 3,970 years ago, the star γ in the constellation Draco fulfilled this office; 12,000 years hence, it will fall to the lot of a brilliant star of the 1st magnitude — *Vega* (a Lyræ) — which is $24° 52'$ from the pole.

Connected with this subject, a curious circumstance was noticed in the recent researches that have been made in Egypt. Of the 9 pyramids which still remain standing at Gizeh, 6 have openings facing the north, leading to straight passages, which descend at an inclination, varying from 26° to 27°, the direction of these passages being, in all cases, parallel to the meridian: now, if we suppose a person to be stationed at the bottom of any one of these passages, and to look up it, as he would through the tube of a telescope, his eye will be directed to a point on the meridian, 26° or 27° above the plane of the horizon; and this is precisely the altitude at which the star γ Draconis must have passed the lower meridian, at the place in question, 3,970 years before the present time. Now, the date of the construction of these pyramids (the great pyramid in 2123 B.C.) corresponds exactly to that epoch, and it cannot be doubted that the peculiar direction given to these passages must have had reference to the position of γ Draconis, the then Pole-star.[1]

The stars, on account of their various degrees of brilliancy, have been distributed into classes or orders. Those which appear largest are called stars of the 1st magnitude; next to these come stars of the 2nd magnitude, and so on to the 6th, which are the smallest visible to the naked eye. This distribution having been made long before the invention of telescopes, such stars as cannot be seen without the assistance of these instruments are called *telescopic*, and are classed in

---

[1] *Phil. Mag.* vol. xxiv. pp. 481–4. 1844. Pytheas of Marseilles, 330 B.C., first noticed that the so-called Pole-star was not situated exactly at the Pole.

magnitudes varying from the 7th to the 18th or 20th; these latter, of course, being only visible in the most powerful instruments hitherto constructed; nor does there seem the least reason to assign a limit to this progressive diminution, inasmuch as past experience has shown that every successive improvement in the construction of telescopes brings to light new objects; and there is no reason why the past should not hold good for the future.

From the earliest ages of antiquity it has been the custom to arrange the stars into groups or constellations, for the purpose of more readily distinguishing them; each one having some special figure to which the configuration of its stars may be supposed to bear some resemblance, which, in the majority of instances, is purely imaginary. Modern astronomers have continued this arrangement chiefly on account of the confusion that would arise were any change made. We often find that one constellation will contain an isolated portion of another, just as one county will sometimes wholly surround a parish belonging to another. Stars, too, often occur under different names.[1] Many catalogue-stars never existed, but owe their creation to mistakes of observers. Constellations are recognised by some and not by others; while the same names are repeated in different parts of the heavens; such are a few of the anomalies of the present system.[2]

The constellations will again be referred to in a subsequent chapter.

Concerning the comparative brilliancy of the stars, we know, for certain, but little. Sir W. Herschel gives the following table of the light emitted by stars of different magnitudes, as deduced from his own observations; an average star of the 6th magnitude being taken as unity.

6th magnitude	= 1	. .	3rd magnitude	=	12
5th ,,	= 2	. .	2nd ,,	=	25
4th ,,	= 6	. .	1st ,,	=	100

[1] Baily, in the *Brit. Assoc. Cat.* of Stars, p. 75, gives a list.
[2] Vide remarks by Baily in the *B.A.C.*, p. 52, *et seq.*

Sir J. Herschel ascertained that the light of *Sirius* (the brightest of all the fixed stars) is about 324 times that of an average star of the 6th magnitude. From direct photometrical experiments, Dr. Wollaston found that the light of the Sun, as received by us, exceeded by 20,000,000,000 times that of Sirius; consequently, in order that the Sun might appear to us no brighter than Sirius, it must be removed from us not less than 13,433,000,000,000 miles; a distance utterly beyond the limits of human comprehension.

Why stars appear to us of different degrees of brilliancy we know not; but it is most likely for one or other of the following reasons: — Either (1) that the stars are all of the same size, but situated at different distances; or (2) that they are of different sizes, but at the same distances. If we suppose the first to be the true hypothesis, as it probably is, and taking the light of a star of each magnitude to be half that of the magnitude next above it, we find that the star of the 16th magnitude cannot be less than 362 times the distance of a star of the 1st magnitude [1]; and " as it has been considered probable from recondite investigations, that the average distance of a star of the 1st magnitude from the Earth is 986,000 radii of our annual orbit," it follows that a 16th magnitude star is distant from us 33,908,540,000,000,000 miles, a distance which light, with a velocity of 192,000 miles per second, would occupy 6611 years in traversing! Although, of course, these stupendous numbers are only approximations founded on calculation, yet the actual distances of 9 stars have been carefully and satisfactorily ascertained within the last few years.

The determination of the distance of the stars is effected by ascertaining by instrumental observations the amount of their annual parallax, or displacement in the heavens. The non-detection of stellar parallax afforded for a long time, a much resorted to, and certainly to some extent plausible, argument against the soundness of the Copernican theory of the universe. Since it happens, that with few exceptions, the stars

---

[1] Sir J. Herschel.

generally do not suffer from parallax, and since also the fact of the orbital motion of the Earth round the Sun rests on the most undoubted evidence, it follows that the general absence of parallax can only be ascribed to the fact that the stars must be placed at such distance from us, that, comparatively, the Earth's orbit, of a diameter of 190,000,000 miles, is something utterly insignificant, — a mere point, when considered in reference to the distances of the stars themselves.

It might be supposed, that where the character and laws of parallax are so clearly understood as they are, the discovery of its existence could present no great difficulty. Nevertheless, nothing in the whole range of astronomical research has more baffled the efforts of observers than this question. This has arisen altogether from the extreme minuteness of its magnitude. It is quite certain that the parallax does not amount to so much as 1″ in the case of any of the numerous stars which have been as yet submitted to the course of observation which is necessary to discover the parallax. Now, since in the determination of the exact uranographical position of a star there are a multitude of disturbing effects to be taken into account and eliminated, such as precession, nutation, aberration, refraction, and others, besides the proper motion of the star, which will be explained hereafter; and since, besides the errors of observation, the quantities of these are subject to more or less uncertainty, it will astonish no one to be told that they may entail, upon the final result of the calculation, an error of 1″; and, if they do, it is vain to expect to discover such a residual phenomenon as parallax, the entire amount of which is less than 1″.

If in any case the parallax could be determined, the distance of the stars could be immediately inferred. For, if this value of the parallax be expressed in seconds, or in decimals of a second, and if r denote the semidiameter of the Earth's orbit, d the distance of the star, and p the parallax, we shall have —

$$d = r \times \frac{206265}{p}.$$

If, therefore, $p = 1''$, the distance of the star would be 206,265 times the distance of the Sun, and since it may be considered satisfactorily proved, that no star which has ever yet been brought under observation has a parallax greater than this, it may be affirmed that the nearest star in the universe to the solar system is at a distance from it, *at least*, 206,265 times greater than that of the Sun.

Let us consider more attentively the import of this conclusion. The distance of the Sun, expressed in round numbers (which are sufficient for our present purpose), is 95 millions of miles. If this be multiplied by 206,265, we shall obtain, — not indeed the distance of the nearest of the fixed stars, — but the *minor limit* of that distance, that is to say, a distance within which the star cannot lie. This limit, expressed in miles, is

$$d = 206,265 \times 95,000,000 = 19,595,175,000,000 \text{ miles,}$$

or nearly 20 *billions of miles.*

In the contemplation of such numbers the imagination is lost, and no other clear conception remains, except of the mere arithmetical expression of the result of the computation. Astronomers themselves, accustomed as they are to deal with stupendous numbers, are compelled to seek for units of proportionate magnitude to bring the arithmetical expression of the quantities within moderate limits. The motion of light supplies one of the most convenient möduli for this purpose, and has, by common consent, been adopted as the unit in all computations whose object is to gauge the universe. It is known that light moves at the rate of 192,000 miles per second. If, then, the distance d above computed be divided by 192,000, the quotient will be the time, expressed in seconds, which light takes to move over that distance. But since even this will be an unwieldy number, it may be reduced to minutes, hours, days, or even to years.

In this manner we find that, if any star have a parallax of $1''$, it must be at such a distance from our system, that light would take 3·234 years, or 3 years and 85 days, to come from it to the Earth.

If the space through which light moves in a year be taken, therefore, as the unit of stellar distance, and p be the parallax expressed in seconds, or decimals of a second, we shall have

$$d = \frac{3 \cdot 234}{p}.$$

It will easily be imagined that astronomers have diligently directed their observations to the discovery of some change of apparent position, however small, produced upon the stars by the Earth's motion. As the stars most likely to be affected by the motion of the Earth are those which are nearest to the system, and therefore probably those which are brightest and largest, it has been to such chiefly, that this kind of observation has been directed; and since it was certain that, if any observable effect be produced by the Earth's motion at all, it must be extremely small, the nicest and most delicate means of observation were those alone from which the discovery could be expected.

One of the earlier expedients adopted for the solution of this problem, was the erection of a telescope, of great length and power, in a position permanently fixed, attached, for example, to the side of a pier of solid masonry, erected upon a foundation of rock. This instrument was screwed into such a position that particular stars, as they crossed the meridian, would necessarily pass within its field of view. Micrometric wires were, in the usual manner, placed in its eye-piece, so that the exact point at which the stars passed the meridian each night, could be observed and recorded with the greatest precision. The instrument being thus fixed and immovable, the transits of the stars were noted each night, and the exact places where they passed the meridian recorded. This kind of observation was carried on through the year; and if the Earth's change of position, by reason of its annual motion, should produce any effect upon the apparent position of the stars, it was anticipated that such effect would be discovered by these means. After, however, making all allowance for the usual causes which affect the apparent position of the stars, no change of position was discovered which could be assigned to the Earth's motion.[1]

The number of stars *known* to possess a sensible parallax amount only to 9, particulars of which are given in the following table.

Star.	Parallax.	Distance.		Observer.
		Sun's dist. =1.	Annual motion of light = 1.	
α Centauri .	0·913″	225920	3·54217	Henderson.
61 Cygni .	0·348	592715	9·29310	Bessel.
α Lyræ .	0·261	790287	12·39080	Struve.
Sirius .	0·230	896804	14·06087	Henderson.
1830 Groombridge	0·226	912677	14·30973	C. H. Peters.
ι Ursæ Maj.	0·133	1550864	24·31579	C. H. Peters.
Arcturus .	0·127	1624134	25·46456	C. H. Peters.
Polaris .	0·067	3078582	48·26866	C. H. Peters.
Capella .	0·046	4484021	70·30435	C. H. Peters.

Stars are distinguished from one another in various ways; the ancients were in the habit of indicating the locality of a star by its position in the constellation to which it belonged; thus Aldebaran was called *Oculus Tauri;* this custom was also followed by the Arabians, and, indeed, many of the names applied by them are still retained in a corrupted form; thus Betelgueze (α Orionis) is a corruption of *ibt-al-jauza*, signifying "the giant's shoulder." Bayer, the German astronomer, was the first to improve upon the old plan, which he did by publishing, in 1604, a celestial atlas, in which the stars of each constellation were distinguished by the letters of the Greek alphabet; the brightest receiving the name of α, the next β, and so on. Bayer's letters are still in common use; the name of the constellation in the genitive case being put after each; thus Procyon is termed α Canis Minoris; Vega, α Lyræ; Arcturus, α Boötis. It should be remarked that, owing to

[1] The preceding remarks on Stellar Parallax are from the *Museum of Science.*

carelessness on the part of Bayer, this alphabetical arrangement does not, in all cases, accurately represent the relative
brilliancy of the different stars in a constellation.   Flamsteed,
the first Astronomer Royal at Greenwich, affixed *numbers* to
the stars observed by him.

The subject of the photometry of stars, though one of much
importance, has received but little attention from the practical
astronomer.   The common method of classifying stars into
arbitrary magnitudes is both vague in theory and contradictory
in practice.   It is vague, inasmuch as the place of a star in
the scale of magnitudes conveys but little definite idea of its
own actual brilliancy ; and it is contradictory in a remarkable
degree, inasmuch as the same star has in numberless instances
a different magnitude assigned to it by different authorities.[1]

The practice of giving names to the stars is an early one,
and probably originated with the Chaldæans.   Intimations of
this custom may be found in the Holy Scriptures.

"Which maketh *Ash, Kesil,* and *Kimah,* and the chambers of the
south." [2]

" Canst thou bind the sweet influences of *Kimah,* or loose the bands
of *Kesil?*   Canst thou bring forth *Mazzaroth* in his season? Or canst
thou guide *Ash* with his sons?" [3]

" Seek him that maketh *Kimah* and *Kesil.*" [4]

The invention of the zodiac has been ascribed to the Egyp-

---

[1] Sir J. Herschel, *Res. of Ast. Obs.* p. 304.
[2] Job ix. 9.   [3] Ibid., xxxviii. 31, 32.
[4] Amos v. 8.   It is not known for certain what stars or constellations
are referred to in these verses, but the rendering given in our translation
rests on insufficient authority.   It is probable, however, that Mazzaroth
may mean the circle of the zodiac, but the others are doubtful.   Park-
hurst (*Lexicon*) thinks that the applying of the Greek names of certain
constellations to the Hebrew originals, as is done in the Authorised
Version, here and elsewhere, and by the LXX previously, is only fancy.
Barnes, however, (*Notes on Job*) derives *Kimah* from a root signifying a heap, and applicable to the Pleiades, and *Kesil* from another
root signifying *to be strong,* and thus applicable to that constellation
known as *the strong man* corresponding, as may be conjectured, to what
the Greeks called *Orion.* (See *Class. Dict.*)   It should be added that
what is commonly considered to be the root of Kimah means *to be hot,*
and of Kesil, *to be stiff.*

tians. Dupuis especially has advocated this opinion, and thinks that the constellations in question had reference to the division of the seasons, and to the agriculture at the time of their invention. He supposes Cancer to represent the retrogradation of the Sun at the solstice, Libra, the equality of the day and night at the equinoxes. This idea is undoubtedly supported by several curious coincidences; for instance, the inundation of the Nile, which takes place after the summer solstice, would happen when the Sun was in Aquarius and Pisces, and Virgo, usually represented as a woman holding an ear of corn, would coincide with the time of the Egyptian harvest.

The insuperable objection to this theory is the excessive antiquity which it assigns to the zodiac (not less than 15,000 years). As this is historically inadmissible, and directly opposed to Divine Revelation, Dupuis gets over the difficulty by supposing the names to have been given, not to the constellations in conjunction, but to those in opposition to the Sun. This only requires the constellations to have been invented B.C. 2500±; in this form the idea is adopted by Laplace and others as correct.[1] It has been asserted that the Jews were acquainted with the zodiac, and that in Gen. i. 14, the uses of the heavenly bodies, to divide the seasons, years, and days, are set forth. Whilst on this subject it may be mentioned that in Gen. i. 5, we find that "the evening and the morning were the first day;" that is to say, the course of this revolution was from evening — place of sunset — or West, to morning — place of sunrise — or East; thereby clearly pointing out to us that the axial rotation of the Earth is from West to East, a fact which we also know from other sources.

Seneca attributes the sub-division of the heavens into constellations to the Greeks, 1400 or 1500 years B.C.[2] It may be mentioned as a somewhat singular fact, that the Iroquois, a North American Indian tribe, should have applied the name of the Bear to the group Ursa Major, in common with the earliest Asiatic nations, so remote from them, more especially as it bears no resemblance whatever to that animal.

The present system of constellations, though on the whole

---

[1] *Hist. of Ast.* L.U.K. p. 16.   [2] *Quæst. Nat.* lib. vii. cap? 25.

most valuable, presents many anomalies, which require reforming. Thus Aries should no longer have a horn in Pisces, and
a leg in Cetus ; nor should 13 Argûs pass through the Unicorn's
flank into the Little Dog: 51 Camelopardi might with pro-
priety be extracted from the eye of Auriga; and the ribs of
Aquarius released from 46 Capricorni, &c.

With reference to the present mode of identifying stars by
letters, it may be remarked that though the idea was carried
out practically by Bayer [1], as mentioned above, yet Piccolomini,
who was born at Vienna in 1508, and died there in 1578, did
the same thing ; the letter system is defective in this respect,
that in large constellations the alphabet is so soon used up;
indeed, as Mr. Baily remarks, La Caille has, in the constellation Argo alone, besides the Greek alphabet, employed the
whole of the Roman alphabet, both in small and capital letters,
each of them more than 3 times; in fact he has used nearly
180 letters in that constellation alone. " Thus we have 3 stars
marked a, and 7 marked A ; 6 marked d, and 5 marked D, and
so on with several others."

As greater pains came to be taken in the observation of the
heavens, the number of stars enumerated, increased; a fact
which of course we should be led to expect. The following table
exemplifies this. It shows the number of stars reckoned in
the following constellations by 5 different catalogue makers
living at different epochs.

	Ptolemy.	Tycho Brahe.	Hevelius.	Flamsteed.	Bode.
Aries . .	18	21	27	66	148
Ursa Major	35	56	73	87	338
Boötes .	23	28	52	54	319
Leo . .	35	40	50	95	337
Virgo . .	32	39	50	110	411
Taurus .	44	43	51	141	394
Orion . .	38	62	62	78	304

---

[1] Bayer was something of an astrologer. In the first edition of the
Uranometria he has marked many objects supposed to have some kind
of influence over mundane affairs.

In the great trapezium of Orion upwards of 2000 stars have been counted.

The ideas of the ancients on the fixed stars were very " misty." Anaximenes (550 B.C.) thought the stars were for ornament, and nailed, as it were, like studs in the crystalline sphere. Pythagoras pronounced each star to be a distinct world with its own land, water, and air. The Stoics, Epicureans, and indeed almost all the ancient schools of philosophy, held, that the stars were celestial fires nourished by the caloric or igneous matter which they considered perpetually streamed out from the centre of the universe. Anaxagoras (450 B.C.) considered that the stars were stones whirled upwards from the Earth, by the rapid motions of the ambient ether, the inflammable properties of which setting them on fire, thus caused them to appear as stars. Callimachus describes the circumpolar stars as feeding on air; and Lucretius, pondering on the subject, and not doubting the fact, asks " Unde æther sidera pascit ? " Stars were at one time looked upon as the *spiraculæ* or breathing holes of the universe.

Sir John Herschel's remarks on the stars are very forcible. He says : " The stars are the land-marks of the universe ; and amidst the endless and complicated fluctuations of our system, seem placed by its Creator, as guides and records not merely to elevate our minds by the contemplation of what is vast, but to teach us to direct our actions by reference to what is immutable in his works. It is indeed hardly possible to over-appreciate their value in this point of view. Every well-determined star, from the moment its place is registered, becomes to the astronomer, the geographer, the navigator, the surveyor, a point of departure which can never deceive or fail him, the same for ever and in all places, of a delicacy so extreme as to be a test for every instrument yet invented by man, yet equally adapted for the most ordinary purposes; as available for regulating a town clock, as for conducting an army to the Indies; as effective for mapping down the intricacies of a petty barony, as for adjusting the boundaries of transatlantic empires. When once its place has been thoroughly ascertained and carefully recorded, the brazen circle, on which that useful work was done, may moulder, the marble pillar totter on its base, and the astronomer only survive in the gratitude of his posterity : but the record remains, and trans-fuses all its own exactness into every determination which takes it for a ground-work, giving to inferior instruments, nay, even to temporary contrivances, and to the observations of a few weeks or days, all the precision attained originally at the cost of so much time, labour, and expense." [1]

---

[1] *Mem.*, R.A.S. vol. iii. p. 125.

CHAPTER II.

DOUBLE STARS, ETC.

*But few known until Sir W. Herschel commenced his Search for them.—
Labours of Sir J. Herschel and F. G. W. Struve.—Examples.—Optical
Double Stars.—Binary Stars.—Discovered by Sir W. Herschel.—
Examples.—List of Optical Doubles.—Coloured Stars.—Examples.—
Generalisations from Struve's Catalogue.—Stars changing Colour.—
Triple Stars.—Quadruple Stars.—Multiple Stars.*

ALTHOUGH to the unaided eye all the stars appear single, yet
in numerous instances the application of suitable optical assistance shows that many consist in reality of two stars, placed in
juxtaposition so close together that they appear as one. These
are termed *double stars*.[1] Only 4 of these objects were known,
until Sir W. Herschel, by means of the powerful telescopes constructed by himself, discovered a large number never before
suspected. He observed and catalogued, altogether, about 500,
which subsequent observers, especially F. G. W. Struve, and
Sir J. Herschel, have augmented to near 6000.

The following have been selected by Sir J. Herschel[2] from
Struve's catalogue, as remarkable examples of each class, well
adapted for observation by amateurs, who may be disposed to
try by them the efficiency of telescopes.

---

[1] The first application of this term was by Ptolemy, who called *ν*
Sagittarii, διπλοῦς.

[2] *Outlines of Ast.* p. 609.

0″ to 1″.	1″ to 2″.	2″ to 4″.	4″ to 8″.
γ Coronæ Bor.	γ Circini.	α Piscium.	α Crucis.
γ Centauri.	δ Cygni.	β Hydræ.	α Herculis.
γ Lupi.	ε Chamæleontis	γ Ceti.	α Gemin.
ε Arietis.	ζ Boötis.	γ Leonis.	δ Gemin.
ζ Herculis.	ι Cassiopeiæ.	γ Cor. Aus.	ζ Cor. Bor.
η Coronæ.	ι 2 Cancri.	γ Virginis.	θ Phænicis.
η Herculis.	ξ Ursæ Maj.	δ Serpentis.	κ Cephei.
λ Cassiopeiæ.	π Aquilæ.	ε Boötis.	λ Orionis.
λ Ophiuchi.	σ Coro. Bor.	ε Draconis.	μ Cygni.
π Lupi.	2 Camelopar.	ε Hydræ.	ξ Boötis.
η Ophiuchi.	32 Orionis.	ζ Aquarii.	ξ Cephei.
φ Draconis.	52 Orionis.	ζ Orionis.	π Boötis.
φ Ursæ Maj.		ι Leonis.	ρ Capricor.
χ Aquilæ.		ι Trianguli.	ν Argûs.
ω Leonis.		κ Leporis.	ω Aurigæ.
Atlas Pleiadum		μ Draconis.	μ Eridani.
4 Aquarii.		μ Canis.	70 Ophiuchi.
42 Comæ.		ρ Herculis.	12 Eridani.
52 Arietis.		σ Cassiopeiæ.	32 Eridani.
66 Piscium.		44 Boötis.	44 Herculis.

8″ to 12″.	12″ to 16″.	16″ to 24″.	24″ to 32″.
β Orionis.	α Centauri.	α Can. Ven.	δ Herculis.
γ Arietis.	β Cephei.	ε Normæ.	η Lyræ.
γ Delphini.	β Scorpii.	ζ Piscium.	ι Cancri.
ζ Antliæ Pn.	γ Volantis.	θ Serpentis.	κ Herculis.
η Cassiopeiæ.	η Lupi.	κ Cor. Aus.	κ Cephei.
θ Eridani.	ζ Ursa Maj.	κ Tauri.	ψ Draconis.
ι Orionis.	κ Boötis.	24 Comæ.	κ Cygni.
f Eridani.	8 Monocerotis.	41 Draconis.	23 Orionis.
2 Can. Ven.	61 Cygni.	61 Ophiuchi.	

If two stars lie very nearly in the same line of vision, whatever may be their distances, they will form an *optical* double star, or one whose components are only apparently, but not really joined together. Herschel, considering it extremely unlikely that this was the case, undertook, in 1778, carefully to investigate the subject. Long observation showed him that the above hypothesis was not the true one, but that the components were at the same distance from the Sun ; and not only that, but, that in many instances, the smaller body described an orbit round the larger star. This great discovery, previously conjectured both by Lambert, and Mitchell[1], was first published by Herschel in 1803, in 2 memorable papers, read before the Royal Society[2], in which he announced the existence of sidereal systems, consisting of two stars revolving about each other in regular elliptic orbits, and constituting what may .be termed *binary stars*, to distinguish them from double stars, generally so called, and in which no periodic change is discoverable.

The following are some of the more remarkable objects of this class[3] :—

Name of Star.	Period, Years.	Dist.	ε of Orbit.[4]
		"	
ζ Herculis . .	36	1·2	0·44
η Coronæ · .	43	...	...
ξ Ursæ Majoris .	58	3·8	0·42
ζ Cancri . .	58	0·9	...
α Centauri . .	78	12·1	0·71
70 Ophiuchi . .	88	4·4	0·47
Castor . .	253	8·1	0·76
6 Coronæ . .	287	3·7	0·76
61 Cygni . .	452	15·4	...
γ Virginis . .	180	...	...
γ Leonis . .	1200	...	...

[1] *Phil. Trans.* vol. lvii. p. 234, *et seq.* 1767.

[2] *Phil. Trans.* vol. xcliii. p. 339, *et seq.* 1803; vol. xcliv. p. 353, *et seq.* 1804.

[3] Arago, *Pop. Ast.* vol.-i. p. 302. Eng. Ed.

[4] The first to compute elliptic elements, for the orbit of any binary

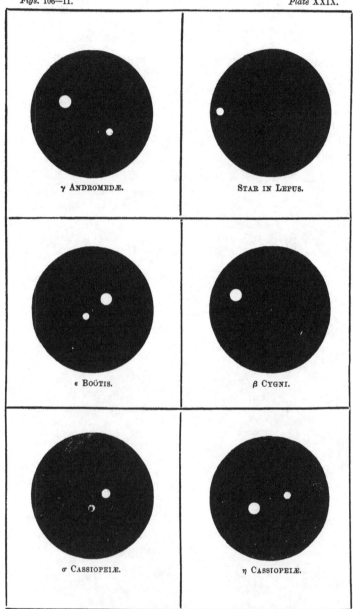

COLOURED STARS.

Of 653 stars in Struve's 8 orders there are only 48 which are probably optically double. Of the wider ones none have so changed in position as to enable any orbit to be approximated, whence it is concluded that even where they have a physical connection, the period of revolution cannot be less than 20,000 years. The following are a few of the more remarkable optically double stars :—

				Mag.		Dist. "
α Lyræ .	.	.	.	1	11	43
α Tauri	.	.	.	1	12	108
α Aquilæ	.	.	.	1½	10	152
β Geminorum	.	.		2	12	208

Many of the double stars exhibit the curious and beautiful phenomenon of complementary colours. In such instances the larger star is usually of a ruddy or orange hue, the smaller one being blue or green. When the complementary colours are remarked in a double star, whose components are of very unequal size, we may attribute the circumstance only to the effect of contrast; yet it can hardly be doubted, that in many cases, the light of the stars is actually of different colours. Single stars of a fiery red, or deep orange colour, are not uncommon; but there is no instance of an isolated blue or green star; those colours are apparently confined to the compound stars, of which the following are good examples :—

Name of Star.	R. A.			Decl.		Mag. of Compon.		Col. of larger.	Col. of smaller.
	h.	m.	s.	°	′				
γ Andromedæ . .	1	54	79	+41	39	3¾	5½	Orange .	Sea Green.
η Cassiopeiæ . .	0	40	38	57	5	4	7½	Yellow .	Purple.
α Piscium . . .	1	54	48	2	5	5	6	Pale Green	Blue.
ι Boötis	14	38	52	27	40	3	7	Pale Orange	Sea Green.
ζ Coronæ	15	34	6	37	5	5	6	White . .	Light Purple.
α Herculis . .	17	8	15	17	33	3½	5½	Orange .	Emerald Green.
β Cygni	19	25	4	27	40	3	7	Yellow .	Sapphire Blue.
σ Cassiopeiæ . .	23	51	55	54	58	6	8	Greenish .	Bright Blue.

star, was Savary, who investigated the elements of ξ Ursæ Majoris. (*Conn. des Temps.* 1830.)

The following are some generalisations from Struve's catalogue.[1] Of 596 bright double stars there were :—

> 375 pairs of the same colour and intensity.
> 101 pairs of the same colour but different intensity.
> 120 pairs of totally different colours.

Amongst those of the same colour the white greatly pre-
dominated, and of 476 specimens of that species there were :—

> 295 pairs, both white.
> 118 pairs, yellowish or reddish.
> 63 pairs, both bluish.

The number of reddish stars is double that of the bluish
tinge; and that of the white stars is $2\frac{1}{2}$ times as great as the
number of red ones. The combination of a blue companion
with a coloured primary happens :—

> 53 times with a white principal star.
> 52 times with a light yellow.
> 52 times with a yellow or red.
> 16 times with a green.

The question of stars changing colour must be answered in
the affirmative, though the examples yet known are few.
Ptolemy and Seneca expressly declare that at their time
Sirius was of a reddish hue, whereas now, as is well-known,
it is of a brilliant white. It would also seem that γ Leonis
and γ Delphini have changed since they were first observed
by Sir W. Herschel. He says[2] that they were perfectly white,
whereas now the larger components of each are both yellow,
and the smaller both green. The same observer also states
that 11 stars in Leo have changed their *lustre* since the time
of Flamsteed.

When very powerful telescopes are directed upon some
stars, which in smaller ones are only seen either singly or as
doubles, they appear to consist of three or more stars grouped

---

[1] Quoted in Smyth's *Cycle of Cel. Obj.* vol. i. p. 301. See the orig.
[2] Quoted in Smyth's *Cycle*, vol. i. p. 303. We have been unable to
find the original.

together, and thence are termed triples, quadruples, multiples,
&c.   The following are examples:—

TRIPLE STARS.

Name.	R.A. h. m. s.	Decl. ° ′	Mag. of Compon.
ψ Cassiopeiæ	1 16 5	+ 67 27·5	4½, 9, 11.
γ Andromedæ	1 55 19	+ 41 39·4	3½, 5½,
11 Monocerotis	6 22 2	− 6 56·7	6½, 7, 8.
12 Lyncis	6 33 50	+ 59 34·6	6, 6½, 7½.
ζ Cancri	8 4 10	+ 18 4·0	6, 7, 7½.
ξ Libræ	14 49 10	− 10 50·5	

QUADRUPLE STARS.

Name	R.A.	Decl.	Mag. of Compon.
π² Canis Majoris	6 48 58	− 20 13·8	6, 9½, 10, 11.
β Geminorum	7 36 45	+ 28 21·6	2, 12½, 11½, 12.
β Lyræ	18 44 54	+ 34 13·1	3, 8, 8½, 9.
178 P. XX Delphini	20 24 32	+ 10 41·0	7½, 8, 16, 9
8² Lacertæ	22 29 38	+ 38 48·4	6½, 6½, 10, 10.

MULTIPLE STARS.

Name	R.A.	Decl.	Mag. of Compon.
σ Orionis	5 31 43	− 2 41·0	4, 11, 8, 7, 8½, 9. 8.
ε′ Lyræ	18 39 42	+ 39 31·5	5, 6½, 5, 5½, 8, 13, 13.
β Capricorni	20 13 8	− 15 13·2	3½, 7.
β Equulei	21 15 56	+ 6 12·9	5½, 13, 16, 14.

CHAPTER III.

Variable Stars. — o Ceti. — Algol. — List of Variables. — Temporary Stars. — Notices of Stars which have disappeared. — Proper Motion. -- Motion of the System through Space. — Summary by W. Struve. — Proper Motion first suspected by Halley. — Wright's Hypothesis of a Central Sun. — Revived by Mädler. — Stars which are probably Centres of Systems.

THERE are many stars which exhibit periodic changes of brilliancy quite unaccounted for by any causes with which we are acquainted; these are termed *variable* stars. One of the most remarkable is *o* Ceti (*Mira*), which was first observed by David Fabricius in 1596.[1] It appears about 12 times in 11 years; remains at its greatest brightness about a fortnight, when it sometimes equals in brilliancy a star of the 2nd magnitude; decreases during about 3 months till it becomes totally invisible; it remains so for about 5 months, and then gradually recovers its brilliancy during the remaining 3 months of its period. Its maximum brightness is not always the same, nor does it always increase or diminish by the same gradations. According to Hevelius, nothing was seen of it between October, 1672, and December, 1676. Algol (β Persei) is another well-known instance of this kind. For about 2d. 13h. it shines as an ordinary star of the 2nd magnitude. In about 3½ hours it is reduced to the 4th magnitude, and thus remains about 20 minutes; it then rapidly increases to the 2nd, and continues so for another period of 2d. 23h., when similar changes recur. The exact period in which all these variations are performed is 2d. 20h. 48m. 55s. This remarkable law of variation seems to suggest that some opaque body revolves

---

[1] Kepler, *De Stellâ Novâ*, cap. xxiii. p. 115.

round the star, as surmised by Goodricke, who re-discovered
the variability of this object in 1782.[1]

The following are some of the more prominent periodic stars
visible to the naked eye[2] :—

Name.	Period.	Changes.		R.A.			Decl.	
	d.	From	to	h.	m.	s.	°	′
β Persei (*Algol*)	2·86	2·5	4	3	0	24	+ 40	24
δ Cephei	5·36	3·7	4·8	22	23	58	57	41
β Lyræ	12·91	3·5	4·5	18	44	55	33	12
η Aquilæ	7·18	3·6	4·4	19	45	9	0	38
α Herculis	67 ±	3·1	3·7	17	8	15	+ 17	33
o Ceti	331·3	2	12	2	12	17	− 3	37
υ Hydræ	440	4	10	13	22	4	22	33
η Argûs	Irr.	1	4	10	39	38	− 58	56

Somewhat similar in character to what we have just been
considering are the *temporary stars.* The first on record was
observed by Hipparchus, 125 B. C. ; the disappearance of which
is said to have led that astronomer to compile the well-known
star-catalogue bearing his name.[3] Brilliant stars appeared
between the constellations Cassiopeia and Cepheus, in the
years 949, 1264, and 1572. The last was a very remarkable
one, and we are indebted to Tycho Brahe for the observations
we possess of it: he was returning home on November 11,
when his attention was drawn by some peasants to a star,
which he was certain did not exist half an hour previously.
It was then equal in brightness to Sirius, and gradually increased till it became so bright as to be seen in the daytime ;
its brilliancy then diminished, and it at length disappeared in
March, 1674.[4] Passing over the stars of 1604[5] and 1670[6],

---

[1] See Goodricke's Memoir in *Phil. Trans.* vol. lxxiii. p. 474, *et seq.*
1783.

[2] A complete catalogue will be found at the end of the volume.

[3] According to Pliny.   [4] *Progymnasmata.*

[5] *De Stellâ Novâ.*

[6] *Phil. Trans.* vol. v. p. 2087, *et seq.* 1670; also vol. vi. p. 2197, *et
seq.* 1671.

we find that the last on record was that seen by Hind in
1848. It was in the constellation of Ophiuchus, of the 5th
magnitude, and easily seen by the naked eye on April 28,
and was observed as long as the season would permit.[1] It
has been conjectured, and with great plausibility, that the
stars of 945, 1264, and 1572, are identical, being apparitions
of a variable star, of long period.

Numerous instances are on record of stars formerly known,
being now nowhere to be found[2], and *vice versâ* of new stars
appearing, not formerly noticed. There were 4 in Hercules, 1 in
Cancer, 1 in Perseus, 1 in Pisces, 1 in Hydra, 1 in Orion, and
2 in Coma Ber., which have apparently disappeared. Several
stars in the catalogue of Ptolemy do not appear in that of
Ulugh-Beigh; 6 of these were near Piscis Australis, and as 4
were of the 3rd magnitude, Baily concludes that they were
visible in Ptolemy's time, but disappeared before the time of
Ulugh-Beigh. Some discrepancies have, no doubt, arisen from
mistaken entries, yet there are other instances in later times
quite certain. Thus 55 Herculis, mag. 5, was observed by
Sir W. Herschel in 1781 and 1782, but 9 years afterwards it
could not be found, and has not been seen since. In May
1828 Sir J. Herschel missed one of De Zach's stars in Virgo.
Montanari observed, in 1670, that " there are now wanting in
the heavens 2 stars of the 2nd magnitude, in the stern and yard
of the ship Argo. I and others observed them in the year
1664 upon the occasion of the comet that appeared that year.
When they first disappeared I know not; only I am sure
that on April 10, 1668, there was not the least glimpse of
them to be seen."[3]

It is probable that all the above stars belong to the temporary class.

To the naked eye the stars appear to preserve the same
positions relatively to one another from year to year; hence
they have been called *fixed stars ;* but, as we have already

---

[1] *Month. Not.* R.A.S., vol. viii. p. 146. See also vol. xxi. p. 232.
[2] Sir W. Herschel, *Phil. Trans.*, vol. lxxiii. p. 250–53. 1783.
[3] *Phil. Trans.*, vol. vi. p. 2202. 1671.

mentioned, this is not wholly true, inasmuch that delicate micrometrical observations show that many stars are endued with a *proper motion* of their own, through space. Motions which require whole centuries to accumulate before they produce changes of arrangements such as the naked eye can detect, are yet too trifling, as far as practical application go, to lead us to speak of them in common parlance as otherwise than fixed. Too little is yet known of their amount and directions, to permit of their being referred to any definite laws. Various attempts, commenced by Mayer, have been made to render the changes of the apparent positions of the stars compatible with some assumed motion of the Sun through space; Sir W. Herschel, in 1783[1], thought that the direction of the solar motion was to a point in the constellation Hercules, R.A., 17h. 5m.; Declination, +26° 17',—near the star λ.

This supposition has been confirmed in a remarkable manner by more recent observers. W. Struve, from a careful examination of the result of the researches of MM. Argelander, O. Struve, and Peters, has summed up as follows:—" *The motion of the solar system in space is directed to a point in the celestial sphere, situated on the right line, which joins the 2 stars of the 3rd magnitude, π aud μ Herculis, at ¼ of the apparent distance between these stars measured from π Herculis. The velocity of the motion is such that the Sun, with the whole cortége of bodies depending on him, advances annually in the direction indicated, through a space equal to* 1·623 *radii of the terrestrial orbit, or* 154,000,000 *of miles.*"[2]

Amongst those stars whose annual proper motion is considerable, may be mentioned :—

Name.			Proper Motion.
			"
2151 Argûs	.	.	7·871
ε Indi	.	.	7·740
1830 Groombridge	.	.	6·974
61 Cygni	.	.	5·123

---

[1] *Phil. Trans.*, vol. lxxiii. p. 247, *et seq.* 1783.
[2] *Etudes d'Astronomie Stell.*, p. 108.

Proper motion appears to have been first suspected by
Halley, in 1718[1], in the case of the stars Aldebaran, Sirius,
and Arcturus; shortly afterwards the idea was supported by
telescopic observation.   It was shown that the latitude of
Arcturus had undergone a change of 5 in 152 years, although
η Boötis, in the same neighbourhood, had not exhibited
any perceptible displacement.   Halley almost exclusively
devoted his attention to variations in latitude; but J. Cassini
ascertained the existence of similar variations in longitude[2];
and it was afterwards remarked by Fontenelle that "There
is a star in the Eagle (α) which, if all things continue in
their present course, will have to the west of it, after the
lapse of a great number of ages, another star which at present
appears to the east."

The consideration of this subject naturally leads us to say
a few words on the Central Sun Hypothesis, first started by
Wright in 1750, and actively revived within the last few
years by Madler, of Dorpat.   This theory simply supposes
the existence of some central point around which the Sun,
with its vast attendant *cortége* of planets and comets, revolves
in the course of millions of years.   Mädler thinks he has
sufficient ground for believing that this point is situated in or
near the Pleiades, or more exactly at the star Alcyone (η
Tauri).   A distinguished living writer very sensibly remarks,
"It is manifest that all such speculations are far in advance
of practical astronomy, and therefore they must be regarded
as premature, however probable the supposition on which
they are based, or however skilfully they may be connected
with the actual observation of astronomers."[3]   Vague ideas
of the motion of the solar system around some common centre
are to be found in Lucretius[4]; it was thought that but for
such motion all celestial objects must have collapsed and
formed a chaos.

There are some stars which seem to be in a great measure
out of the reach of the attractive force of other stars, whence

---

[1] *Phil. Trans.*, vol. xxx. pp. 736–8.   1718.

[2] *Mém. Acad. des Sciences*, 1738, p. 337, *et seq.*

[3] Grant, *Hist. Phys. Ast.*, p. 558.   [4] *De Rer. Nat.*, lib. i.

Sir W. Herschel is disposed to consider that they are probably centres of extensive systems like our own. Among them, with probably many others, are :—

Vega (α Lyræ).	Bellatrix (γ Orionis).
Capella (α Aurigæ).	Menkab (α Ceti).
Arcturus (α Boötis).	Schedir (α Cassiopeiæ).
Sirius (α Canis Maj.).	Algorab (δ Corvi).
Canopus (α Argûs).	Propus (ı Gem.)
Markab (α Pegasi).	

CHAPTER IV.

CLUSTERS AND NEBULÆ.

Arranged in Three Classes. — Five Kinds of Nebulæ. — The Pleiades. — The Hyades. — Mentioned by Homer. — Præsepe. — Opinion of Aratus and Theophrastus. — Coma Berenicis. — List of Clusters. — Annular Nebulæ. — Elliptic Nebulæ. — Spiral Nebulæ. — Planetary Nebulæ. — Nebulous Stars. — List of Irregular Clusters. — Notes to the Objects in the List. — The Nubeculæ Major and Minor. — List of Nebulæ in Herschel's Catalogue.

On examining the heavens on a clear evening, when the Moon is not shining, we find here and there groups of stars which seem to be compressed together in such a manner as to present a dull cloud-like appearance : these are termed *clusters* and *nebulæ*, and are usually classed as follows : —

1. Irregular groups, visible more or less to the naked eye.
2. Clusters resolvable into separate stars with the aid of a telescope.
3. Nebulæ, for the most part irresolvable.

The latter are subdivided into : —

      i.   Annular Nebulæ.
      ii. Elliptic Nebulæ.
      iii. Spiral Nebulæ.
      iv. Planetary Nebulæ.
      v. Nebulous Stars.

Of the 1st class, there are several examples to be found, with all of which the reader is probably more or less familiar. The cluster of the *Pleiades* in Taurus is probably best known.[1] When examined *directly*, few persons can see more than 6 stars, but by turning the eye *sideways*, we discover there are

---

[1] The Pleiades and Hyades are among the few stars mentioned by Homer (*Odyssey*, lib. v. ver. 270).

Fig. 112.

Plate XXX.

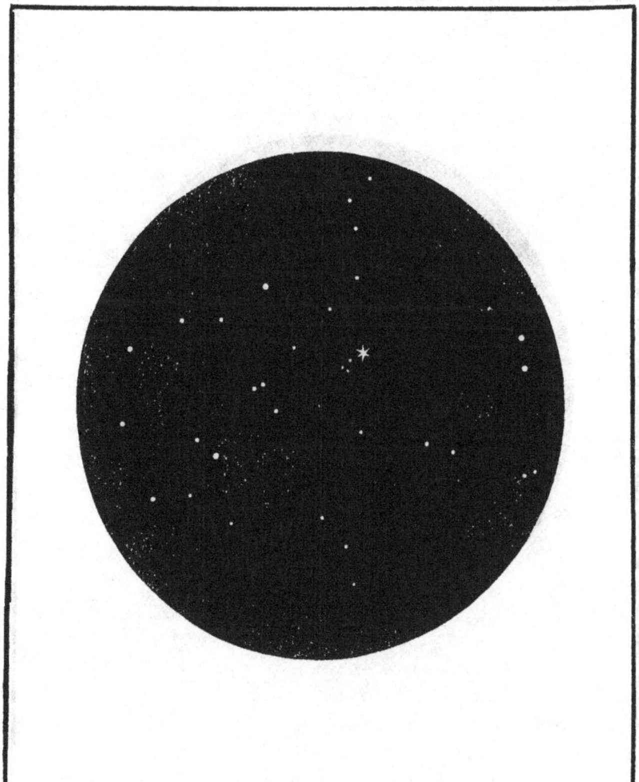

THE PLEIADES, IN TAURUS.

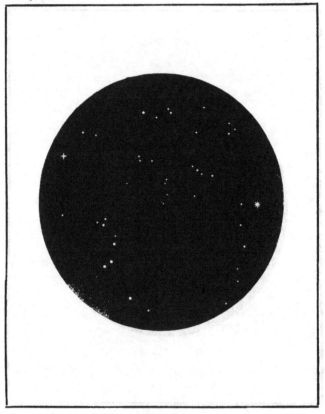

THE HYADES, IN TAURUS.

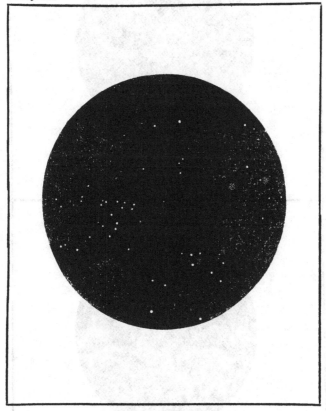

PRÆSEPE, IN CANCER.

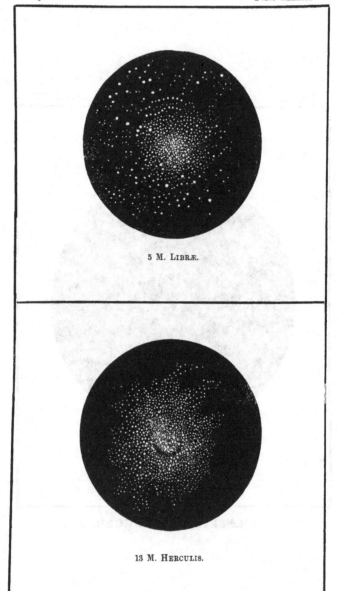

5 M. LIBRÆ.

13 M. HERCULIS.

CLUSTERS.

3 M. Canum Venaticorum.

11 M. Antinoï.

CLUSTERS.

Figs. 119, 120.   Plate XXXV.

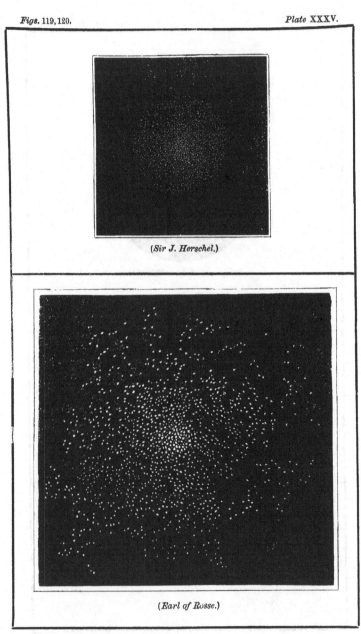

(Sir J. Herschel.)

(Earl of Rosse.)

THE CLUSTER, 2 M. AQUARII.

many more, and between 50 and 60 are visible in a telescope.[1]
The most brilliant star in the group is *Alcyone*, or η Tauri, of
the 3rd magnitude; next in order come *Electra* and *Atlas* of
the 4th; *Maia* and *Taygeta* of the 5th; *Pleione* and *Celeno*,
which are between the 6th and 7th; *Asterope* between the
7th and 8th; and finally, a great number of smaller stars.
The *Hyades* are another group in Taurus, near Aldebaran, of
a somewhat similar character to the cluster near λ Orionis.
Præsepe, or the "Beehive," in Cancer, is one of the finest
objects of this kind for a small telescope; it is an aggregation
of little stars, which has long borne the name of a nebula, its
components not being visible to the naked eye; indeed, before
the invention of the telescope, it must have been the only
recognised one. Aratus[2] and Theophrastus[3] tell us that its
dimness and disappearance during the progressive condensation
of the atmosphere were regarded as the first signs of approaching rain. *Coma Berenicis* has fewer stars, but they are of a
larger size, and more diffused.

From the 2nd class we have selected the following as the
most interesting : —

	Name.	R.A.			Decl.	
		h.	m.	s.	°	′
1.	33 H. VI. Persei	2	9	20	+ 56	28
2	35 M. Geminorum	6	0	33	+ 24	15
3.	3 ,, Canum Venat.	13	35	40	+ 29	4
4.	5 ,, Libræ	15	11	26	+ 2	45
5.	13 ,, Herculis	16	36	40	+ 36	43
6.	92 ,, Herculis	17	12	51	+ 43	17
7.	11 ,, Antinoï	18	43	36	— 6	26
8.	15 ,, Pegasi	21	23	11	+ 11	32
9.	2 ,, Aquarii	21	26	12	— 1	27
10.	52 ,, Cephei	23	18	2	+ 60	49
11.	30 H. VI. Cassiopeiæ	23	50	6	+ 55	56

---

[1] The following are some of the different estimations :—

Kepler .	.	.	32	Hook .	.	.	78
La Hire	.	.	64	De Rheita .	.		118

[2] *Diosemeia*, verse 160. See Lamb's Translation, p. 70, where the
passage is very prettily rendered into English verse.

[3] *De Signis Pluviarum*, p. 419. Heinsius's Ed. Lugd. Batavor.

We now pass on to another order of stars, which present themselves much less clearly to our eyes than the brilliant clusters enumerated above; we mean the *nebulæ* properly so called. Some of them are resolvable in large telescopes, but the greater number defy the utmost efforts made to separate them into their component stars. They are usually faint misty objects, many of them not unlike comets or specks of fog. We have seen that they may be subdivided into 5 classes, which we shall now proceed to consider.

Of *annular* nebulæ the heavens afford only 4 examples; but the most remarkable instance occurs in Lyræ, R.A. 18h. 48m. 21s.; Decl., + 32° 51′ (Messier's 57th). It is situated about midway between the stars β and γ, and may be seen with a telescope of moderate power. Sir J. Herschel, in his description of it, says, "It is small, and particularly well defined, so as, in fact, to have much more the appearance of a flat, oval, solid ring than of a nebula. The axes of the ellipse are to each other in the proportion of about 4 to 5. Its light is not quite uniform, but has a somewhat curdled appearance, particularly at the exterior edge; the central opening is not entirely dark, but is filled up with a faint hazy light, uniformly spread over it, like a fine gauze stretched over a hoop." [1]

Elliptic nebulæ, of greater or less eccentricity, are not uncommon; the well-known "Great nebula in Andromeda," R.A. 0h. 35m. 9s.; Decl. + 40° 30′ (Messier's 31st.) is an object of this kind. Several elliptic nebulæ are remarkable as having double stars at or near each of their foci; an instance occurs in the constellation Sagittarius, R.A. 18h. 8m. 48s.; Decl. —19° 55′. Other elliptic nebulæ may be found as follows: —

		h.	R.A. m.	s.		Decl. °	′
1.	...	0	39	12	...	— 26	35
2.	...	13	47		...	— 39	9
3.	...	18	25		...	+ 64	53

The discovery of *spiral*, or *whirlpool* nebulæ, is due to the Earl of Rosse. The best known is in the constellation Canes

[1] *Treat. on Ast.*

(Sir J. Herschel.)   *(Earl of Rosse.)*

THE ANNULAR NEBULA, 57 M. LYRÆ.

(Sir J. Herschel.)   *(Earl of Rosse.)*

THE PLANETARY NEBULA, 97 M. URSÆ MAJORIS.

PLANETARY NEBULA IN VIRGO.   NEBULOUS STAR
R.A. 13h. 30m.; Decl.—17° 10′.   ORIONIS.
(Sir J. Herschel.)   *(Earl of Rosse.)*

VARIOUS NEBULÆ.

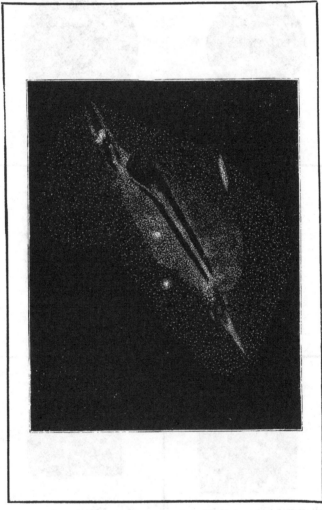

THE GREAT NEBULA IN ANDROMEDA.
DRAWN BY G. P. BOND.

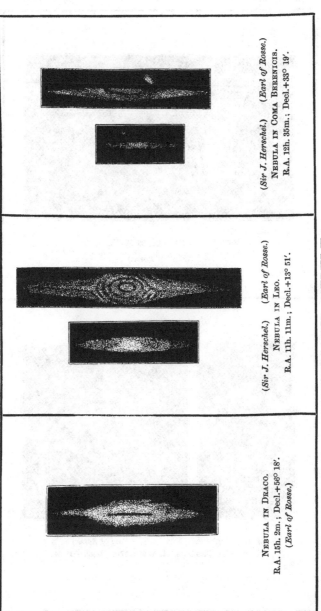

Plate XXXVIII.

Figs. 128—132.

NEBULA IN DRACO.
R.A. 15h. 2m.; Decl.+56° 18'.
(*Earl of Rosse.*)

(*Sir J. Herschel.*)   (*Earl of Rosse.*)
NEBULA IN LEO.
R.A. 11h. 11m.; Decl.+13° 51'.

(*Sir J. Herschel.*)   (*Earl of Rosse.*)
NEBULA IN COMA BERENICIS.
R.A. 12h. 35m.; Decl.+38° 19'.

ELONGATED NEBULÆ.

THE SPIRAL NEBULA, 99 M. VIRGINIS.

(*Earl of Rosse.*)

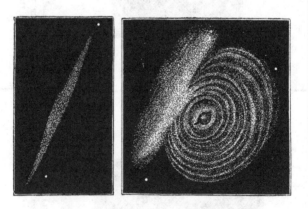

(*Sir J. Herschel.*)   (*Earl of Rosse.*)

NEBULA IN PEGASUS.   R.A. 22h. 57m.; Decl.+11° 34′.

SPIRAL NEBULÆ.

Figs. 136, 137.   Plate XL.

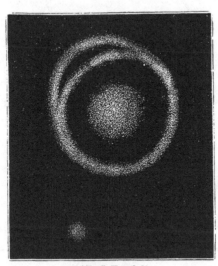

(*Sir J. Herschel.*)

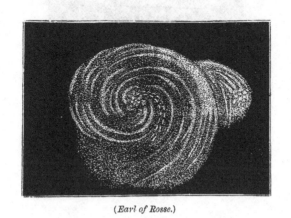

(*Earl of Rosse.*)

**THE SPIRAL NEBULA,
51 M. CANUM VENATICORUM.**

(*Sir J. Herschel.*)

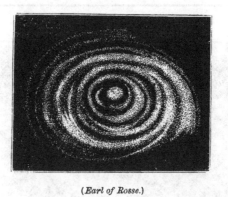

(*Earl of Rosse.*)

THE SPIRAL NEBULA IN LEO.
R.A. 9h. 24m.; Decl.+22° 7′.

Venatici, R.A. 13h. 23m. 56s.; Decl. + 47° 54′ (Messier's 51st.) To Sir J. Herschel "it presented the appearance of a large and brilliant globular cluster surrounded by a ring, at a considerable distance, in which inequalities of brightness were remarked;" but Lord Rosse's telescope entirely changed the aspect of this object into a magnificent spiral with unequal folds, the extremities of which are terminated by thick granular and rounded knots. There is, also, another fine spiral nebula, in Virgo, R.A., 12h. 10m. 30s.; Decl. + 15° 17′ (Messier's 99th).

Planetary nebulæ received their name from Sir W. Herschel, on account of their resembling in form the planets of our system. They are either circular or slightly elliptical; some have well-defined outlines, in others the edges appear hazy; they are uniformly bright all over without any trace of a nucleus. A good example is to be found near β Ursæ Majoris, R.A. 11h. 6m. 34s.; Decl. + 55° 46′ (Messier's 97th). It was discovered by Méchain in 1781, and is described as "a very singular object, circular and uniform, and after a long inspection looks like a condensed mass of attenuated light, seemingly of the size of Jupiter."

Other planetary nebulæ may be found as follows : —

		R.A.			Decl.	
		h.	m.	s.	°	′
1.	...	7	35	39	... — 17	53
2.	...	10	17	58	... — 17	56

Nebulous stars are so called from their being surrounded by a faint nebulosity, usually of a circular form, and sometimes several minutes in diameter. Hind remarks that the nebulosity, in some cases, is well defined; but, in others, quite the reverse. He says, moreover, "the stars thus attended have nothing in their appearance to distinguish them from others entirely destitute of such appendages; nor does the nebulous matter, in which they are situated, offer the slightest indications of resolvability into stars, with any telescopes hitherto constructed." The following stars are instances of this kind : —

	Name.	R.A.	Decl.
		h. m. s.	o
1.	55 Andromedæ	1 44 54	+ 40 2·2
2.	69 ♄. IV. Tauri	4 0 20	+ 30 23·8
3.	ι Orionis	5 28 35	− 6 0·3
4.	ε Orionis	5 29 6	− 1 17·7
5.	45 ♄. IV. Geminorum	7 20 54	+ 21 11·6
6.	8 Canum Venat.	12 27 5	+ 42 7·1

Besides the clusters and nebulæ belonging to the foregoing classes, there are others, for the most part of irregular
form and large dimensions, which it is desirable to class by
themselves. We may here include : —

	Name.	R.A.	Decl.	
		h. m. s.	o ′	
1.	47 Toucani	0 17	− 72 51	
2.	1 M. Tauri	5 26 3	+ 21 55·2	
3.	θ Orionis	5 28	− 5 29	
4.	30 Doradûs	5 39	− 69 8	
5.	η Argûs	10 39	− 58 56	
6.	ω Centauri	13 18	− 46 44	
7.	41 ♄. IV. Sagittarii	17 52	− 23 1	(1830)
8.	17 M. Clypei Sobieskii	18 12 31	− 16 15·5	
9.	27 M. Vulpeculæ	19 53 31	+ 22 20·2	
10.	14 ♄. V. Cygni	20 49	+ 31 23	(1830)

1. This is described by Sir J. Herschel as "a superb
globular cluster, immediately preceding the centre of the
Nubecula minor ; it is very visible to the naked eye, and one
of the finest objects of this kind in the heavens, and consists
of a very condensed mass of stars, of a pale rose colour, con-
centrically enclosed in a much less condensed globe of white
ones, 15′ or 20′ in diameter."

2. Frequently called the " Crab nebula in Taurus." Has
an elliptic outline in most instruments, but in Lord Rosse's
reflector " it is transformed into a closely-crowded cluster,

_____

[1] *Res. Ast. Obs.*, p. 18.

Fig. 140.

Plate XLII.

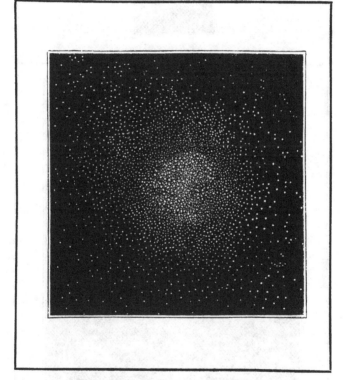

THE CLUSTER, 47 TOUCANI.
DRAWN BY SIR J. HERSCHEL.

(*Sir J. Herschel.*)

(*Earl of Rosse.*)

THE CRAB NEBULA, IN TAURUS.

Fig. 143.

Plate XLIV.

THE GREAT NEBULA IN ORION.
DRAWN BY W. C. BOND.

Plate XLV.

Fig. 144.

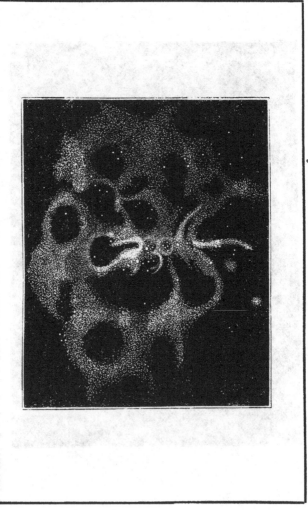

THE NEBULA, 30 DORADÛS.

DRAWN BY SIR J. HERSCHEL.

Plate XLVI.

Fig. 145.

THE NEBULA SURROUNDING η ARGÛS.

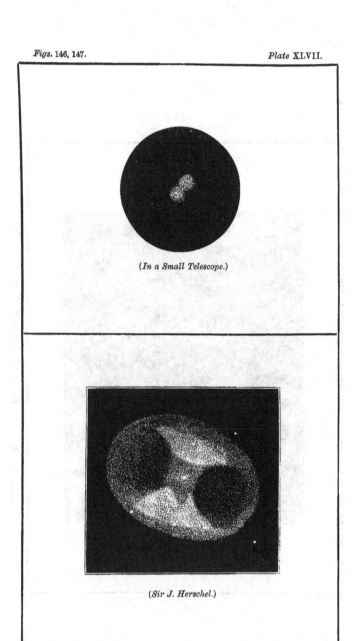

(*In a Small Telescope.*)

(*Sir J. Herschel.*)

THE DUMB BELL NEBULA IN VULPECULA.

with branches streaming off from the oval boundary, like
claws, so as to give it an appearance that, in a measure
justifies the name by which it is distinguished."

3. Is the " great nebula in the sword handle of Orion," surrounding the multiple star θ in that constellation. It was
discovered by G. Huyghens, about the middle of the 17th
century. " In its more prominent details may be traced some
slight resemblance to the wing of a bird. In the brightest
portion are four conspicuous stars forming a trapezium. The
nebulosity in the immediate vicinity of these stars is flocculent,
and of a greenish-white tinge; about half a degree northward
of the trapezium are 2 stars involved in a bright branching
nebula of singular form, and southward is the star ι Orionis,
also situated in a nebula. Careful examination with powerful
telescopes has traced out a continuity of nebulous light
between the great nebula and both these objects, and there
can be but little doubt that the nebulous region extends
northwards as far as ε in the belt of Orion, which is involved
in a strong nebulosity, as well as several smaller stars in the
immediate neighbourhood." [1]

4. Is a singular nebula, faintly visible to the naked eye,
situated within the limits of the *Nubecula major;* it was
noticed by La Caille as resembling the nucleus of a comet.

5. Is a very large nebula surrounding the star η *Argûs,*
and occupying a space equal to about 5 times the area of the
Moon. Sir J. Herschel, who carefully examined this object
when at the Cape of Good Hope in 1833, says that, " viewed
with an 18-inch reflector, no part of this nebula shows any
sign of resolution into stars, nor in the brightest and most
condensed portion, adjacent to the singular oval vacancy in
the middle of the figure, is there any of that curdled appearance, or that tendency to break up into bright spots, with
intervening darker portions, which characterise the nebula of
Orion, and indicate its resolvability. It is not easy for
language to convey a full impression of the beauty and

[1] See Struve, *Month. Not.* R.A.S., vol. xvii. pp. 225-30. Bond, *Mem.
Amer. Acad.*, vol. iii. N.S. p. 87. J. Herschel, *Res. Ast. Obs.*, pp. 25-32.

sublimity of the spectacle which the nebula offers as it enters
the field of the telescope, ushered in as it is by so glorious and
innumerable a procession of stars, to which it forms a sort of
climax." [1]

6. Is visible to the naked eye, and presents the appearance
of a comet without a tail; its brilliancy is about equal to a
star of the 4½ magnitude, but, "viewed in a telescope, it is
like a globe fully 20' in diameter, very gradually increasing
in brightness to the centre, and composed of innumerable stars
of the 13th and 15th magnitude." [2]

8. Usually termed the "Horse-shoe nebula," from its
resemblance to the shoe of a horse, or a capital Greek
omega (Ω). [3]

9. Is a curious object, near the 5th magnitude star 14 *Vulpe-
culæ*; it is shaped like a double-headed shot, or dumb-bell,
and is generally known as the "Dumb-bell nebula." In a
small telescope it appears like two roundish nebulosities,
nearly in contact; in a very powerful telescope, however, "it
has an elliptical outline of faint light enclosing the two chief
masses."

In the southern hemisphere, and not far from the pole, are
the Magellanic clouds, or *nubeculæ major* and *minor*, so called
from their cloud-like appearance. The former is situated in
the constellation Dorado, and the latter in Toucan. They are
of a somewhat oval shape, and both visible to the naked eye
when the Moon is not shining; but the smaller disappears in
strong moonlight. Sir John Herschel, when at the Cape,
examined these remarkable objects with his large telescope,
and describes them as consisting of swarms of stars, clusters,
and nebulæ of every description. The larger one covers an
area of about 42 square degrees, and the smaller of 10 square
degrees. [4]

The nebulæ are not very uniformly distributed in the
heavens, but congregate especially to a zone, crossing at right
angles the Milky Way. They are exceedingly abundant in

[1] *Res. Ast. Obs.*, pp. 32–47.   [2] Ibid. p. 21.
[3] Ibid. p. 7.   [4] Ibid. pp. 143–164.

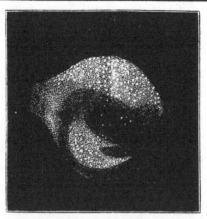

(*Earl of Rosse :* 3-Foot Reflector.)

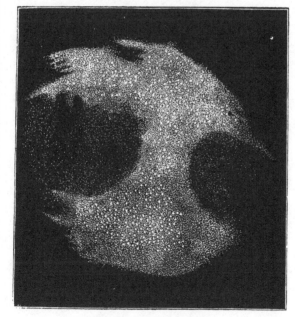

(*Earl of Rosse :* 6-Foot Reflector.)

THE DUMB BELL NEBULA IN VULPECULA.

the constellation Virgo.   Sir J. Herschel's catalogue of 1833 [1]
contains 2306 of these objects, which are thus distributed
through the different hours of R. A.

I.	Hour	89	Neb.	IX.	Hour	72	Neb.	XVII.	Hour	32	Neb.
II.	„	109	„	X.	„	110	„	XVIII.	„	18	„
III.	„	89	„	XI.	„	153	„	XIX.	„	34	„
IV.	„	24	„	XII.	„	271	„	XX.	„	37	„
V.	„	36	„	XIII.	„	441	„	XXI.	„	36	„
VI.	„	32	„	XIV.	„	214	„	XXII.	„	45'	„
VII.	„	56	„	XV.	„	153	„	XXIII.	„	60	„
VIII.	„	55	„	XVI.	„	42	„	XXIV.	„	98	„

[1] *Phil. Trans.*, vol. cxxiii. pp. 359-505. 1833.

CHAPTER V.

THE MILKY WAY.

Its Course amongst the Stars described by Sir J. Herschel. — The "Coal Sack" in the Southern Hemisphere. — Remarks by Sir W. Herschel as to the prodigious Number of Stars in the Milky Way. — Computation by Sir J. Herschel of the Total Number of Stars visible in an 18-inch Reflector. — Terms applied to the Milky Way by the Greeks. — By the Romans. — By our early Ancestors.

FOREMOST amongst the great clusters with which we are acquainted stands the Milky Way, which has pre-eminently occupied the attention of philosophers from the earliest ages of antiquity.

The course of the Milky Way amongst the constellations is thus sketched by Sir J. Herschel, whose description we shall quote with a few verbal alterations.[1]

Neglecting occasional deviations, and following the line of its greatest brightness, as well as its varying breadth and intensity will permit, its course conforms nearly to that of a great circle inclined at an angle of about 63° to the equinoctial, and cutting that circle in R.A. 0h. 47m., and 12h. 47m.; so that its northern and southern poles respectively are situated in R.A. 12h. 47m., Decl. N. 27° and R.A. 0h. 47m., Decl. S. 27°. Throughout the region where it is so remarkably subdivided, this great circle holds an intermediate situation between the 2 great streams; with a nearer approximation, however, to the brighter and continuous stream, than to the fainter and interrupted one. If we trace its course in order of right ascension, we find it traversing the constellation Cassiopeia, its brighter part passing about 2° north of the star δ of that

---

[1] *Outlines of Ast.*, p. 569.

constellation, *i. e.* in about 62° of north declination. Passing
thence between γ and ε Cassiopeiæ, it sends off a branch to
the south preceding side, towards α Persei, very conspicuous
as far as that star, prolonged faintly towards ε of the same
constellation, and possibly traceable towards the Hyades and
Pleiades as remote outliers. The main stream, however,
(which is here very faint) passes on through Auriga, over the
3 remarkable stars, ε, ζ, η, of that constellation, preceding
Capella (α Aurigæ), called the Hædi, preceding Capella, between the feet of Gemini and the horns of the Bull (where it
intersects the ecliptic, nearly in the solstitial colure), and thence
over the Club of Orion to the neck of Monoceros, intersecting
the equinoctial in R.A. 6h. 54m. Up to this point, from the
offset in Perseus, its light is feeble and indefinite, but thence-
forward it receives a gradual accession of brightness, and when
it passes through the shoulder of Monoceros, and over the
head of Canis Major, it presents a broad, moderately bright,
very uniform, and to the naked eye, slender stream up to the
point where it enters the prow of the ship Argo, nearly in the
southern tropic. Here it again subdivides (about the star *m*
Puppis), sending off a narrow and winding branch on the preceding side as far as γ Argûs, where it terminates abruptly.
The main stream pursues its southward course to the 33rd
parallel of south declination, where it diffuses itself broadly
and again subdivides, opening out into a wide fan-like ex-
panse, nearly 20° in breadth, formed of interlacing branches,
all of which terminate abruptly, in a line drawn nearly through
λ and γ Argûs.

At this place the continuity of the Milky Way is interrupted
by a wide gap, and when it recommences on the opposite side
it is by a somewhat similar fan-shaped assemblage of branches
which converge upon the bright star η Argûs. Thence it
crosses the hind feet of the Centaur, forming a curious and
sharply defined semi-circular concavity of small radius, and
enters the Cross by a very bright neck of isthmus, of not more
than 3° or 4° in breadth, being the narrowest portion of the
Milky Way. After this it immediately expands into a broad
and bright mass, enclosing the stars α and β Crucis, and β

Centauri, and extending almost up to α of the latter constellation. In the midst of this bright mass, surrounded by it on
all sides, and occupying about half its breadth, occurs a singular dark pear-shaped vacancy, so conspicuous and remarkable as to attract the notice of the most superficial gazer, and
to have acquired, amongst the early southern navigators, the
uncouth but expressive appellation of the *Coal Sack*. In this
vacancy, which is about 8° in length, and 5° in breadth, only
one very small star visible to the naked eye occurs, though it
is far from devoid of telescopic stars, so that its striking blackness is simply due to the effect of contrast with the brilliant
ground with which it is on all sides surrounded. This is the
place of the nearest approach of the Milky Way to the South
Pole. Throughout all this region its brightness is very striking,
and when compared with that of its more northern course,
already traced, conveys strongly the impression of greater
proximity, and would almost lead to a belief that our situation
as spectators is separated on all sides by a considerable interval from the dense body of stars composing the Galaxy, which
in this view of the subject would come to be considered as a
flat ring of immense and irregular breadth and thickness,
within which we are eccentrically situated, nearer to the
southern than to the northern part of its circuit.

At α Centauri, the Milky Way again subdivides, sending off
a great branch of nearly half its breadth, but which thins
off rapidly at an angle of about 20° with its general direction
towards the preceding side to η and d Lupi, beyond which it
loses itself in a narrow and faint streamlet. The main stream
passes on, increasing in breadth to γ Normæ, where it makes
an abrupt elbow, and again subdivides into one principal and
continuous stream of very irregular breadth, and brightens on
the following side, and a complicated system of interlaced
streaks and masses on the precedings, which cover the tail of
Scorpio, and terminates in a vast and faint effusion over the
whole extensive region occupied by the preceding leg of
Ophiuchus, extending northwards to a parallel of 13° of south
declination, beyond which it cannot be traced; a wide interval
of 14° free from all appearance of nebulous light, separating it

from the great branch on the north side of the equinoctial, of which it is usually represented as a continuation.

Returning to the point of separation of this great branch from the main stream, let us now pursue the course of the latter. Making an abrupt bend to the following side, it passes over the stars ι Aræ, θ and ι Scorpii, and γ Tubi to γ Sagittarii, when it suddenly collects into a vivid oval mass about 6° in length and 4° in breadth, so excessively rich in stars that a very moderate calculation makes their number exceed 100,000. Northward of this mass, this stream crosses the ecliptic in longitude about 276°, and proceeding along the bow of Sagittarius into Antinoüs has its course supplied by 3 deep concavities, separated from each other by remarkable protuberances, of which the larger and brighter (situated between Flamsteed's stars 3 and 6 Aquilæ) forms the most conspicuous path in the southern portion of the Milky Way, visible in our latitudes.

Crossing the equinoctial at the 19th hour of Right Ascension, it next runs in an irregular, patchy, and winding stream through Aquila, Sagitta, and Vulpecula up to Cygnus; at ε of which constellation its continuity is interrupted, and a very confused and irregular region commences, marked by a broad dark vacuity, not unlike the southern Coal Sack, occupying the space between ε, α, and γ Cygni, which serves as a kind of centre for the divergence of 3 great streams; 1 which we have already traced; a 2nd, the continuation of the 1st (across the interval) from α northward, between Lacerta and the head of Cepheus to the point in Cassiopeia whence we set out; and a 3rd branches off from γ Cygni, very vivid and conspicuous, running off in a southern direction through β Cygni, and δ Aquilæ almost to the equinoctial, when it loses itself in a region thinly sprinkled with stars, where in some maps the modern constellation Taurus Poniatowskii is placed.

This is the branch which, if continued across the equinoctial, might be disposed to unite with the great southern effusion in Ophiuchus, already noticed. A considerable offset, or protuberant appendage, is also thrown off by the northern stream from the head of Cepheus directly towards the pole, occupying the greater part of the quartile formed by α, β, ι, and δ of that constellation.

It is impossible to give any idea of the enormous number of stars in the Milky Way, but Sir W. Herschel has recorded some facts that will assist us. That observer states that on one occasion he estimated that 116,000 stars passed through the field of his telescope in ¼ hour[1]; and again that on Aug. 22, 1792, he saw 258,000 stars pass in 41m.[2] The surprising character of this result will be more adequately appreciated when compared with the number of stars that are visible to the naked eye. The common estimation gives between 3000 and 4000, though Struve augments the number to 6000 for persons of very acute vision.[3]

Sir John Herschel has computed that the total number of stars visible in an 18-inch reflector cannot be less than $5\frac{1}{4}$ millions, and probably many more.[4]

Fig. 25.

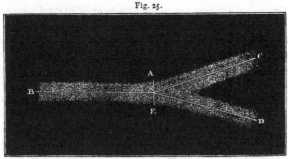

Herschel's Stratum Theory.

A brief reference must here be made to what is commonly known as Sir W. Herschel's theory of the Milky Way. He conjectured that the stars were not indifferently scattered though the heavens, but were rather arranged in a certain

---

[1] *Phil. Trans.*, vol. lxxv. p. 244.   1785.

[2] Ibid. vol. lxxxv. p. 70.   1795.

[3] *Etudes d'Astronomie Stellaire*, p. 61.

[4] *Res. of Ast. Obs. &c.*, p. 381. For more on this subject, see *Outlines of Ast.*

definite stratum, the thickness of which, as compared with its length and breadth, was inconsiderable; and that the Sun occupies a place somewhere about the middle of its thickness, and near the point where it subdivides into 2 principal streams, inclined to each other at a small angle. It is clear then, that to an eye viewing the stratum from S, the apparent density of the stars would be least in the direction S A, or S E, and greatest in the direction S B, S C, S D, and this corresponds generally to the observed facts.[1] " Such is the view of the construction of the starry firmament taken by Sir William Herschel[2], whose powerful telescope has effected a complete analysis of the wonderful zone, and demonstrated the fact of its entirely consisting of stars."

By the Greeks the Milky Way was termed the Γαλαξίας or κύκλος γαλακτικός, and by the Romans the *Circulus lacteus* or *Orbis lacteus;* from our ancestors it received the name of *Jacob's Ladder, the Way to St. James's, Watling Street,* &c. The diversity of the ancient names was equalled only by the diversity of opinions that prevailed as to what it was. Meliodorus considered that it was the original course of the Sun, but that it was abandoned by him after the bloody banquet of Thyestes; others thought that it pointed out the place of Phaëthon's accident; whilst a 3rd class thought it was caused by the ears of corn dropped by Isis in her flight from Typhon. Aristotle considered it was caused by gaseous exhalations from the Earth, which were set on fire in the sky. Theophrastus declared it to be no other than the soldering together of 2 hemispheres; and finally Diodorus conceived it to be a dense celestial fire, showing itself through the clefts of the starting and dividing semi-globes.

The speculations of Democritus[3] and Pythagoras were to the effect, that the Galaxy was nothing more or less than a vast assemblage of stars. Ovid speaks of it as a high road " whose ground-work is of stars." Manilius uses similar

---

[1] Hind in *Atlas of Astronomy.*

[2] Thomas Wright of Durham (see his *Theory of the Universe,* London, 1751,) first started this idea in 1734.

[3] Plutarch, *De Placit.,* lib. iii. cap. 1.

language.[1] It is singular that Ptolemy has in none of his writings expressed any opinion on it. Our own ancestors supported the star theory.

In Milton we find mention of that —

> "broad and ample road,
> Whose dust is gold, and pavement, stars."

---

[1] *De Sphæro*, lib. i. cap. 9.

CHAPTER VI.

THE CONSTELLATIONS.

List of those formed by Ptolemy.—Subsequent Additions.—Remarks by Herschel, &c.—Catalogue of the Constellations with the Position of, and Stars in, each.

WE have already referred to the constellations; in this chapter we shall catalogue them.

Ptolemy enumerates 48 constellations: 21 northern, 12 zodiacal, and 15 southern ones, as follows:—

Northern.

1. Ursa Minor.	The Little Bear.
2. Ursa Major.	The Great Bear.
3. Draco.	The Dragon.
4. Cepheus.	
5. Böotes or *Arctophylax.*	The Bear Keeper.
6. Corona Borealis.	The Northern Crown.
7. Hercules, *Engonasin.*	Hercules kneeling.
8. Lyra.	The Harp.
9. Cygnus, *Gallina.*	The Swan.
10. Cassiopeia.	The Lady in her Chair.
11. Perseus.	
12. Auriga.	The Charioteer.
13. Serpentarius.	The Serpent Bearer.
14. Serpens.	The Serpent.
15. Sagitta.	The Arrow.
16. Aquila, *Vultur volans.*	The Eagle.
17. Delphinus.	The Dolphin.
18. Equuleus.	The Little Horse.
19. Pegasus, *Equus.*	The Winged Horse.
20. Andromeda.	The Chained Lady.
21. Triangulum.	The Triangle.

Zodiacal.

1. Aries.	The Ram.
2. Taurus.	The Bull.
3. Gemini.	The Twins.
4. Cancer.	The Crab.
5. Leo.	The Lion.
6. Virgo.	The Virgin.
7. Libra, *Chelæ.*	The Balance.
8. Scorpio.	The Scorpion.
9. Sagittarius.	The Archer.
10. Capricornus.	The Goat.
11. Aquarius.	The Water Bearer.
12. Pisces.	The Fishes.

Southern.

1. Cetus.	The Whale.
2. Orion.	
3. Eridanus, *Fluvius.*	Eridanus, The River.
4. Lepus.	The Hare.
5. Canis Major.	The Great Dog.
6. Canis Minor.	The Little Dog.
7. Argo Navis.	The Ship " Argo."
8. Hydra.	The Snake.
9. Crater.	The Cup.
10. Corvus.	The Crow.
11. Centaurus.	The Centaur.
12. Lupus.	The Wolf.
13. Ara.	The Altar.
14. Corona Australis.	The Southern Crown.
15. Piscis Australis.	The Southern Fish.

Tycho Brahe, about 1603, added —

1. Coma Berenicis.	The Hair of Berenice.
2. Antinoüs.	

(Both Northern Constellations.)

Bayer, in 1604, added —

1. Pavo.	The Peacock.
2. Toucan.	The American Goose.
3. Grus.	The Crane.
4. Phœnix.	The Phœnix.

5. Dorado, *Xiphias.*   The Sword Fish.
6. Piscis Volans.   The Flying Fish.
7. Hydrus.   The Water Snake.
8. Chamæleon.   The Chameleon.
9. Apis.   The Bee.
10. Avis Indica.   The Bird of Paradise.
11. Triangulum Australe.   The Southern Triangle.
12. Indus.   The Indian.

(All Southern.)

Hevelius, in 1690, added —

1. Camelopardalis.   The Cameleopard.
2. Canes Venatici, *Asterion et Chara.*   The Hunting Dogs.
3. Vulpecula et Anser.   The Fox and the Goose.
4. Lacerta.   The Lizard.
5. Leo Minor.   The Little Lion.
6. Lynx.   The Lynx.
7. Clypeus, or Scutum, Sobieskii.   The Shield of Sobieski.
8. Triangulum Minor.   The Little Triangle.
9. Cerberus.   The Cerberus.

(All Northern: and)

10. Monoceros.   The Unicorn.
11. Sextans Uraniæ.   The Sextant of Urania.

(Southern Constellations.)

Royer, in 1679, added —

1. Columba Noachi.   The Dove of Noah.
2. Crux Australis.   The Southern Cross.
3. Nubis Major.   The Great Cloud.
4. Nubis Minor.   The Little Cloud.
5. Fleur-de-Lis.   The Lily.

(All Southern Constellations.)

Halley, about the same period, added —

1. Robur Caroli.   Charles's Oak.

(A Southern Constellation.)

Flamsteed's maps also contain —

1. Mons Mænalus.   The Mountain Mænalus.
2. Cor Caroli.   Charles's Heart.

(Both Northern Constellations.)

La Caille, in 1752, added —

1. Apparatus Sculptoris.	The Apparatus of the Sculptor.
2. Fornax Chemica.	The Chemical Furnace.
3. Horologium.	The Clock.
4. Reticulus Rhoimboidalis.	The Rhoimboidal Net.
5. Cæla Sculptoris.	The Sculptor's Tools.
6. Equuleus Pictoris.	The Painter's Easel.
7. Pixis Nautica.	The Mariner's Compass.
8. Antlia Pneumatica.	The Air Pump.
9. Octans.	The Octant.
10. Circinus.	The Compasses.
11. Norma, or Quadra Euclidis.	Euclid's Square
12. Telescopium.	The Telescope.
13. Microscopium.	The Microscope.
14. Mons Mensæ.	The Table Mountain.

(All Southern Constellations.)

Le Monnier, in 1776, added —

1. Tarandus.	The Rein Deer.
2. Solitarius.	The Solitaire.

(The former in the northern, the latter in the southern hemisphere.)

In the same year Lalande placed Messier's name in the heavens, by forming a constellation in his honour, near Tarandus.

Poczobut, in 1777, added —

Taurus Poniatowskii.	The Bull of Poniatowski.

(Between Aquila and Serpentarius.)

Hell formed in Eridanus —

Psalterium Georgianum.	George's Harp.

And finally in Bode's maps we meet with —

1. Honores Frederici.	The Honours of Frederick.
2. Sceptrum Brandenburgicum.	The Sceptre of Brandenburg.
3. Telescopium Herschellii.	Herschel's Telescope.
4. Globus Aërostaticus.	The Balloon.
5. Quadrans Muralis.	The Mural Quadrant.
6. Lochium Funis.	The Log Line.
7. Machina Electrica.	The Electrical Machine.
8. Officina Typographica.	The Printing Press.
9. Felis.	The Cat.

Making in all 109 constellations. This number by no means exhausts the list of those which have been proposed by different persons. It does however exhaust our patience, and we shall give no more. A writer in the *English Cyclopædia* very pertinently remarks: "In fact, half-a-century ago, no astronomer seemed comfortable in his position till he had ornamented some little cluster of stars of his own picking with a name of his own making."

Sir J. Herschel says: — "The constellations seem to have been almost purposely named and delineated to cause as much confusion and inconvenience as possible. Innumerable snakes twine through long and contorted areas of the heavens, where no memory can follow them; bears, lions, and fishes, small and large, northern and southern, confuse all nomenclature, &c."

Many of the above smaller constellations are very properly rejected by modern uranographers, and in the list which we append, we shall insert those asterisms only which are generally acknowledged in the present day.

The column headed "Co-ordinates" may be thus explained. Project a line through the given parallel of R.A., and another through the given parallel of Declination, and their point of intersection will fall on a central part of the constellation, a celestial globe or map being employed.

THE NORTHERN CONSTELLATIONS.

No.	Name.	Co-ordinates.		Stars of Mag.					
		R. A. h. m.	Decl. °	i.	ii.	iii.	iv.	v.	Total.
1	Andromeda . .	1 0	35	1	1	2	6	8	18
2	Aquila	19 30	10	1	...	7	3	22	33
3	Auriga	6 0	42	1	1	1	5	27	35
4	Boötes	14 35	30	1	...	8	7	19	35
5	Camelopardus .	5 45	68	...	...	...	5	31	36

THE NORTHERN CONSTELLATIONS — *continued*.

No.	Name.	Co-ordinates.			Stars of Mag.					
		R. A.		Decl.	i.	ii.	iii.	iv.	v.	Total.
		h.	m.	o						
6	Canes Venatici .	13	0	40	...	1	...	1	13	15
7	Cassiopeia . . .	1	10	60	...	1	4	4	37	46
8	Cepheus . . .	21	40	65	...	...	3	4	37	44
9	Clypeus Sobieskii	18	10	15 S.	...	...	...	...	4	4
10	Coma Berenicis .	12	40	26	...	...	...	5	15	20
11	Corona Borealis .	15	40	30	...	1	...	3	15	19
12	Cygnus	20	20	42	...	1	5	10	51	67
13	Delphinus . . .	20	40	15	...	...	1	5	4	10
14	Draco	17	20	66	...	2	3	7	68	80
15	Equuleus . . .	21	0	6	...	...	...	2	3	5
16	Hercules . .	16	45	27	...	1	6	12	46	65
17	Lacerta . . .	20	20	44	...	...	...	2	11	13
18	Leo Minor . .	10	5	36	...	...	...	5	10	15
19	Lynx	7	55	50	...	...	...	4	24	28
20	Lyra	18	40	35	1	...	2	1	14	18
21	Pegasus . . .	22	25	15	...	4	2	5	32	43
22	Perseus et Caput Medusæ.	3	30	47	...	2	4	7	27	40
23	Sagitta	19	40	18	...	...	...	3	2	5
24	Serpens . . .	15	40	10	...	1	5	8	9	23
25	Taurus Ponia-towskii.	17	50	5	...	...	...	4	2	6
26	Triangulum . .	2	0	32	...	...	1	1	3	5
27	Ursa Major . .	10	40	58	1	3	8	7	34	53
28	Ursa Minor . .	15	0	78	...	1	3	4	15	23
29	Vulpecula et Anser	20	0	25	...	...	...	2	21	23
29	Grand Total				6	20	65	132	604	827

THE ZODIACAL CONSTELLATIONS.

No.	Name.		Co-ordinates.		Stars of Mag.					
		Symbol.	R. A. h. m.	Decl. °	i.	ii.	iii.	iv.	v.	Total.
1	Aries	♈	2 30	18 N.	...	1	2	3	11	17
2	Taurus	♉	4 0	18	1	1	4	8	44	58
3	Gemini	♊	7 0	25	1	1	2	7	17	28
4	Cancer	♋	8 40	20	...	...	...	5	10	15
5	Leo	♌	10 20	15	1	3	6	11	26	47
6	Virgo	♍	13 20	3 N.	1	1	3	10	24	39
7	Libra	♎	15 0	15 S.	...	2	1	11	9	23
8	Scorpio	♏	16 15	26	1	1	10	7	15	34
9	Sagittarius	♐	18 55	32	...	...	3	10	25	38
10	Capricornus	♑	21 0	20	...	...	2	4	16	22
11	Aquarius	♒	22 0	9 S.	...	...	3	4	18	25
12	Pisces	♓	0 20	10 N.	...	...	1	3	14	18
12	Grand Total				5	10	37	83	229	364

THE SOUTHERN CONSTELLATIONS.

No.	Name.	Co-ordinates.		Stars of Mag.					
		R. A. h. m.	Decl. °	i.	ii.	iii.	iv.	v.	Total.
1	Antlia Pneumatica.	10 0	35	...	...	...	1	6	7
2	Sculptor (App. Sculpt.).	0 20	32	...	...	...	...	13	13
3	Apis . . .	15 20	76	...	...	...	1	6	7
4	Ara	17 0	54	...	...	3	4	8	15
5	Argo	7 40	50	2	4	9	14	104	133

THE SOUTHERN CONSTELLATIONS — *continued.*

No.	Name.	Co-ordinates.		Stars of Mag.					
		R. A. h.　m.	Decl. °	i.	ii.	iii.	iv.	v.	Total.
6	Cæla Sculptoris	4　40	42	...	...	...	1	5	6
7	Canis Major　.	6　45	24	1	2	4	7	13	27
8	Canis Minor　.	7　25	5 N.	1	...	1	...	4	6
9	Centaurus . .	13　0	48	2	1	5	7	39	54
10	Cetus　. . .	2　0	12	...	3	4	9	16	32
11	Chamæleon　.	10　50	78	...	...	...	...	17	17
12	Circinus　. .	15　0	64	...	...	...	1	1	2
13	Columba Noachi	5　25	35	...	1	1	3	10	15
14	Corona　Aus- tralis.	18　30	40	...	...	...	1	6	7
15	Corvus . . .	12　20	18	...	1	...	2	5	8
16	Crater　. . .	11　20	15	...	...	1	2	6	9
17	Crux　. . .	12　15	60	1	2	1	2	4	10
18	Dorado . . .	4　40	62	...	...	1	3	13	17
19	Pictor　(Eq. Pict.).	5　25	55	...	...	...	3	14	17
20	Eridanus　. .	3　40	30	1	1	7	18	37	64
21	Fornax (Forn. Chem.).	2　20	30	...	...	...	...	6	6
22	Grus	22　20	47	...	1	1	2	7	11
23	Horologium　.	3　15	57	...	...	...	...	11	11
24	Hydra . . .	10　0	10	...	1	...	7	41	49
25	Hydrus . . .	2　40	70	...	...	3	1	21	25
26	Indus　. . .	21　0	55	...	...	1	1	13	15
27	Lepus　. . .	5　25	20	...	...	1	9	8	18
28	Lupus　. . .	15　25	45	...	...	3	8	23	34
29	Monoceros . .	7　0	2	...	...	...	3	9	12
30	Mons Mensæ .	5　20	75	...	...	...	1	8	9

THE SOUTHERN CONSTELLATIONS — *continued.*

No.	Name.	Co-ordinates.			Stars of Mag.					
		R. A. h.	m.	Decl. °	i.	ii.	iii.	iv.	v.	Total.
31	Microscopium	20	40	37	...	...	...	1	4	5
32	Musca Australis	12	25	68	...	...	...	4	3	7
33	Norma . . .	16	0	45	...	...	...	...	12	12
34	Octans . . .	21	0	80	...	...	3	1	12	16
35	Ophiuchus . .	17	0	0	...	1	5	11	23	40
36	Orion . . .	5	30	0	2	4	2	7	22	37
37	Pavo	19	20	68	...	1	2	5	19	27
38	Phœnix . . .	1	0	50	...	2	1	4	25	32
39	Piscis Australis	21	40	32	1	...	...	3	12	16
40	Piscis Volans .	7	40	68	...	...	...	1	8	9
41	Reticulus Rhomboidalis.	4	0	62	...	...	1	1	9	11
42	Sextans . . .	10	0	0	...	...	...	...	3	3
43	Telescopium .	18	40	53	...	...	...	4	4	8
44	Toucan . . .	23	45	66	...	...	1	3	17	21
45	Triangulum Australe.	15	40	65	...	1	2	1	7	11
45	Grand Total				11	26	63	157	654	91

SUMMARY.

	No. of Cons.	Stars of Mag.					
		i.	ii.	iii.	iv.	v.	Total.
Northern . .	29	6	20	65	132·	604	827
Zodiacal . .	12	5	10	37	83	229	364
Southern . .	45	11	26	63	157	654	911
	86	22	56	165	372	1487	2102

Argelander's numbers differ slightly. He gives the following : —

1st Mag.	=	20	4th Mag.	=	425	7th Mag.	=	13,000
2nd „	=	65	5th „	=	1100	8th „	=	40,000
3rd „	=	190	6th „	=	3200	9th „	=	142,000

The following is the number contained in Ptolemy's Catalogue : it must be observed that his range of observation at
Alexandria was somewhat limited.

1st.	2nd.	3rd.	4th.	5th.	6th.
15	45	208	474	217	49

Making a total of 1008.

BOOK VIII.

ASTRONOMICAL INSTRUMENTS.

CHAPTER I.

TELESCOPES.

Telescopes of two Kinds. — Reflecting Telescopes. — The Gregorian Reflector.—The Newtonian Reflector.—The Cassegrainian Reflector.—The Herschelian Reflector.—Nasmyth's Telescope.—Refracting Telescopes. —Spherical Aberration.— Chromatic Aberration. — Theory of Achromatic Combinations.—Dollond's Discovery.—Galileo's Principle.—Eye-Pieces.—The Positive Eye-piece.—The Negative Eye-piece.—Formulæ for calculating the Focal Lengths of equivalent Lenses.—Micrometers. — The Reticulated Micrometer. — The Parallel-Wire Micrometer. — Telescope Tubes.—Description of the Manufacture of Veneered Tubes. —Clarke's Zinc and Brown Paper Tubes.

THE most important instrument used for astronomical observations is the telescope [1]; the different kinds and applications of which, we purpose now briefly to describe. Telescopes are of 2 kinds—reflecting and refracting: in the former kind an image of the object to be viewed is produced by a concave reflector, in the latter by a converging lens.

Four principal varieties of reflectors are in general use : —

1. The Gregorian, invented by James Gregory, of Aberdeen, in 1663.

2. The Newtonian, by Sir I. Newton, in 1669.

3. The Cassegrainian, by Cassegrain, in 1672.

---

[1] τῆλε, at a distance, and σκόπειν, to see.

P 4

4. The Herschelian, by Sir W. Herschel, towards the close of the last century.

The *Gregorian Telescope* consists of a large concave metal speculum, in the centre of which a circular aperture is pierced. A 2nd concave speculum, with its concave surface turned in the

Fig. 26.

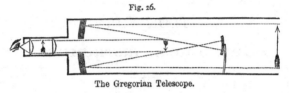

The Gregorian Telescope.

other direction, is placed in the axis of the tube at a distance from the larger speculum greater than its focal length. A smaller tube, carrying an eye-piece, is placed at the extremity of the larger tube, at the end where the speculum is.

The *Cassegrainian Telescope* is similar in all respects to the Gregorian, except that the smaller speculum is convex instead

Fig. 27.

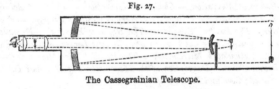

The Cassegrainian Telescope.

of concave, and that it is placed in the tube at a distance from the larger speculum *less than* its focal length.

In the *Newtonian Telescope* a large concave reflector is

Fig. 28.

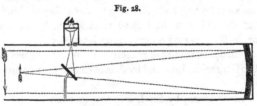

The Newtonian Telescope.

placed at one end of the tube. At a distance from the larger mirror, less than at its focal length, is placed at an angle of

45° to the optic axis of the telescope, a plane reflector, by which the rays proceeding from the object are turned to the side of the larger tube where there is a smaller one in which the eye-piece for viewing them is placed.

In all these telescopes the central rays are lost, because in the Gregorian and Cassegrainian arrangements the central portion of the mirror is cut away, and in the Newtonian, the central rays are intercepted by the plane mirror.

In the *Herschelian Telescope* the large speculum is not fixed in the tube with its diameter at right angles to the axis of the tube, but is slightly inclined thereto; by this means the image of the object observed is brought to the interior edge of the tube, where it is directly examined by the eye-piece,

Fig. 29.

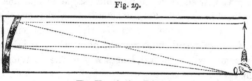

The Herschelian Telescope.

instead of through the medium of the 2nd reflector. The advantage of this plan is, not only that observations can be made with greater ease, but a large saving of light is effected by dispensing with the 2nd reflector. With an instrument constructed in this way, Sir W. Herschel was enabled to make use of, for observations on the fixed stars, magnifying powers ranging up to 6450.

Mr. James Nasmyth, the well-known astronomer and engineer, a few years ago invented a telescope which combines the principles of Cassegrain and Newton. The rays reflected from the great speculum are received either upon a small speculum or a prism placed in the axis of the tube, between the focus and the great speculum. By this they are reflected at right angles, and the image is formed in a tube inserted in one of the trunnions upon which the instrument turns. The image is then viewed in the usual way. The advantage derived from this arrangement is, that the great tube can be

moved to any extent in altitude, whilst the lateral tube, placed
in the trunnions, is fixed, and thus the observer can survey any
vertical circle without changing his own position.

The inventor has one erected at his own residence, the tube
of which is 28 ft. long, and 4 ft. 6 in. in diameter. The
azimuthal movement is given by a turn-table, similar to that
used on railways for turning locomotive engines.

We come now to speak of the 2nd, and by far the most
important class of telescopes now in use; those formed on
the refracting principle. Such an instrument in its simplest
form consists merely of a double convex lens, which forms an
image of the object to be viewed (and hence termed the *object-
glass*), and a 2nd and smaller double convex lens (called
the *eye*-glass), used as a simple microscope to examine the
image formed by the 1st. For astronomical purposes the object-
glass is commonly double, sometimes triple, for the purpose of
getting rid of *spherical* and *chromatic aberration*; and the eye-
glass is composed of several lenses which when put together
form the *eye-piece*.

For a proper account of the theory of chromatic and spherical
aberration the reader is referred to some treatise on optics:
we shall merely state that chromatic aberration is caused by
the unequal refrangibility of the different colours which together form white light; so that when an observer views an
object through a lens he will not see the image perfectly
defined and colourless; but it will appear more or less surrounded with colour. Spherical aberration depends on the
figure of the lens, there being no curvature such that all the
rays of light coming from any point of an object are *exactly*
united in a common focus. In order to get rid of the first
source of indistinctness, the telescope makers of by-gone days
constructed lenses of great focal length (some of them were as
much as 300 feet long); by this means the chromatic dispersion was diminished in comparison with the size of the
image formed. Dollond, by an examination of the different
kinds of glass then in use, found that some specimens dispersed
colours, under a given mean refraction, much more than
others: this discovery led to the construction of the com-

pound object-glass, which consists in general of a convex lens of crown glass of a greenish tinge, and a concave lens of flint glass of a white tint, placed in contact. By a proper arrangement of their focal lengths, any 2 selected parts of the spectra formed by these lenses, can be united, and the chromatic dispersion is thus to a great extent got rid of, without destroying the refractive power of the object-glass. When a ray of light falls upon such an arrangement of lenses, it is acted upon by each component: the crown glass renders it convergent and decomposes it; the concave lens neutralises the decomposition, and an uncoloured image is formed at the focus. By this discovery, which was made in the year 1758, the inconveniences of the old aërial telescope were done away with, and to an Englishman the world owes that most beautiful instrument the "Achromatic Refractor," so many fine examples of which are now to met with, both in this country and abroad.

The refracting telescope invented by Galileo consists of a double convex object-glass, having for the eye-glass a double

Fig. 30.

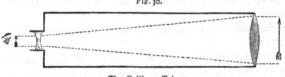

The Galilean Telescope.

concave lens placed *in front* of the image formed by the object-glass.

The common opera-glass is a telescope on the Galilean principle. Supposing the reader to have possessed himself of a refracting telescope, we strongly recommend him never to remove the object-glass, for the purpose of picking it to pieces; they always are or ought to be put in proper adjustment by the maker, and the amateur is recommended to let them remain so: the case, however, is different with the eyepiece; occasions will often arise when it will be desirable to take that to pieces for cleaning the interior, and therefore we will now describe somewhat in detail this important appendage to a telescope.

The eye-pieces commonly in use are the positive or Ramsden's, and the negative or Huyghenian, having been first used by Ramsden and Huyghens respectively.

A positive eye-piece consists of 2 plano-convex lenses placed with the convex sides towards each other, of which the inner-

Fig. 31.

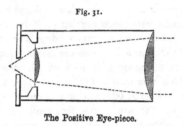

The Positive Eye-piece.

most is called the *field-glass*, and the outermost, the *eye-glass*. The focal lengths are equal to each other; and the field-glass should be so far within the focus of the eye-glass, that particles of dust upon the former cannot be seen when looking through the latter.

To find the single lens equivalent to an eye-piece of this description.

Divide the product of the focal lengths of the component lenses by their sum, minus the distance between them.

Thus if the focal length of each lens be 1·5 inch, and the distance between them, 1 inch, the length of the equivalent single lens will be —

$$\frac{1 \cdot 5 \times 1 \cdot 5}{3 - 1} = \frac{2 \cdot 25}{2} = 1 \cdot 125 \text{ inch.}$$

The positive eye-piece having its focus beyond the field-glass is suited for use with micrometers, and other instruments having wires in the focus of the object-glass; a case to which the negative eye-piece, in consequence of its having, as we shall see, the focus between the glasses, is not suited.

A negative eye-piece consists also of 2 plano-convex lenses, the convex sides of both being in this case turned towards the object-glass. The ratio of the focal lengths of the lenses is usually 3 to 1, the latter representing the eye-glass. In order that the combination may be achromatic, it is indispensable that the distance between the lenses be equal to half the sum of their focal lengths.

Fig. 32.

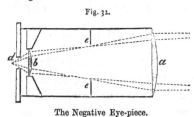

The Negative Eye-piece.

a being the field-glass, a stop, *cc*, to limit the field of view, is placed in the focus of the eye-glass *b*, and the eye-hole *d* is of such magnitude and distance from the eye-glass that the emergent pencils just find a passage through it. The passage of rays proceeding from an achromatic object-glass is shown in the figure, where it will be seen, that after refraction by the field-glass, they come to a focus at *c*, at which place the image of the object is found. The rays again diverge, and by passing through the eye-glass *b* are, in turn, converged towards the point *d*, where they enter the eye, and form an inverted image on the retina.

To find the single lens equivalent to an eye-piece of this description.

Divide twice the product of the focal lengths of the component lenses by their sum.

Thus, if the focal length of the field lens be 3, and of the eye-lens be 1, the length of the equivalent lens will be —

$$\frac{3 \times 1 \times 2}{4} = \frac{6}{4} = 1\tfrac{1}{2}.$$

Negative eye-pieces are almost universally used for all astronomical purposes, except in those cases where they are inadmissible, and which have already been pointed out, when speaking of the positive eye-piece.

The following should be borne in mind —

1. *That in an astronomical refracting telescope the distance between the object-glass and the eye-piece is the sum of their focal lengths, and the magnifying power is the ratio of their focal lengths.*

Thus, let the focal length of the object-glass of a telescope be 10 feet, and of its eye-glass $\frac{1}{4}$ inch. Find its magnifying power.

Here 10 ft. = 120 inches.

Then as $\frac{1}{4}$: 120 :: 1 : x

: 480.

2. *That in a Galilean Telescope, the distance between the glasses is the difference of their focal lengths, and the magnifying power is the ratio of their focal lengths.*

Thus, let the focal length of the object-glass of a telescope be 6 inches, and of its eye-glass $\frac{3}{4}$ inch. Find its magnifying power.

As $\frac{3}{4}$: 6 : 1 : x

: 8.

A micrometer is used for measuring small distances, such as the diameter of a comet, the distance between 2 double stars, &c.

The simplest form is that known as the Reticulated Micrometer. It consists of an eye-piece of low power, having

Fig. 33.

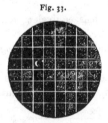

Arrangement of the Wires in the Reticulated Micrometer.

stretched across it a number of wires at right angles to, and at an equal and known distance from each other.

All that the observer has to do, is to put in the eye-piece, illuminate the field with the small lantern, shown in Fig. 39,

and notice how many divisions cover the object, whose size
it is desired to measure, &c.; knowing previously the value
of each division; the solution then becomes a simple matter of
arithmetic.

The other contrivance in most general use is the Parallel-
wire Micrometer, which consists of 2 sliding frames across
which parallel wires are stretched. These frames are moveable in a direction at right angles to that of the wires by

Fig. 34.

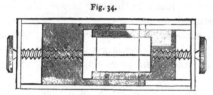

The Parallel-wire Micrometer.

means of very fine-threaded screws. This apparatus is placed
in the focus of the object-glass of the telescope, so that the eye
viewing the object under examination is enabled (at night by
suitable illumination) to see clearly the wires. Supposing it
is desired to know the diameter of a comet, seen in the field
of the telescope, all the observer has to do, is to bring the 2
wires together on one side of the comet, and then to turn the
screw until the wire moved is brought to coincide with the
other margin of the comet. The number of turns and parts of
a turn necessary to effect this, is a measure of the angular
diameter of the object. The observer now requires to ascertain the value of each turn of the screw. This he does by
directing the telescope to some terrestrial object, for example, a
foot rule, placed at a distance of 100 yards, and observing how
many turns are required to encompass its image by the 2
wires. Knowing by calculation the angle subtended by the
foot rule at the distance at which it is placed, it follows that
if a is the angle subtended, n the number of turns of the
micrometer screw, then v, the value of each turn, is given by
the equation

$$v = \frac{a}{n}.$$

For observations of the spots on the Sun, it is necessary to adopt some expedient by which the power of the Sun's light may be overcome; this is done by employing coloured glasses of different tints.   This method answers very well, but where expense is not an object, the observer is recommended to supply his telescope with a special solar eye-piece, of the kind invented by W. R. Dawes, Esq.   In large instruments the heat of the concentrated rays of the Sun may be got rid of by the use of a prism; in this case, however, a diagonal eye-piece is required.

For telescopes up to 5 feet in length the tubes are usually made of brass; but for sizes larger than that, in consequence of its expense, other materials are employed, plate iron, or wood, for instance.

The following is a description of a tube of a very strong and convenient style of manufacture, formed of veneers of mahogany.

" This tube was formed upon a core of dry, well-seasoned deal planks, and turned down in the lathe to the following dimensions : —

Length of core	.	. 9 ft.	0 in.
Diameter, large end	.	0	$7\frac{7}{8}$
Do. small end	.	0	$5\frac{7}{8}$

" The core was well soaped to prevent the possibility of any portion of the glue adhering to it, and thereby rendering the removal of the tube difficult when finished.   Upon the core a sheet of thick brown paper was wrapped, its edges being pasted together to serve as a foundation for the first veneer. The core was then brushed over with glue, and a veneer of mahogany then laid round it; a moveable caul made of double canvass was then laid round the veneer, and drawn up tightly by means of screws.

" This caul exactly fitted the taper of the tube, and its longitudinal edges were securely fastened to 2 strips of wood. In one of these strips, at intervals of about 5 inches, were inserted common bed screws, moving freely through it, and screwing into corresponding nuts in the other strip.   By this means the caul was screwed up tightly round the veneer, thereby ensuring a close contact between it and the core.

" The core was then placed over a stove pipe, which extended along its whole length, and was slowly turned on its axis until it became sufficiently hot to melt the glue, which thus became equalised, while the superabundant quantity was squeezed out; a portion even permeating the veneer, owing to the pressure of the caul. The core was then set aside for 2 or 3 days to dry.

" Every successive veneer, prior to being laid on, was lined with a piece of thin calico, to prevent the veneer from splitting while being turned round the core. This calico was removed when the veneer was dry; after which the surface was prepared for another veneer, by being levelled over and freed from inequalities by the veneer plane. Each veneer was in a single piece, the joints being placed at alternate sides of the tube, and the grain of the wood reversed at every layer. A tube was thus formed of 8 thicknesses, the 7 under ones being of Spanish mahogany having stains or other faults which rendered them unfit for *fine* cabinet-work, and therefore of moderate cost (about 9*d.* per foot super). The 8th was of the finest Spanish mahogany, and cost (on account of its great size) about 1*s.* 8*d.* per foot super. The thickness of the tube when finished was $\frac{3}{8}$ of an inch. It was French polished on the outside."[1]

Mr. Alvan Clarke, of Boston, U.S., makes excellent tubes of sheet zinc, and brown paper, painted and varnished on the outside. Mr. Dawes possesses one carrying an 8-inch object-glass, which, although only $\frac{2}{10}$ of an inch thick, is remarkably hard and solid to the touch.

[1] Brodie, *Month. Not.* R.A.S., vol. xvii. p. 33.

CHAPTER II.

TELESCOPE STANDS.

Importance of having a good Stand.—"Pillar-and-Claw" Stand.—The "Finder."—Vertical and Horizontal Rack Motions.—Steadying Rods. —Cooke's Mounting.—Varley's Stand.

THE manner in which a telescope is mounted is by no means an unimportant matter. It frequently happens that a good glass is nearly or quite useless because it is provided with no

Fig. 35.

Telescope Mounted on a Pillar-and-Claw Stand.

proper stand on which it may be placed. Far better is it for observers, desirous of deriving some benefit from the instrument they purpose obtaining, to get one of smaller size and power, thoroughly well mounted, than to devote their resources to the purchase of some immense glass which is only to be hung up by ropes and spars, under which condition it would fail to possess, in the slightest degree, those most indispensable qualifications, steadiness and facility of movement.

The simplest form of stand is that known as the "pillar-and-claw," and is suited for telescopes of from 24 to 48 inches focal length. By its use the observer is enabled to impart to his tube 2 motions, one in altitude, the other in azimuth. The former or vertical motion is obtained by a joint at the top of the pillar; the latter or horizontal motion by a conical axis carefully fitted into the capital of the pillar by a nut and screw.

Fig. 36.

Telescope Mounted on Pillar-and-Claw Stand, with Finder and Vertical Rack Motion.

The telescope is brought to a focus by a rack and pinion. Sometimes, especially in telescopes furnished with a long range of powers, a tube sliding easily within the tube to which the rack and pinion is attached, and called a *tail-piece*, is employed for first getting an approximate focus.

Supposing several powers be supplied to such an instrument, the observer is recommended to make use of the lowest to find the object he is in quest of : by a little care and practice he will be enabled to change the eye-piece for one of higher power, all the while keeping the object within the field of view.

The first addition to such an instrument as we have just described is the finder, which is a small achromatic telescope mounted on the outside of the tube of the large one, the optical axis of both being coincident, and used for the purpose which its name implies. The eye-piece is of very low power, and consequently the field of view is extensive, and being furnished with 2 wires crossing each other at right angles, it will happen (supposing both to be in good adjustment) that when an object is seen at the intersection of the cross wires in the finder, it will also be in the centre of the field of the large telescope. Perfect parallelism between the axes of the 2 telescopes is obtained by the use of the screws in the collar which holds the eye-end of the finder, the method of using which will be obvious from inspection.

When it is desired to have greater precision, for the purpose of moving the telescope, than could be given by the hand, vertical and horizontal rack motions are applied.

The former consists of 2 or 3 tubes which slide one within the other, the largest being attached by a point to the base of the pillar, and the smallest being secured to the eye-end of the telescope. The 2 larger tubes slide freely, but can be fixed in any position by a clamp ; the smallest is moveable by a rack and pinion.

In cases where the horizontal rack motion is applied, the construction of the pillar differs somewhat from that described at the commencement of this chapter. It consists, as in the former case, of an outside and an inside cone; but instead of the telescope being attached to the inside cone, and that dropping into the outside one, the arrangement is reversed, and the telescope is fastened to the outside one, which drops *upon* the inner one. Upon the lower end of this fixed inside cone, a ring is made to move stiffly, and in the edge of this ring teeth are cut, into which an endless screw, attached to the outer

Fig. 37.

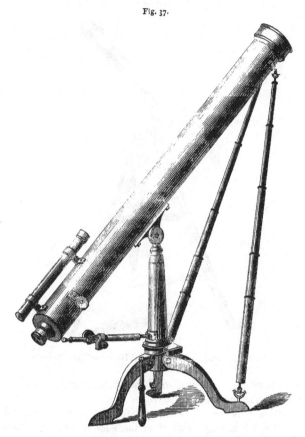

Telescope Mounted on a Pillar-and-Claw Stand, with Finder, Vertical and Horizontal Rack Motions, and Steadying Rods.

cone, works. When therefore the observer wishes to impart a rapid horizontal motion, he has only to apply to the telescope a force sufficient to cause the outside cone, together with the ring, to revolve upon the inside one; but when a slow motion is desired, it suffices that the endless screw be turned, in which

Fig. 38.

Telescope on Stand, with Vertical and Horizontal Rack Motions.

case the ring, in consequence of the friction, remains attached to the inside and immoveable cone.

Motion may be conveniently imparted in the horizontal direction by having attached to the endless screw a Hook's joint in a handle.

To all telescopes, of a greater length than 3 feet, and mounted on a pillar and claw stand, steadying rods are a very desirable addition. They are generally 2 in number, and consist of 4 or more tubes sliding one within another, their ends terminating in universal points, fixed to the object-end of the telescope, and the 2 front "claw" legs, respectively. (Fig. 37.)

Fig. 38 represents a telescope and mounting by Mr. Cooke, of York, a maker whose instruments give general satisfaction to those who use them, and whom from experience we cordially recommend.

A contrivance, known as *Varley's Stand,* is sometimes used for telescopes longer than 4 feet: but the regular equatorial stand is recommended for instruments too large to be conveniently placed on one of the "pillar-and-claw" construction.

CHAPTER III.

THE EQUATORIAL.

Brief Epitome of the Facts connected with the apparent Rotation of the Celestial Sphere. — Principle of the Equatorial Instrument. — Two Forms in general Use.—Description of Sisson's Form, and of the different Accessories to the Instrument generally.—Description of Fraunhofer's Form of Equatorial.—In what its Superiority consists.—The Adjustments of the Equatorial.— Six in Number.— Method of performing them.—Method of observing with the Instrument, reading the Circles, &c.—Examples.

THE reader doubtless is aware that the celestial sphere has an apparent motion of rotation round certain imaginary points in the heavens termed the poles, one only of which is visible from any given point on the Earth, the equator excepted ; that the altitude of the pole, or angular elevation above the horizon, is equal to the latitude of the place of observation ; and also that every star describes, apparently, a circular path around the pole of the heavens, increasing in magnitude with the increase of the angular distance of the star from the pole, up to a distance of 90° or a quadrant, after which it again diminishes towards the opposite pole.

If now we were to incline the pillar-and-claw stand already described, in such a manner that the vertical axis should point towards the pole, and thus be parallel to the axis on which the sphere is supposed to revolve, — in point of fact give the pillar an inclination equal to the latitude of the place where we are,— it would clearly follow, that a single motion of the telescope, viz. one of rotation, about that inclined axis, would cause the line of sight (called also the optical axis) to trace upon the sphere a circle corresponding to those in which the heavenly bodies appear to move, these circles increasing or diminishing as, by moving the telescope upon the stand or horizontal pivot, we increase or diminish the angle between the line of sight and the first or inclined axis; just as the circles which the

Fig. 39.

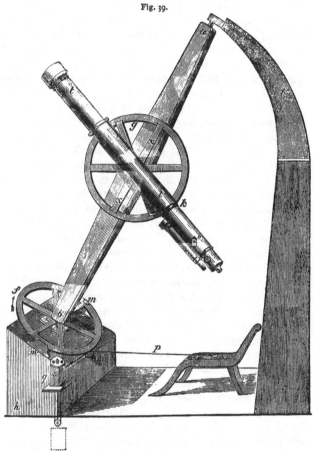

The English Equatorial.

heavenly bodies themselves describe, increase or diminish according as the polar distance increases or diminishes.

Q

A pillar-and-claw stand placed in the above described position constitutes a simple equatorial, though in the construction of one specially designed to serve that purpose, numerous alterations are introduced. Equatorials are all constructed on similar principles, but vary in the manner in which those principles are arranged. The forms commonly met with are the English or Sisson's, and the German or Fraunhofer's; the latter is for many reasons to be preferred, but the general construction of the former being most readily comprehensible, will be described first. Fig. 39 is a general representation of an equatorial; $a\,b$, represents the polar axis; it is directed to the pole of the heavens, and is supported in this position by the stone piers h, i, the curved portion of the latter, i, being generally of cast-iron. This axis terminates in cylindrical pivots, which rest in Y's; and one of the Y's, usually the lower one, is provided with means for altering, for the purpose of adjustment, the direction of the polar axis. The declination axis (not shown, but which the declination circle, g, is attached to) passes through the polar axis, and rests upon collars, and as it is necessary that the 2 axes should be at right angles to each other, the collar at the end opposite to which the telescope is fastened, is adjustible by screws. The collar, in which the telescope end revolves, is held by pivots, which allow a lateral motion through a small arc, in order to prevent any strain being given when the adjustment at the other end is performed. $t\,t$ is the telescope, fixed at right angles to the declination axis, and therefore parallel to the polar axis, great care being taken that the fastenings are perfectly rigid. The eye end is furnished with means for the adjustment of the line of sight.

This is done by means of a transit eye-piece in which a system of cross wires are arranged, the line of sight passing through the intersection of the middle vertical with the horizontal wire, the whole system being removeable from right to left, by screws placed for the purpose.

The angle between the line of sight and the polar axis is measured in the declination circle g, divided into degrees and fractions of a degree, and capable of being read off to minutes and seconds by 2 verniers placed at the end of the index plate *x*, and carried round with the telescope. When the line of sight is parallel with the polar axis, and consequently, if the latter is in adjustment, pointing to the pole, the index arrow on each vernier should point to zero, and in order that they may do so, means for adjustment are generally applied to the verniers themselves. A clamp and tangent screw (not shown) near to *k* give the observer the power of fixing the telescope or moving it through very small arcs.

The angle through which the plane containing the line of sight and polar axis revolves is measured on the circle *f*, called the hour circle, which is fixed to the polar axis [1], and divided to show portions of time, hours, minutes, and by means of verniers (marked *m*) seconds, the hours being marked from I to XXIV. When the declination axis is horizontal, the zero arrows on the verniers should point to XII and XXIV, facilities for bringing about that coincidence being provided. The hour circle has a milled edge in which the threads of an endless screw, which forms part of the clamp at *n*, are arranged to work so as to give a slow motion in right ascension, which may either be imparted by the observer himself, through the medium of a rod *p*, terminating in an universal joint, or by means of clockwork at *q*, if a uniform motion is desired for the purpose, for example, of following an object. In order that the 2 actions may subsist independently one of the other, it is usual to attach the clock to one end of the tangent screw, and a rod for the use of the observer to the other. The screw is so mounted that when it is required to turn the telescope through a large arc, it can be thrown out of gear.

There are various expedients resorted to in practice, in connexion with the clockwork, and the method of making use of it, &c., which need not here be specified in detail.

---

[1] In some equatorials of the largest class, as, for instance, in the Northumberland, at Cambridge, the hour circle turns on the lower pivot, and a different method of procedure is adopted.

Certain micrometers having wires in their eye-pieces, it is necessary, in order that, when these wires are required to be

Fig. 40.

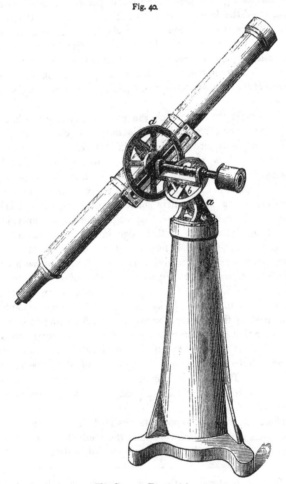

The German Equatorial.

made use of, they may be visible, that we should possess the power of illuminating them. This is done by means of a small lantern l, the light passing into the interior of the tube through an aperture, the size of which may be varied at pleasure.

A weight is placed at the further end of the declination axis to counterbalance the telescope; and a spirit level to set the declination axis horizontal, is also arranged that it can be placed in position when required.

The German, or Fraunhofer's equatorial, so called from its inventor, the celebrated optician of Munich, is represented in its simplest form in the annexed engraving. (Fig. 40.)

a is the polar axis, and b the attached hour circle, c the declination axis, and d its attached circle. Appliances for the adjustment of the polar and declination axes, and the verniers of the 2 circles are provided, though not shown in the figure.

There are 2 principal reasons why this form of mounting is preferable to the one just described. 1°, The telescope will reach every part of the heavens without interruption; whereas it will be noticed that the upper support required for the polar axis of Sisson necessarily interferes with the view of the objects at and below the pole. 2°, That this stand requires but 1 pier easily erected, instead of 2, the proper placing of which occasions much labour and trouble, in order to ensure their being brought within the limits of the adjustment.

We shall now proceed to give the adjustments which the equatorial requires.

They are 6 in number. For correct observation it is necessary —

1st. That the polar axis be placed at the altitude of the pole.

2nd. That the polar axis be placed in the meridian.

3rd. That the polar and declination axes be at right angles to each other.

4th. That the optical axis, or line of sight of the telescope, be at right angles to the declination axis.

5th. That the indices of the declination circle point to 90° when the telescope points to the pole.

6th. That the index of the hour circle point to 0 h. when

the telescope is placed in the meridian, that is to say, when
the declination axis is horizontal.

1st. To bring the polar axis to the altitude of the pole.

Put on the transit eye-piece.

Select from a standard catalogue some star which at the
time in question happens to be on, or very near the meridian.
Set the index of the declination circle to the declination of the
star given in the catalogue, correcting the same for refraction :
if the star does not run along the horizontal wire, the end of
the polar axis must be elevated or depressed as the case may
require.

2nd. To place the polar axis in the meridian. Direct the
telescope on some known star, 6 or 7 hours from or past the
meridian, and note its declination by the index of the circle.
Apply to this reading the proper refraction correction, and
compare the result with the declination of the star as given in
a standard catalogue. Supposing that the star is east of the
meridian, and its observed declination exceed that given in
the catalogue, the lower end of the polar axis will be to the
west of its true place, and must be moved accordingly. Supposing the observed declination is less than that given in the
catalogue, the lower end of the polar axis is too much to the
east, and must be shifted.

Should the star observed be west of, or past the meridian,
the effects of the erroneous position of the polar axis will be
reversed, and the adjustments must be reversed also. It is
desirable that these 2 adjustments should be repeated until the
results are satisfactory.

3rd. To set the declination and polar axis at right angles to
each other.

Place a striding spirit level upon the cylindrical points on
which the former turns, and by moving the hour circle bring
the bubble to the middle of its run. The declination axis will
then be horizontal.[1]

Read the hour circle, turn the polar axis half round, again
bring the declination axis into a horizontal position, and again

---

[1] This supposes the level to be rigorously true.

Plate XLIX.

Fig. 150.

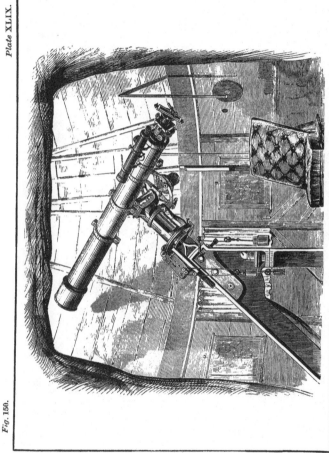

A 7½-INCH EQUATORIAL.

read the hour circle. If the readings are the same (or, where the circle is graduated to 24h., differ exactly by 12h.) in both positions, the declination axis is in adjustment. But if not, place the hour circle half way between the positions it has and ought to have, and make the declination axis horizontal for the 3rd time by means of the screws which adjust it.

4th. To place the optical axis exactly at right angles to the declination axis.

Put in a transit eye-piece, and observe the time of the passage of some star over the centre wire, and read off the hour circle. Turn the polar axis half round, observe a 2nd passage, note the time, and again read off the hour circle. If the interval of the time between the 2 observations, shown by the clock and shown by the circle coincide, all is right, if not it is evident that one of the transits has been observed too early, and the other too late, on account of the erroneous position of the wires. If the time elapsed *exceed* by 4s., suppose, the difference of the hour angles, the first transit has taken place 2s. too soon, and the second, 2s. too late. Turn the instrument again upon the star, and observe how far it appears to travel in 2s.; then if the instrument is in the 1st position, move the wires this quantity in R. A. with the star, and *vice versâ*, if the instrument be in the 2nd position. With a micrometer, or with a mark, this adjustment may be accurately performed at once. It is desirable that the star chosen be situated not far from the equator, in consequence of an apparent angular motion of the sphere in that point being so much faster than in the pole or midway between.

5th. To make the indices of the declination circle read 90° when the line of sight and polar axis are parallel. Turn the instrument on any object in the heavens, and read the declination circle. Turn the instrument half round, again direct the telescope to the object, and again read the circle. Half the difference of the readings will be an index error, which may be corrected by moving the verniers upon the index bar, or kept as a *constant of correction* to be applied with the proper sign to the observations. This latter course is recommended when the account of the error is not too great to affect the ready finding of an object.

6th. To make the indices of the hour circle point to o h. when the declination axis is horizontal.

This is simply done by moving the verniers so that the arrows shall point to 12h. and 24h. respectively, the instrument being placed in the meridian.

It is also requisite that the polar axis should be at right angles to the plane of the hour-circle. This, however, need hardly be called an adjustment, as it ought to be attended to by the maker.

Every one of these adjustments should be effected (the second excepted), with the telescope as near as possible in the meridian, this being the most favourable position of the instrument, as ordinarily constructed, for symmetry and strength; moreover the correction for refraction is applied with greatest facility under these circumstances.

In performing them, the most important thing for the observer to possess is —

PATIENCE.

The equatorial being now in adjustment is ready for use. A few remarks on this subject may not be out of place. Supposing it be required to find a certain star whose R. A. is 16h. 40m. and declination + 45° 47′, and that the time shown by the sidereal clock is 12h. 16m. As the R. A. of the star is greater than the sidereal hour on the meridian, the star sought for, has not yet come to the meridian. Subtracting 12h. 16m. from 16h. 40m. we have 4h. 24m. as the hour angle. Turn the telescope to the east, and set it to the reading, 12h. less 4h. 24m., or 7h. 36m. of the hour circle; then setting the declination circle to 45° 47′ north, the object sought will be seen in the centre of the field. With a little practice in this way the observer will soon be able to fix his telescope upon an object, a small allowance being made to the circle reading for the effect of refraction.

Let us take now the converse of this proposition. Suppose the observer suddenly picks up an unknown comet, and he desires to obtain a record of its place at some given time. If he is content with an approximation to the truth, he may proceed thus. Let the comet be placed by eye in the centre of the field, the time noted, and the circles read off, and he will then possess all the required data.

For instance, suppose at 13h. 25m. sidereal time a comet is seen in the field of the telescope, the hour angle of which is 4h. 17m. west, and the declination circle — 9° 35', what is the comet's position?

Seeing that in this case the object is 4h. 17m. past the meridian, and the hour on the meridian, when the observation was made, was 13h. 25m., it is clear that the R. A. of the comet is 13h. 25m. less 4h. 17m., or 9h. 6m.; and the declination 9° 35' south.

Supposing the observer not to be content with an approximate position, a micrometer must be called into requisition, and a *star of comparison* selected. The mode of procedure is this: the difference between the Right Ascension and the Declination of the comet and star is ascertained by measurement, and the position of the star being known from a standard catalogue, the position of the comet is readily ascertained.

For instance, suppose that the R. A. of a standard star is 16h. 16m. 35·4s., and its declination +47° 15' 37", and by a micrometer it is found that a comet is preceding the star in question by 2·7s., and is north of it 4' 21" in declination, what is the comet's position?

	R. A.				Decl.		
	h. m. s.				° ' "		
Star ...	16 16 35·4	...		...	+47 15 37
Subtract ...	2·7	...	Add	...	4 21
True R.A.	16 16 32·7			True Decl.	+47 19 58

In practice, index errors and correction for refraction must be scrupulously taken into account, when precision is required, as it always ought to be.

CHAPTER IV.

THE TRANSIT INSTRUMENT.

Its Importance.—Description of the Portable Transit.—Adjustments of the Transit. — Four in Number.—Method of performing them.— Example of the Manner of recording Transit Observations of Stars. —Of the Sun.—Remarks on Observations of the Moon.—Of the larger Planets.—Mode of completing imperfect·Sets.—The Uses to which the Transit Instrument is applied.

By far the most important of what may be termed the miscellaneous astronomical instruments is the Transit, or Transit-circle, the smaller and less perfect kinds being chiefly used for taking the time, and the larger for measuring the positions of stars, &c., for forming catalogues.

We shall only describe the small, or portable, transit.

The instrument consists of three principal parts, the telescope, the stand, and the circle : *a b*, is a telescope of a large field and low power, the tube of which is in 2 parts, connected by a cubical centre-piece, into which, at right angles to the optical axis, are fitted the larger ends of 2 cones, *c, c*, which form the horizontal axis of the telescope, the smaller ends of each cone are accurately ground to 2 perfectly equal cylinders, or pivots. These pivots rest on Y's, angular bearings, which surmount the 2 side standards, *e* and *w*, of which *e* may be called the eastern, and *w* the western. One of the Y's is fixed in a horizontal groove, so that, by means of a screw, a small azimuthal motion may be imparted to the instrument; in like manner a small motion in altitude may be obtained by turning the foot screw *g*. On one end of the axis is fixed, so that it may revolve with it, a declination circle, *d*, divided to degrees and read by verniers to minutes, &c. Over this is fixed a level, *f.* The other cone is hollow, in order that light coming from the lamp, *h*, may pass to the centre-piece, where there is a plane pierced mirror inclined at an angle of 45°, which reflects the light to the wires placed in the principal focus. There are usually 4 or 6 wires; in the

Fig. 41.

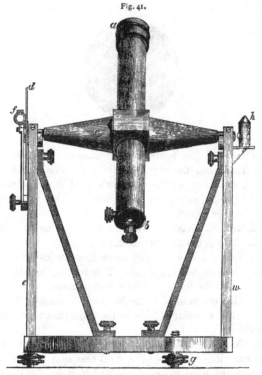

The Portable Transit Instrument.

former case, 1 is placed horizontally, and 3 vertically; in the latter, 1 horizontally, and 5 vertically. The lamp is furnished with a sliding diaphragm, by which the quantity of light allowed to pass out may be increased or diminished as may be required.

A striding level is furnished with the instrument, to ascertain the level of the horizontal axis when required.

It is needless to say that it is of paramount importance that all the parts be perfectly rigid, and free from the slightest flexure, which would immediately vitiate all the observations.

Fig. 42.

Arrangement of the Wires in a Transit Instrument.

The Transit adjustments are 4 in number. It is necessary,

1st. That when the wires and the object are both in focus, a lateral movement of the observer's head does not cause a similar apparent movement of the wires. This is called the adjustment for *parallax*.

2nd. That the axis on which the telescope moves be horizontal. This is the adjustment in *level*.

3rd. That the line of sight move in a vertical circle perpendicular to the horizontal axis. This is sometimes called the adjustment of the *line of collimation in azimuth*.

4th. That the vertical circle be made to coincide with the plane of the meridian. This is the adjustment for *error in azimuth*.

1st. To adjust for parallax. First focus the wires by means of the moveable eye-piece tube, until they appear well defined, then turn the telescope on a star, and if it can be distinctly seen, and the star when bisected by the horizontal wire, remains so when the eye is quickly moved into another position, this adjustment is complete ; if not, move in or out, as may be required, the larger sliding tube, until the proper focus both of the star and wires can be obtained. This adjustment frequently occasions some trouble, but when once done seldom requires repeating.

Fig. 151.   *Plate* L.

A 3-INCH TRANSIT INSTRUMENT.

2nd. To level the axis in which the telescope moves. Place upon the pivots the ends of a long striding level, and bring the air bubble to the centre of its run by turning the foot-screw. Turn the level end for end, and if the bubble retains its middle position the axis is horizontal, but if not it must be brought back half, by the foot-screw, and the other half by turning the small adjusting-screw at one end of the level. Repeat this operation until the results are satisfactory.

3rd. To make the line of sight move in a vertical circle, at right angles to the horizontal axis.

Turn the telescope on some distant, small, and well-defined terrestrial object, and bisect it with the centre wire, giving an azimuthal motion by means of the screw, if necessary. Elevate or depress the telescope to see whether the object still remains bisected in every part by the middle wire; if not, loosen the screws which hold the eye end of the telescope in its place, and turn the end round very carefully until the error is removed. Lift up the whole instrument bodily from the Y's and reverse, end to end: if the object is still bisected by the centre wire, the collimation in azimuth is perfect; but if not, move the centre of the cross wires half way towards the object by turning the small screws which hold the wire plate, and if this half distance has been correctly estimated, the operation of adjustment will be finished. Again bisect the object by the centre of the cross wires by turning the azimuthal screw, and repeat the operation till the object is bisected by the centre of the cross wires in both positions of the instrument, and the adjustment will be known to be perfect.

4th. To make the vertical circle described by the telescope when moving in its horizontal axis, coincide with the plane of the meridian.

This may be effected in several ways. The vertical circle bisects the zenith, and therefore passes through one part of the meridian. If, then, we can make it also bisect another part in the meridian, it follows it will wholly move in that plane. Compute from tables to be found in the *Nautical Almanac* the time of the meridian passage of Polaris, and at the computed time look at the star by the middle vertical wire, and the instrument will be then adjusted to the plane of the meridian.

The following is another and better method : —

Observe the interval elapsing between meridian passage shown by the instrument at the upper and lower culmination. Then if the interval between the inferior and superior passages be equal to the interval between the superior and inferior, the adjustment to the meridian is perfect; but if the interval between the inferior and superior passages be less than the interval between the superior and inferior, the circle described by the line of sight deviates to the eastward of the true meridian from the zenith to the north point of the horizon, and to the westward from the zenith to the south; while if the intervals between the inferior and superior passages be the greater, the deviation is in the contrary direction.

Let Δ be the difference between the observed interval and 12 hours, or half the difference of the 2 intervals in seconds, π, the polar distance of Polaris, and λ the latitude of the place; then x, representing the deviation of the meridian in time, will be found by the logarithmic formula —

$$\log. x = \log. \frac{\Delta}{2} + \log. \sec. \lambda + \log. \tan. \pi - 20.$$

EXAMPLE.

Place of observation, Greenwich, $\lambda = 51^\circ\ 28'\ 38\cdot2''$.
Polar distance of Polaris, $\pi = 1^\circ\ 25'\ 45\cdot8''$.
Difference of intervals from 12 hours ; 1m. 12s. = 72s.

$\frac{\Delta}{2} = 36$	... log.	$= 1\cdot5563025$
$\lambda = 51^\circ\ 28'\ 38\cdot2''$ ...	log. sec.	$= 10\cdot2056070$
$\pi = 1^\circ\ 25'\ 45\cdot8''$ ...	log. tan.	$= \underline{8\cdot3970448}$
$x = 1\cdot44$s.	... log.	$= 20\cdot1589543$

The value of a revolution of the azimuthal screw must next be determined. Note the sidereal time of the passage across the centre wire of an equatorial star on one day ; turn the screw through 1 revolution, and then note the time of the second passage; the interval between these 2 passages will be the value in time of 1 revolution of the screw. Suppose the difference thus observed to amount to 2 seconds, then the value of 1 complete revolution of the screw will amount to 2 seconds; this must be reduced to the horizon, by increasing it in the ratio of co-sine of latitude to radius, and may then be thus applied to correct the error of deviation as found above.

The instrument being in adjustment, the following is an example of the reduction of an observation: —

Transit of a^2 Capricorni over the meridian at Uckfield, June 30, 1861: —

		m.	s.
Wire I.	...		48·5
II.	...		20·2
III.	...	10	52·0
IV.	...		9·0
V.	...		14·0
			143·7
			·2
			28·74
			+24
			52·74

	h.	m.	s.
Computed time of passage	20	10	24·37
Observed ,,	20	10	52·74
			28·37 clock fast.

If the computed time is less than the observed, the clock is fast, and *vice versâ*.

In taking observations of the Sun, the time of the transit of its centre is the time required, but as it would be impossible to estimate this accurately, the time of each limb coming in contact with each wire is noted, and thus a mean of all these gives the required result.

EXAMPLE.

			☉ 1st Limb.			☉ 2nd Limb.
Wire I. ...			20·4 ...			28·7
II. ...			38·7 ...			47·2
III. ...	11	58	57·0 ...	12	1	5·7
IV. ...			15·5 ...			24·0
V. ...			33·7 ...			42·3
			165·3			147·9
			·2			·2
			33·06			29·58
			+24			−24
			57·06			5·58

Combining these results for the two limbs : —

$$
\begin{array}{rrr}
11 & 58 & 57·06 \\
12 & 1 & 5·58 \\
\hline
2)24 & 0 & 2·64 \\
\hline
12 & 0 & 1·32
\end{array}
$$

= the mean value, for the time of the meridian passage of the Sun's centre.

In taking transits of the Moon, only the bright limb can be observed [1]; the time of the centre transit can however be deduced by the use of tables.

In observing transits of the larger planets, it is recommended that one limb be observed at the 1st, 3rd, and 5th wires, and the other at the 2nd and 4th; then the mean of these observations will give the passage of the centre.

In practice it will frequently happen that owing to clouds, or other causes, the transit of an object across each wire cannot be noted. In this case, provided the star's declination be known, the imperfect set can be reduced without difficulty to the centre wire; and thus the instant of meridian passage ascertained.

If, however, observations at corresponding wires, as for instance, the 1st and 5th, or the 2nd and 4th, have been ob-

---

[1] Sometimes when the Moon is very near its full phase, both limbs are observed, a small correction for defective illumination being applied.

tained, and the angular interval between the wires is exactly the same, the time for the centre wire can be obtained by simply taking a mean.   On the other hand supposing, as will commonly be the case, that the wires are not equidistant, a formula of reduction must be made use of.   Let R represent the reduction required, ϵ, the equatorial interval in seconds of arc, and π, the polar distance of the object; then,

$$R = \epsilon \, \text{cosec. } \pi.$$
or
$$\log. R = \log. \epsilon + \log. \text{cosec. } \pi.$$

This formula must be applied to each wire observed, and a mean of all the results will give a fair value for the centre passage.

It is obviously necessary, however, that the equatorial intervals should be carefully known.   This must be done beforehand in the following manner, and the resulting values of the intervals preserved as a *constant*.

Observe the time occupied by some star (very nearly in the equator) whose declination is very accurately known, in passing from one of the side wires to the centre.   Multiply this interval by the sine of the star's polar distance.

This operation must be repeated for every wire, and on several stars for each; a very accurate determination can thus be arrived at.   For large instruments capable of giving results within 0·05″, a further correction is necessary for objects within 10° of the pole.   In this case the following is the formula for the whole reduction :—

$$R' = \tfrac{1}{15} \sin.^{-1} (\text{cosec. } \pi \sin. 15 \, \epsilon).$$

The large transits in use in first class observations, and commonly termed " transit circles," are used for determining with the greatest accuracy the Right Ascension of heavenly bodies.   For such purposes the portable transit is unsuited, except for approximations, its chief use being to ascertain the time and the latitude.   The method of performing the former problem has already been explained.   The latter is effected by placing the instrument in the prime vertical, or at right

angles to the meridian. The process involves spherical trigonometry, and is unsuited to these pages.[1]

With reference to the former problem it may be well to
mention, what probably most people know, that the observed
time of the Sun's centre passing the meridian is only the
instant of *apparent noon;* the time of *mean noon,* in use in the
civil reckoning of time, must be deduced from the apparent
by means of an equation-of-time table.[2]

The time shown by a sidereal clock when any celestial
object crosses the meridian, should coincide with its Tabular
Right Ascension. The difference then between the time
shown by the clock, and the R. A. as tabulated, is the clock-
error, as before explained.[3]

---

[1] Two valuable papers on these applications of the Transit Instrument
will be found in the *Memoirs* R. A. S., vol. xxviii. p. 235, *et seq.*

[2] See above, page 221.

[3] For a further elucidation of the details connected with the subject
of the preceding chapter, the reader is referred to the *English Cyclopædia,*
Arts and Sciences div., art. "Transit," where will be found, incomparably, the best treatise on the Transit Instrument extant. It was
written, we believe, by the late Rev. R. Sheepshanks, M.A.

355

CHAPTER V.

OTHER ASTRONOMICAL INSTRUMENTS.

The Altazimuth.— The Mural Circle.—Borda's Repeating Circle.— The Zenith Sector.— The Reflex Zenith Tube.— The Sextant. —The Box Sextant. — Troughton's Reflecting Circle. — The Dip Sector. — The Floating Collimator.—Airy's Orbit Sweeper.

OF the other instruments used for astronomical purposes the following are the chief:—

1. The Altazimuth.
2. The Mural Circle.
3. Borda's Repeating Circle.
4. The Zenith Sector.
5. The Reflex Zenith Tube.
6. The Sextant.
7. The Box Sextant.
8. Troughton's Reflecting Circle.
9. The Dip Sector.
10. The Floating Collimator.
11. The Orbit Sweeper.

The instruments comprised in the above list are, for the most part, not required for the purposes of the amateur astronomer. We shall therefore content ourselves with a very brief mention of each, adding, however, references for the use of those who may desire to know more about them.

The *Altazimuth*, as its name implies, is for the measurement of altitudes and azimuths. It may be considered as a modification of the ordinary transit instrument, the telescope, circle, and stand of which are capable of motion round a vertical pivot. The altazimuth may therefore be used for meridional or extra-meridional observations indifferently, and when of a portable size, may be looked upon as a theodolite of a superior construction.[1]

The *Mural Circle* consists of a graduated circle furnished with a suitable telescope, and very firmly affixed to a wall (*murus*) in the plane of the meridian. It is used for determining, with great accuracy, meridian altitudes and zenith distances, from which may be found declinations and polar distances.[2]

Borda's Repeating Circle, so called from the name of its inventor, is employed for the measurement of angular distances, both of celestial and terrestrial objects. The principle consists in repeating the readings of an angle, and thus eliminating almost wholly the errors due to defective graduation. This instrument was invented in France, somewhere between 1780 and 1790, in which country it was much used. In England, however, it was never popular, firstly, because when invented, the graduation of English instruments was so much superior to those of foreign make, as to render it less needed; and secondly, because of the calculation required.[3]

The *Zenith Sector*, used for determining the zenith point, and whether or not the Earth's orbit afforded any sensible parallax. Invented by Hook, about the year 1669, and chiefly used now in geodetical operations.[4]

The *Reflex Zenith Tube*, used for observations of the star γ Draconis. Invented by the Astronomer Royal some years ago.[5]

The *Sextant*, sometimes called from its inventor *Hadley's sextant*, is a graduated arc occupying about the sixth part of the circumference of a circle. It is employed for measuring

---

[1] Pearson, *Pract. Ast.*, vol. ii. pp. 413–436, 457, 472; Heather, *Mathematical Instruments*, pp. 153–9; Simms, *Treatise on Instruments*, pp. 92–112.

[2] Pearson, *Pract. Ast.*, vol. ii. pp. 472–488; *English Cyclopædia*, art. "Astronomical Circle."

[3] Brewster's *Cyclopædia*, art. "Circle;" *English Cyclopædia*, art. "Repeating Circle;" *Memoirs* R. A. S., vol. i. p. 33; Pearson, *Pract. Ast.*, vol. ii. pp. 578–580; Breen, *Pract. Ast.*, pp. 381–384.

[4] Pearson, *Pract. Ast.*, vol. ii. pp. 531–549.

[5] *Greenwich Observations*, 1854.

angular distances especially of celestial objects, and is of
great importance to the navigator and traveller for obtaining
local time, latitude, longitude, &c.

The principle of the sextant depends on the practical
application of the following theorem in optics : *That when a
ray of light, proceeding in a plane at right angles to each of
two plane mirrors, which are inclined to each other, at any
angle whatever, is necessarily reflected at the plane surfaces of
each of the mirrors, the total deviation of the ray is double the
angle of inclination of the mirrors.*[1]

The *Box Sextant* is similar in principle to the preceding,
but is more portable, and used rather for surveying than
astronomical purposes, though in conjunction with an artificial
horizon, it becomes valuable for obtaining solar time.[2]

Troughton's Reflecting Circle is a different adaptation of the
same principle involved in the two kinds of sextants. It consists of a complete graduated circle, having the telescope and
reflector on one side of the circle, whilst the graduations and
verniers are on the other.[3]

The *Dip Sector*, another instrument of Troughton's invention,
is used for determining the dip of the horizon. The principle
is similar to that of the sextant.[4]

The *Floating Collimator* is used for determining the position
of a fixed point. It is an instrument of great practical value,
and was invented by the late Captain Kater.[5]

Airy's Orbit Sweeper. Since these pages were first prepared for the press, the Astronomer Royal has published a

---

[1] *Phil. Trans.*, vol. xxxvii. p. 147. 1731; Pearson, *Pract. Ast.*, vol. ii.
pp. 572–577; *Nautical Magazine*, vol. i. p. 351; *English Cyclopædia*, art.
"Sextant;" Galbraith and Haughton, *Optics*, pp. 63–66; Heather,
Math. Instr., pp. 137–141; Simms, *Treat. on Instr.*, pp. 49–56; Young,
Nautical Astronomy, pp. 172–7.

[2] Heather, *Math. Instr.*, pp. 117–122; Simms, *Treat. on Instr.*, pp. 61–3.

[3] Pearson, *Pract. Ast.*, vol. ii. pp. 586–592; Simms, *Treat. on Instr.*,
pp. 57–60.

[4] Heather, *Math. Instr.*, pp. 141–4; Simms, *Treat. on Instr.*, pp. 65–8.

[5] *Phil. Trans.*, vol. cxv. pp. 147–178. 1825; Pearson, *Pract. Ast.*,
vol. ii. pp. 446–457; Herschel, *Outlines of Ast.*, p. 107.

description of a new contrivance under the above name which
he thinks will possibly meet an acknowledged difficulty.

The case for application is that of a comet or planet, which
by previous calculation is known to be pursuing a certain, and
tolerably definite, track through the heavens, but whose actual
position at any given time is unknown.   To sweep for such
an object, an unmounted telescope is of little or no use, and
an equatorial is hardly more suitable, unless the path of the
object be continuously through the same parallel of declination

Fig. 43.

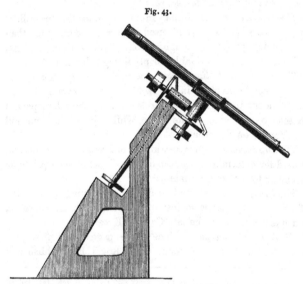

Airy's Orbit Sweeper.

eastward and westwards, or through the same hour of Right
Ascension, northwards and southwards, a condition which it
is scarcely necessary to mention, never subsists, except it be
for a very limited period of time.   The apparent course of
celestial objects, whether planets or stars, being almost always
inclined.

It is to follow by one motion this inclined path that Mr.
Airy's new instrument is designed.   It resembles a German

equatorial, the polar axis of which is of a greater length than usual, and which works, at its upper end for some distance, in a tubular bearing. The declination, or cross axis, carries at one end a counterbalance, and at the other, not as in the regular equatorial, the telescope, but a small trunk in which a second and smaller cross axis turns; to one end of this is attached a counterbalance, and then to the other the telescope.

"By giving a proper position in rotation to the first cross-axis, the inclination of the second cross-axis to an astronomical meridian may be made any whatever, and therefore the inclination of the circle in which the telescope will sweep may be made any whatever; and it may be made to coincide with the definite line drawn on the celestial sphere in which the comet is to be sought."

The inventor thinks that an instrument of this kind would be found of great service to lunar photographers, as the difficulty of following the Moon with an equatorial is well known.[1]

The most complete treatise on astronomical instruments ever published is undoubtedly that of Dr. Pearson, to which the reader is referred for a full account of the various matters touched upon in the preceding pages.

---

[1] *Month. Not.* R. A. S., vol. xxi. p. 159. A model was exhibited at the Society's meeting on March 8, 1861.

CHAPTER VI.

HISTORY OF THE TELESCOPE.

Early History lost in obscurity.—Vitello.—Roger Bacon.—Dr. Dee.—Digges.— Borelli's Endeavour to find out who was the Inventor.—His Verdict in favour of Jansen and Lippersheim of Middleburg.—Statements by Boreel.—Galileo's Invention.—Scheiner's use of Two double convex Lenses.—Lenses of long Focus used towards the close of the 17th Century.—Invention of Reflectors.—Labours of Newton.—Of Halley.—Of Bradley and Molyneux.—Of Mudge.—Of Sir W. Herschel.—Of the Earl of Rosse.—Of Mr. Lassell.—Improvements in Refracting Telescopes.—Labours of Hall.—Of Euler.—Of the Dollonds.—The largest Refractor yet made.

THE early origin of the telescope, like that of most other important inventions, is lost in obscurity, and it is now impossible to determine who was the first maker. It is certain that some time prior to the end of the 13th century lenses were in common use for assisting in procuring distinctness of vision. A certain Vitello, a native of Poland, seems to have done something in this line; and Roger Bacon, in one of his works, employs expressions which show that even in his time (he died in 1292), spectacles were known.[1]

Seeing that this was the case, it is almost certain that some combination of 2 or more lenses must have been made in the interval which elapsed between Bacon's time and the commencement of the 17th century, when telescopes are usually considered to have been invented. Dr. Dee [2] mentions that though some skill is required to ascer-

---

[1] *Opus Majus*, Part iii. cap. iv. p. 357, Ed. S. Jebb, fol. : London, 1733.
[2] Preface to Euclid's *Elements*, 1570.

tain the strength of an enemy's force, yet that the commander of an army might wonderfully help himself by the aid of "perspective glasses," a phrase which must refer to some kind of optical instrument then in use.   The well-known old philosophical writer, Digges, states that "by concave and convex mirrors of circular [spherical] and parabolic forms, or by paines of them placed at due angles, and using the aid of transparent glasses which may break, or unite, the images produced by the reflection of the mirrors, there may be represented a whole region ; also any part of it may be augmented, so that a small object may be discerned as plainly as if it were close to the observer, though it may be as far distant as the eye can descrie."[1]

A second edition of Digges's work, edited by his son, was published in 1591, in which the latter affirms that "by proportional mirrors placed at convenient angles, his father could discover things far off; that he could know a man at a distance of 3 miles, and could read the superscriptions on coins deposited in the open fields."   Though these statements are doubtless exaggerations, yet that some kind of optical instruments were known to the writer in question there can be no doubt.

A claim to the invention of the telescope has been put in on behalf of Baptista Porta, who lived between 1545 and 1618; it is probable, however, that all he noticed was, that an object viewed through a convex lens was apparently enlarged in size.

Towards the middle of the 17th century, Borelli, a Dutch mathematician of some repute, published a book[2] containing the result of some researches carried on by him for ascertaining what he could, connected with the invention of the telescope.   He decides, on the whole, in favour of Zachariah Jansen and Hans Lippersheim, two spectacle-makers of Middleburg, Holland.   In a letter written by Jansen's son, the date of the discovery is stated to have been 1590, though

---

[1] *Pantometria,* 1571.
[2] *De Vero Telescopii Inventore,* 4to. : Hagæ, 1665.

another account makes it 1610. In the same work is given a letter, written by M. Boreel (Dutch minister at the Court of St. James's), who mentions that he was acquainted with the younger Jansen, and had often heard of his father as the reputed inventor of the *microscope*, adding that telescopes were first made by Jansen and Lippersheim in 1610, who presented one to Prince Maurice, of Nassau, by whom they were desired to keep secret their discovery, as he thought he might, by means of one of these instruments, obtain advantages over the enemy, Holland being then at war with France.

Boreel further mentions that Adrian Metius and Cornelius Drebbel went to Middleburg and purchased telescopes from Jansen. Descartes's account differs from this. He says [1], that about 30 years previous (to 1637, when his book was published) Metius, who was fond of making burning glasses, by chance placed at the end of a tube 2 lenses, the one thicker in the middle, and the other thinner, than at the edges, and thus was formed the first telescope. It is now impossible to reconcile these discrepancies. Harriot observed the Sun with a telescope in the year 1610, as we learn from his papers, edited and published a few years ago by Professor Rigaud, but there is no evidence to show whether it was of English or foreign construction.

Whatever might have been the exact period at which the telescope was invented, certain it is that the knowledge of it was for some years confined to northern Europe. Galileo knew nothing of it until 1609, when he casually received some information on the subject from a German he met at Venice. He mentions that he then desired a friend at Paris to make certain inquiries for him. On receiving some intelligence to guide him, he was enabled to contrive a telescope on the principle already referred to, magnifying no less than 3 times! He subsequently made one magnifying 30 times. The fruits of this discovery are well known, and include spots on the Sun, the satellites of Jupiter, the phases of Venus, &c.

---

[1] *Dioptrica*, cap. i.: Lugduni Batavorum, 1637.

Though a telescope made on this principle is exceedingly defective for viewing distant objects, on account of the small field it embraces, yet some years elapsed before any improvement was made.

Kepler first pointed out the possibility of forming telescopes of 2 convex lenses[1], but he did not reduce his idea to practice, neither was it done for many years afterwards. Scheiner, in 1650, described an instrument of this kind, adding that he showed one to the Archduke Maximilian 13 years previously[2], and that the images were inverted. About this time De Rheita constructed telescopes of 3 lenses, which he stated gave a better image than 2; he also made binocular telescopes, instruments having 2 tubes placed side by side, and furnished with similar magnifying powers. We have already spoken of chromatic aberration, and of the immense focal length of some of the lenses used for telescopes. This was towards the close of the 17th century. Campani of Bologna, in 1672, made for Louis XIV. a telescope the focal length of whose object-lens was 136 feet; Auzout had one 600 feet, but he was unable to use it, it seems. Huyghens presented one to the Royal Society 123 feet long[3], and which is, we believe, still preserved by that illustrious body. Practical astronomy is indebted to Huyghens for the negative eye-piece, a most valuable invention.

The extravagant lengths which the dioptric telescopes had now reached resulted in attempts being made to see whether an equal magnifying power could not be attained in some other manner.

Mersenne, in 1639, suggested the employment of a spherical reflector for forming an image which might be magnified by means of a lens. Descartes, to whom the proposal was submitted, ridiculed it, and in consequence (we may presume) the idea was dropped. However, in 1663, Gregory renewed it, though it does not appear that he had any previous knowledge of what Mersenne had proposed, using, instead of

---

[1] *Dioptrica.*   [2] *Rosa Ursina,* &c.
[3] *Astroscopia Compendiaria:* Hagæ, 1684.

a spherical, a paraboloidal speculum. He came to London for the purpose of getting an instrument of the kind constructed, but not finding any workman who could do it, he was obliged to relinquish the project.

Shortly after, Newton, finding that it was impossible to overcome the aberration caused by the unequal refrangibility of the different coloured rays of light, gave up the hope of constructing refracting telescopes which were likely to be of any great use, and turned his attention to the manufacture of reflectors. Having in 1669 found an alloy which he thought would be suited for a speculum, he began to cast and grind one with his own hands, and early in 1672 he completed 2 telescopes, a detailed account of which he transmitted to the Royal Society.[1] The radius of the concavity of the one was 13 inches, and its magnifying power, 38. In the same year that Newton finished his telescopes, Cassegrain proposed the arrangement which now bears his name, though it does not appear that he actually constructed one. The first reflecting telescope, the speculum of which was pierced in the centre so as to permit objects to be viewed by directly looking at them, was made by Hook in 1674.[2]

Very little progress was made in improving reflecting telescopes for many years, in consequence of the difficulty in obtaining metal suitable for specula. In 1718, Hadley made 2, each 5 feet long[3]; and Bradley and Molyneux, in 1738, succeeded in making a satisfactory one, and having instructed two London opticians, Scarlet and Hearne, they made some for general sale. Mudge was the next great labourer in this field.[5] He was soon followed, and completely eclipsed, by Dr. (afterwards Sir William) Herschel.[6] In late years the

---

[1] *Phil. Trans.*, vol. vii. p. 4004. 1672.
[2] Birch, *Hist. Roy. Soc.*, vol. iii. p. 122.
[3] *Phil. Trans.*, vol. xxxii. p. 303. 1723.
[4] Smith's *Opticks*, vol. ii. p. 302.
[5] *Phil. Trans.*, vol. lxvii. p. 296. 1777.
[6] *Phil. Trans.*, vol. lxxxv. p. 347. 1795.

Earl of Rosse[1] and Mr. Lassell have both made some very large and perfect reflecting telescopes.

Though the above improvements were progressively made in reflecting telescopes, it must not be supposed that attempts to obtain achromatic combinations of glass lenses were abandoned. In 1729, Mr. Chester More Hall, being of opinion that an examination of the physical constitution of the eye would afford some clue to the best means for forming achromatic combinations of lenses, set to work, and at length succeeded in obtaining the much desired result — an image free from colour. Several persons are said towards the close of the last century to have possessed telescopes made by or under the superintendence of Mr. Hall.[2]

In 1747 Euler came to the same determination as Hall, but did not obtain the same successful results. He proposed to employ a lens compounded of glass and water, but it was a signal failure.[3]

Dollond in 1758 invented the achromatic combination now in use, for which he received from the Royal Society the Copley medal; and in 1765, his son, Peter Dollond, found that spherical aberration could be diminished by using lenses of different kinds of glass.[4]

Since this period great advances have been made in the manufacture of telescopes, more especially in the size and purity of the glass employed. The *largest* refracting telescope in this country is, we believe, the one erected at Markree Castle, Sligo, the residence of E. J. Cooper, Esq., who for many years sat as M.P. in the Conservative interest for that county. The object-glass is 14 inches in diameter, and 25 feet focal length, but we are informed that the mounting is very defective. On the whole, perhaps, the new equatorial at Greenwich is one of the finest ever constructed.

---

[1] *Phil. Trans.*, vol. cxl. p. 499. 1850.
[2] *Gent. Mag.*, vol. lx. part ii. p. 890. 1790.
[3] *Transactions of the Berlin Academy:* 1747.
[4] *Phil. Trans.*, vol. l. p. 733. 1758.

BOOK IX.

A SKETCH OF THE HISTORY OF ASTRONOMY.

——◆——

It is not our intention, in the present Book, to enter into a regular history of astronomy : that would occupy more space than we could afford, more especially as there are several works, now extant in the English language, on the subject; all therefore we shall do, will be to lay before the reader a chronological summary of the rise and progress of the science from the earliest period.[1]

It is difficult to assign any exact date for the origin of astronomy, so ancient and so lost in obscurity is it; we shall therefore not attempt to go further back than —

B. C.

720. Occurrence of an eclipse of the Moon, observed at Babylon, and recorded by Ptolemy.

719. Occurrence of 2 eclipses of the Moon, also observed at Babylon, and recorded by Ptolemy.

600. *Thales* founds the Ionian school.

585. Eclipse of the Sun, said to have been predicted by Thales.

500. *Pythagoras* founds the school of Croton, and suspects the motion of the Earth.

---

[1] The first time any astronomer's name occurs, it is printed in Italic letters, but not subsequently.

B.C.

450. *Empedocles*, author of a poem on the sphere.

432. *Meton* introduces the luni-solar period of 19 years.

424. *Meton* and *Euctemon* observe a solstice at Athens.

370. *Eudoxus*, of Cnidus, introduces into Greece the year of 365¼ days.

330. *Calippus* introduces the cycle of 76 years, as an improvement on Meton's. *Pytheas* measures the latitude of Marseilles, and points out the connexion between the Moon and the tides.

300. *Autolychus*, author of the earliest works on astronomy extant in the Greek language. About this period *Timocharis* and *Aristyllus* make those observations which afterwards enable Hipparchus to discover and to determine the precession of the equinoxes. About this period also flourishes *Euclid*, author of the well-known *Elements* of Geometry.

281. *Aratus*, of Cilicia, author of an interesting poem on astronomy, which has been translated into English verse by Lamb.

280. *Aristarchus*, of Samos, author of a work on the magnitudes and distances of the Sun and Moon.

260. *Manetho*, author of a history, now lost.

240. *Eratosthenes*, of Cyrene, determines, with considerable accuracy, the obliquity of the ecliptic, and also the latitude of Alexandria. Other important observations are attributed to him.

212. *Archimedes*, of Syracuse, observed solstices, and attempted to measure the Sun's diameter; he was, however, more distinguished as a natural philosopher.

160–125. *Hipparchus*, possibly of Bithynia, the most distinguished of the Greek astronomers. He wrote a commentary on Aratus; discovered the precession of the equinoxes; first used Right Ascensions and Declinations, though afterwards abandoned them for latitudes and longitudes; probably invented the stereographic projection of the sphere; determined the mean motion of the Sun and Moon with considerable exactness;

B. C.

        suspected that inequality, afterwards discovered by Ptolemy, and known as the Evection; calculated eclipses, and formed the first regular catalogue of stars, &c. &c. : altogether we may fairly call him the Greek Newton.

70±. *Posidonius*, who attempts to verify Eratosthenes' measure of the Earth. His works are all lost.

50. *Sosigenes*, of Alexandria, in conjunction with Julius Cæsar, plans the Julian reform of the calendar.

A. D.

10. *Manilius*, who wrote a poem on astronomy and astrology.

50. *Seneca*, tutor to the Emperor Nero; he wrote a work on natural philosophy, which contains many astronomical allusions, more especially to comets; the writer surmises they are planets of some kind.

80. *Menelaus*, who wrote treatises on spherical trigonometry, and made observations at Rome and Rhodes.

117 (?). *Theon*, of Smyrna, made observations at Alexandria. He wrote on astronomy.

    Cleomedes wrote on astronomy. It is however uncertain whether he lived before or after Ptolemy; though probably before.

130–150. *Ptolemy*, of Alexandria. A well-known observer and writer, author of the celebrated Μεγάλη Σύνταξις, called by the Arabians *The Almagest*. This work contains, among other things, a review of the labours of Hipparchus; a description of the heavens and the Milky Way; a catalogue of stars; sundry mechanical arguments against the motion of the Earth; notes on the length of the year, &c.

    Ptolemy was an industrious observer: to him we owe the discovery of lunar Evection, and atmospheric refraction. He also propounded a theory of the universe, which bears his name.

173. *Sextus Empiricus* wrote against Chaldæan astrology.

238. *Censorinus* wrote on astrology and chronology.

370. *Julius Firmicus Maternus* wrote on astronomy.

A.D.

383. *Pappus*, of Alexandria, wrote a commentary on Ptolemy,
all of which is lost.

385. *Theon*, of Alexandria, wrote an able commentary on
Ptolemy. He has left some tables, and methods for
constructing almanacs.

415. *Hypatia*, daughter of Theon, the first female on record
celebrated for her scientific attainments. She was
murdered in the above year.

470. *Martianus Capella* wrote a work called the *Satyricon*,
which contains a few astronomical ideas. Amongst
others, that Mercury and Venus revolve round the
Sun.

500. *Thius*, of Athens, who made some occultation observations, &c.

546. *Simplicius* wrote a commentary on a work of Aristotle,
now lost.

550. *Proclus Diadochus* wrote a commentary on Euclid, and
the astrology of Aristotle, and also on some astronomical phenomena.

636. *Isidore*, Archbishop of Hispalis (Seville), who wrote on
astronomy.

640. Destruction of the Alexandrian School of Astronomy
by the Saracens under Omar.

720. *Bede*, who wrote an astronomical work.

762. Rise of astronomy amongst the eastern Saracens, on
the building of Bagdad by the Caliph Al Mansur.
Amongst this nation astronomy made great progress
during the succeeding centuries.

880. *Albategnius* or *Albatani*. The most distinguished astronomer between Hipparchus and Tycho Brahe. He
discovered the motion of the solar Apogee, corrected
the value of precession and the obliquity of the
ecliptic. Formed a catalogue of stars. First used
sines, chords, &c.

950. *Alfraganus* or *Al-Fergadi* and *Thalet Ben Korrah*, both
lived about this time. The first wrote on astronomy;

A. D.

and the second propounded a theory relating to the ecliptic.

1000. *Ebn Yunis* and *Abùl-Wefa,* both lived about this time. The former was an Egyptian astronomer of merit. He has left a work containing tables and observations, which displays considerable knowledge of trigonometry. He was the first to use subsidiary angles. Abùl-Wefa first employed tangents, co-tangents, and secants, and is thought by some to have discovered the lunar inequality known as the variation.

1050. *Michel Psellus.* The last Greek writer on astronomy, of note. *Alphetragius* devised an explanation of the motions of the planets.

1080. *Arsachel,* a Spanish Moor, constructed some tables. *Alhazen* wrote on Refraction, and *Geber*, about this time, introduces the use of the Co-sine, and makes some improvements in spherical trigonometry.

1200. *Abùl Hassan* formed a catalogue of stars, and made some improvements in the practice of dialling.

About this time or earlier the Persians constructed some tables which were translated by a Greek physician named *Chrysococca*, in the 14th century. The best known, however, are those of *Nasireddin,* published in 1270, under the patronage of Hulagu, grandson of Genghis Khan.

1220. *Sacrobosco* (Anglicè *Holywood*), wrote a work on the sphere, based on Ptolemy; he also wrote on the calendar. About this time *Jordanus* wrote on the planosphere.

1230. About this year the *Almagest* is translated into Latin, under the auspices of Frederick II., Emperor of Germany.

1252. *Alphonso X.*, King of Castile, aided, as it is supposed, by certain Arabs and Jews, compiles the *Alphonsine Tables.*

1255. *Roger Bacon* wrote on astronomy.

A.D.

1280. *Cocheou-king* made a number of good observations, and used spherical trigonometry, under the patronage of Kublai, brother of Hulagu.

1433. *Ulugh Beigh*, grandson of Timour or Tamerlane, made numerous observations at Samarkand, and is especially noted for his catalogue of stars. He also gave tables of geographical latitudes and longitudes.

1440. Cardinal *Cusa* wrote on the calendar, and, as some affirm, in favour of the Earth's motion.

1460. *George Purbach* publishes trigonometrical tables, and a planetary theory somewhat à la Ptolemy.

1476. *John Müller*, better known by his Latin name, *Regiomontanus*, wrote an abridgment of the *Almagest*, and formed some extensive trigonometrical tables and almanacs.

1486. *George of Trebizonde*, called *Trapezuntius*, first translated the *Almagest* from Greek into Latin.

1495. *Bianchini* publishes tables.

1504. (d.) *Waltherus*, a pupil of Regiomontanus, made numerous observations.

1521. *Riccius* writes a treatise on astronomy, with especial reference to its history.

1528. (d.) *Werner* gave a more correct value of the precession of the equinoxes. *Fernel* gave a very correct measure of a degree on the meridian. Delambre remarks, it must have been accidental, seeing that his data were very imperfect.

1531. (d.) *Stoffler* wrote on the astrolabe, and published almanacs for 50 years.

1543. Publication of Copernicus' *Re Revolutionibus Orbium Celestium*, in which is propounded his theory of the universe, &c. The illustrious author dies in this year.

1552. (d.) *Apian* studied comets with great diligence. *Munster* wrote on clocks and dials.

1553. (d.) *Rheinhold*, a friend of Copernicus, constructed the *Prutenic Tables*.

A.D.

1555. (d.) *Gemma Frisius.*

1556. Dr. *Dee* publishes a book on geometry.

1558. (d.) *Recorde,* said to have been the first English writer on astronomy and the sphere.

1571. (d.) *Leonard Digges,* author of *the Prognostication Everlasting,* and other works.

1572. Apparition of a new star in Cassiopeia, whose position is determined by *Hagecius,* by measuring the meridian altitude, and noting the time at which the observation was made.

1573. *Thomas Digges* proposes, as a means for determining the positions of celestial objects, the method of equal altitudes.

1576. (d.) *Rheticus,* editor of the *Opus Palatinum.*

1577. (d.) *Nonius,* inventor of an ingenious division of the circle. Apparition of a comet upon which *Tycho Brahe* makes observations for the detection of parallax, in which he fails, thus showing that comets traverse regions more removed from the Earth than the Moon.

1581. About this time, *Galileo* remarks the isochronism of the pendulum.

1582. Tycho Brahe commences making observations, in the island of Huenen, in the Baltic, near Copenhagen.

1592. (d.) The *Landgrave of Hesse Cassel,* a diligent amateur observer.

1594. (d.) *Gerard Mercator,* author of the projection of the sphere which bears his name.

1595. Thomas Digges, the son of Leonard Digges, publishes a work in this year.

1599. (d.) *Rothmann* observed comets. Publication of *Kepler's Mysterium Cosmographicum. Fabricius* discovers the variability of *o* Ceti.

1600. *Jordanus Brunus* burnt to death at Rome for holding certain opinions on the system of the universe.

After the close of the 16th century, observers and observations begin to multiply so, that henceforth we shall find it convenient to tabulate the principal astronomers of note, and then to give, in chronological order, an epitome of their labours. The dates are those of their deaths, but when that is not known, the date of the publication of some work is given in parentheses.

During the 17th century we have the following : —

Tycho Brahe	. 1601	Fontana	. (1646)
Bayer	. (1603)	Longomontanus	. 1647
Scaliger, Jos.	. 1609	Torricelli	. —
Clavius	. 1612	Descartes	. 1650
Calvisius	. 1615	Scheiner	. —
Wright	. —	Wing	. 1651
Fabricius	. 1616	Petavius	. 1652
Napier	. 1617	Crabtree	. —
Gunter	. 1626	Pascal	. 1653
Snellius	. —	Gassendi	. 1655
Malapertius	. 1630	Lubienitz	. (1668)
Vernier	. (1631)	Riccioli	. 1671
Mœstlin	. —	Borelli	. 1679
Kepler	. —	Dörfel	. (1680)
Lansberg	. 1632	Picard	. 1682
Schickhardt	. 1635	Hevelius	. 1687
Horrox	. 1641	Auzout	. 1693
Galileo	. 1642	Mercator, N.	. 1694
Gascoigne	. 1644	Huyghens	. 1695
De Rheita	. (1645)		

A.D.

1603. Publication of *Bayer's* Maps of the Stars.

1604. Kepler succeeds in obtaining an approximate value of the correction for refraction. Apparition of a new star in Serpentarius.

1608. *Hans Lippersheim*, of Middleburg, Holland, invents the refracting telescope, employing a convex object-lens.

1609. Galileo makes a telescope with concave object-lens. Kepler publishes his work on Mars, in which he determines, by Tycho Brahe's observations, the elliptic form of its orbit, and ratio between the areas and the times, thus enunciating his 1st and 2nd laws.

A. D.

1610. Galileo announces the discovery of Jupiter's satellites — of nebulæ — of some phenomena in connection with the appearance of Saturn, afterwards found to proceed from the ring — the phases of Venus,— the diurnal and latitudinal libration of the Moon. Harriot observes spots on the Sun.

1611. Foundation of the Lycean Academy. Galileo observes spots on the Sun.

1614. *Napier* invents logarithms.

1617. *Snellius* measures, by triangulation, an arc of the meridian at Leyden. In consequence, however, of the imperfection of his instruments, the results are not much more certain than the old ones.

1618. Kepler publishes his 3rd law.

1619. Snellius discovers the law of refraction from one medium into another.

1626. *Wendelinus* determines the diminution of the obliquity of the ecliptic; extends Kepler's laws to Jupiter's satellites; and ascertains the Sun's parallax.

1627. Kepler publishes his *Rudolphine Tables*, based on the observations of Tycho Brahe.

1631. *Gassendi* observes the first recorded transit of Mercury over the Sun; and measures the diameter of the planet. *Vernier* describes the instrument which bears his name.

1633. *Norwood* measures an arc of the meridian between London and York, and obtains a more correct value of a degree. *Descartes* promulgates his " System of Vortices." Galileo is forced, by the bigotry of Romish ecclesiastics, to recant his Copernican opinions.

1637. *Horrox* suspects the long inequality in the mean motions of Jupiter and Saturn.

1638. Horrox ascribes the motion of the lunar apsides to the disturbing influence of the Sun, and adduces the oscillations of the conical pendulum as an illustration of the planetary movements.

A.D.

1639. Horrox and *Crabtree* observe the first recorded transit of Venus over the Sun, and the former measures the planet's diameter.

1640. *Gascoigne* applies the telescope to the quadrant and the micrometer to the telescope.

1646. *Fontana* observes the belts of Jupiter.

1647. Publication of *Hevelius's Selenographia*, in which is announced the Moon's libration in longitude.

1650. *Spencer* constructs a telescope with a convex object-glass.

1651. *Shakerley* observes a transit of Mercury at Surat, in the East Indies.

1654. *Huyghens* completes the discovery of Saturn's ring.

1655. Huyghens discovers that satellite of Saturn now known as Titan.

1656. Huyghens publishes his First Treatise on Saturn.

1657. Foundation of the Academia del Cimento at Florence.

1658. Huyghens makes the first pendulum clock.

1659. Huyghens, ignorant of what Gascoigne had previously done, invents a micrometer, and publishes a second treatise on Saturn. *Childrey* writes on the Zodiacal Light.

1660. *Mouton* applies the simple pendulum to observations of differences of Right Ascension, and by this means obtains a very good measurement of the Sun's diameter.

1661. Hevelius, at Dantzic, observes a transit of Mercury.

1662. Foundation of the Royal Society of London. *D. Cassini* begins his researches on refraction. *Malvasia* improves Huyghens' micrometer.

1663. *Gregory* invents the reflecting telescope which now bears his name.[1]

1664. *Hook* detects the rotation of Jupiter on its axis.

---

[1] Gregory possessed another claim to our good opinions: he was a staunch Royalist and supporter of Episcopacy.

A. D.

1665. Cassini determines the time of Jupiter s rotation, and
      publishes the first Tables of the Satellites.
      Hook proposes the reticulated micrometer for the
      measurement of lunar distances. The Brothers *Ball*,
      at Minehead, detect the duplicity of Saturn's ring.

1666. Cassini determines the rotation of Mars and approximates
      to that of Venus. Foundation of the Academy of
      Sciences at Paris. *Auzout*, ignorant of Gascoigne's
      previous labours, applies the micrometer to the tele-
      scope. *Newton* first directs his attention to the
      question of gravitation.

1667. Auzout and *Picard* apply the telescope to the mural
      quadrant, without knowing that Gascoigne had pre-
      viously done the same thing.

1668. Cassini publishes his Second Tables of Jupiter's Satel-
      lites ; and Hevelius, his *Cometographia*.

1669. Newton invents the reflecting telescope which now
      bears his name.

1670. Mouton first uses interpolations in observations.

1671. Picard and *La Hire* publish their degree of the meridian,
      obtained by measuring the arc between Paris and
      Amiens. *Richer*, in a voyage to Cayenne, observes
      the shortening of the seconds pendulum as it is
      brought towards the equator. *Flamsteed* com-
      mences observations at Derby. Cassini begins the
      observations which led to his discovery of the in-
      clination of the Moon's equator, and the coincidence
      of its nodes with those of its orbit. He also discovers
      that satellite of Saturn now known as Iapetus.

1672. Cassini discovers that satellite of Saturn now known as
      Rhea.

1673. Publication of Huyghens' *Horologium Oscillatorium*, in
      which are found the five theories relating to central
      forces. Flamsteed explains the equation of time.

1674. Huyghens, ignorant of what Hook has previously done,
      causes spring watches to be made.

A. D.

1675. *Römer* propounds his discovery relating to the transmission of light, as detected by observations on Jupiter's satellites. Foundation of the Royal Greenwich Observatory. Römer applies the transit instrument for the determination of right ascensions.

1676. Flamsteed commences observations at the Royal Observatory, Greenwich.

1677 *Halley* observes, at St. Helena, a transit of Mercury.

1679. Publication of Halley's Catalogue of Southern Stars. Commencement of the *Connaissance des Temps*.

1680. Flamsteed enunciates the law of the Moon's annual equation. Apparition of a celebrated comet, noticeable on account of its very small perihelion distance, and from its having led Newton to the opinion that comets moved in conic sections.

1681. Publication of *Dörfel's* work on Comets.

1683. Cassini and La Hire discontinue, till 1700, the arc of the meridian, commenced in 1680. Erection of a mural quadrant in the meridian at the Royal Observatory of Paris. Cassini investigates the Zodiacal Light.

1684. Cassini discovers those satellites of Saturn now known as Tethys and Dione.

1687. Publication of Newton's great *Principia*.

1689. Römer uses the transit instrument for taking time.

1690. Huyghens determines, theoretically, the ellipticity of the Earth. Publication of Hevelius's Catalogue of Stars.

1693. Cassini publishes his Third Tables of Jupiter's Satellites, and announces his discoveries on libration. Halley discovers the secular acceleration of the Moon's mean motion.

1694. Newton and Flamsteed commence their correspondence on the subject of the lunar theory, and the theory of refraction.

1700. D. Cassini, aided by J. Cassini, extend southwards the arc, commenced by the former.

The following is a list of the chief astronomers of note during the 18th century : —

Hook	 1703	Bird	. . . .(1766)
Römer	 1710	De L'Isle .	 1768
Cassini, J. D.	. . . 1712	Long	 1770
Leibnitz .	. . . 1716	Harrison .	 1776
La Hire, P.	. . . 1718	Ferguson .	. . . . —
La Hire, G. P. .	. . 1719	Lambert .	 1777
Flamsteed	. . . —	Zanotti .	 1782
Newton .	. . 1727	Wargentin	. . . 1783
Maraldi, J. P. .	. . 1729	Lexell .	. . .(—)
Bianchini	. . . —	D'Alembert	. . . —
Manfredi .	. . . 1739	Euler .	. . . . —
Halley .	. . . 1742	Cassini De Thury	. . 1784
Maclaurin	. . . 1746	Boscovich	. . . 1787
Bernouilli, J.	. . . 1748	Maraldi, J. D. .	1788
Graham .	. . . 1751	Palitzch .	. . . —
Whiston .	. . . 1755	Lepaute, Mdme.	. . 1789
Cassini, James .	. . 1756	Le Gentil	. . . 1792
Fontenelle	. . . —	Bailly	. . . 1793
Ximenes .	. . .(1757)	Saron .	. . . 1794
Simpson, T.	. . . 1760	Mudge .	. . . —
Dollond .	. . . 1761	Du Sèjour	. . . —
Bradley .	. . . 1762	Pingré .	. . . 1796
La Caille .	. . . —	Maraldi, J. P. .	. . 1797
Mayer, T.	. . . —	Borda .	. . . 1799
Bliss .	. . . 1764	Le Monnier .	. . . —
Horrebow	. . . —	Cassini, Count	. .(1800)
Clairaut .	. . . 1765	Ramsden .	. . . —

A. D.

1702. La Hire's researches on the theory of refraction.

1704. Römer commences star observations with a meridian circle.

1705. Halley predicts the return of the comet of 1682 in 1759.

1711. Foundation of the Royal Observatory, Berlin.

1714. J. Cassini discovers the inclination of the 5th satellite of Saturn.

1715. *Taylor's* researches on refraction.

1718. *Bradley* publishes his Tables of Jupiter's Satellites.

A. D.

J. Cassini, and *J. P. Maraldi*, complete, at Dunkirk, the arc, commenced by D. Cassini.

1719. Maraldi's researches on the rotation of Jupiter.

1721. Halley communicates to the Royal Society Newton's table of refractions.

1725. Publication of Flamsteed's *Historia Celestis*. Foundation of the St. Petersburg Observatory. *Harrison* announces the compensation pendulum.

1726. *Bianchini* determines the rotation of Venus. *Graham* invents the mercurial pendulum.

1727. Bradley discovers the aberration of light.

1728. Destruction, by fire, of Copenhagen Observatory, in which were stored the observations of Römer, and *Horrebow*, his successor, all of which are lost.

1729. *Bouger* investigates the theory of refraction.

1731. *Hadley* invents the Sextant.

1732. *J. D. Maraldi* improves the theory of Jupiter's satellites. *Maupertius* introduces into France Newton's theory. *Wright* publishes his Lunar Tables.

1736. Maupertius, and others, measure an arc in Lapland, and Bouger and *La Condamine* in Peru.

1737. *La Caille* and *Cassini* (III.), remeasure the arc of D. Cassini. *Clairaut* improves the theory of the figure of the Earth.

1739. Publication of *Dunthorne's* Lunar Tables.

1740. Publication of J. Cassini's Treatise on Astronomy, in which are given many new tables by himself and his father.

1744. Publication of *Euler's Theoria Motuum*, the first analytical work on the planetary motions.

1745. Bradley discovers the nutation of the Earth's axis. *Bird* commences his improvements in the graduation of instruments.

1746. Publication of Euler's Solar and Lunar Tables, and *Wargentin's* Tables of Jupiter's Satellites.

1747. Researches of Euler, Clairaut, and *D'Alembert* on the theory of the planets. *Mayer* confirms by observation Cassini's theory on the lunar libration.

A.D.

1748. Bouger proposes a divided object-glass micrometer.
      Publication of Euler's Essay on the Motions of
      Jupiter and Saturn.

1749. Investigations by Euler and D'Alembert on precession,
      by D'Alembert on nutation, and by Clairaut on the
      motion of the lunar apogee.   Publication of Hal-
      ley's Tables.

1750. Mayer introduces the use of equations of condition.
      Boscovich measures an arc of the meridian at Rimini.
      Publication of Wright's *Theory of the Universe*, in
      which is propounded that theory of the Milky Way
      afterwards adopted by Sir W. Herschel, and generally
      since.

1751. La Caille goes to the Cape of Good Hope to commence
      a course of observations.

1752. La Caille measures an arc of the meridian at the Cape
      of Good Hope.

1754. Publication of Halley's Solar and Lunar Tables by
      Chappe; also of Clairaut's Lunar Tables.

1755. *Dollond* makes a double object-glass micrometer.
      Mayer first suggests the idea of a repeating circle.
      Occurrence of a transit of Mercury.

1756. Researches of D'Alembert on the figure of the Earth;
      by Euler on the variation of the elements of elliptic
      orbits; and by Clairaut on the perturbation of
      comets.   Mayer's Catalogue of Zodiacal Stars.

1757. Publication of La Caille's *Astronomiæ Fundamenta.*

1758. Publication of La Caille's Solar Tables.   Invention by
      Dollond of the achromatic object-glass.   Researches
      by Clairaut and *Lalande* on the orbit of Halley's
      Comet.

1759. Publication of Halley's Planetary Tables by Lalande.
      Publication of an improved edition of Wargentin's
      Tables of Jupiter's Satellites.

1760. Bird's Standard Scale.

1761. *Maskelyne* at St. Helena.   Transit of Venus.

1762. Researches by Euler and Clairaut on the perturbations
      of comets.

A.D.

1763. Publication of La Caille's Catalogue of Southern Stars.

1764. Lalande confirms the observations of Mayer on the lunar libration. Publication of *La Grange's* prize essay on the same subject, containing the first application of the principle of virtual velocities. *Mason* and *Dixon* begin the measurement of an arc in Pennsylvania.

1765. Harrison obtains, after many vexatious delays, the reward promised by Parliament for the invention of the chronometer. J. D. Maraldi discovers the libratory motion of the nodes of Jupiter's second satellite.

1766. Publication by La Grange, and also by *Bailly*, of a theory of Jupiter's satellites.

1767. Commencement of the *Nautical Almanac.*

1768. *Beccaria* measures an arc of the meridian in Piedmont, and *Liesganig* in Hungary.

1769. Transit of Venus, which was very successfully observed.

1770. Publication of Mayer's Solar and Lunar Tables. Discovery of Lexell's Comet.

1771. Further researches by Bailly on Jupiter's satellites.

1772. Publication by *Bode* of *Titius's* law of planetary distances.

1773. Researches by La Grange on the attraction of spheroids; by *Laplace* on the secular inequalities of the solar system.

1774. Experiments by Maskelyne on the Earth's attraction, on Mount Schehallien.

1780. Publication of Mason's Lunar Tables.

1781. *W. Herschel* discovers the planet Uranus. Publication of *Messier's* Catalogue of Nebulæ. (*Conn. des Temps,* 1784.) Wargentin discovers that the inclination of Jupiter's 4th satellite is variable.

1782. Laplace calculates the elements of the orbit of Uranus, and investigates the attraction of spheroids.

A.D.

1783. Publication of *Nouet's* Tables of Uranus, and Pingré's *Cométographie*.

1784. Researches by Laplace on the stability of the solar system; on the relation between the longitudes of Jupiter's satellites; and on the great inequality of Jupiter and Saturn. *Roy* measures a base on Hounslow Heath for the connection of the observations of Greenwich and Paris.

1786. Publication of Herschel's first catalogue of 1000 nebulæ (in *Phil. Trans.*) La Grange gives the differential equations for the variation of the elliptic elements.

1787. Laplace's theory of Saturn's ring, and explanation of the acceleration of the Moon's mean motion. Herschel discovers 2 satellites of Uranus. *Le Gendre* and Roy complete the connection of the observations of Greenwich and Paris. Commencement of the trigonometrical survey of England. Herschel commences observations with his 40 feet reflector, and discovers those satellites of Uranus now known as Oberon and Titania.

1788. Publication of La Grange's *Mécanique Analytique*. Herschel suspects that the motions of the satellites of Uranus are retrograde.

1789. Herschel determines the rotation of Saturn, discovers the satellites Mimas and Enceladus, and publishes a catalogue of a second 1000 nebulæ in *Phil. Trans.* Publication of Delambre's Tables of Jupiter and Saturn.

1790. Herschel determines the rotation of Saturn's ring, and discovers 2 new satellites of Uranus. Publication of Delambre's Tables of Uranus, and Maskelyne's Catalogue of Stars. *Brinckley* appointed director of the Dublin Observatory.

1792. Commencement of the trigonometrical survey of France. Publication of *Taylor's* Logarithms, Lalande's Improved Planetary Tables, *De Zach's* first Solar Tables, and his Catalogue of Stars.

A.D.

1793. Laplace's researches on the satellites of Jupiter and the figure of the Earth. *Schröter* determines the rotation of Venus.

1795. Herschel's observations on variable stars, and the dismemberment of the Milky Way.

1796. Foundation of the French Institute of Science. Herschel suspects that the rotations of the satellites of Jupiter are of the same duration as their orbital revolutions. *Oriani* investigates the perturbations of Mercury.

1797. Delambre's observations on refraction. Laplace's theory of tides. *Olbers* publishes his method for determining the parabolic elements of a comet's orbit, since generally adopted by German astronomers.

1798. *Cavendish* demonstrates and measures the mutual attraction of metal balls. Herschel announces his discovery of the retrograde motions of the satellites of Uranus.

1799. Commencement of Laplace's *Mécanique Celeste*. Occurrence of a transit of Mercury. *Kramp's* researches on refraction.

The following is a list of the chief astronomers of note during the present century : —

Méchain	.	.	. 1804	Laplace	.	.	. 1827
Lalande, J.	.	.	. 1807	Wollaston, W.	.	.	. 1828
Cavendish	.	.	. 1810	Young	.	.	. 1829
Maskelyne	.	.	. 1811	Fallows	.	.	. 1831
La Grange	.	.	. 1813	Pons	.	.	—
Messier	.	.	. 1817	Oriani	.	.	. 1832
Burckhardt	.	.	—	De Zach	.	.	—
Mudge	.	.	. 1821	Groombridge	.	.	—
Herschel, W.	.	.	. 1822	Le Gendre	.	.	. 1833
Delambre	.	.	—	Harding	.	.	. 1834
Hutton	.	.	. 1823	Troughton	.	.	. 1835
Bode	.	.	. 1826	Kater	.	.	—
Fraunhofer	.	.	—	Brinckley	.	.	—
Piazzi	.	.	—	Pond	.	.	. 1836

Gambart .	.	.	. 1836	Boguslawski	. .	. 1850
Moll	.	.	. 1837	Colby	. .	. 1852
Rigaud	.	.	. 1839	Arago	. .	. 1853
Olbers	.	.	. 1840	Lindenau	. .	. 1854
Poisson	.	.	. —	Petersen .	. .	. —
Bouvard .	.	.	. —	Mauvais .	. .	. —
Littrow .	.	.	. —	Gauss	. .	. 1855
Cacciatore	.	.	. 1841	Sheepshanks	. .	. —
Henderson	.	.	. 1844	Colla	. .	. 1857
Baily	.	.	. —	Raper	. .	. 1858
Bessel	.	.	. 1846	Bond, W. C.	. .	. 1859
Damoiseau	.	.	. —	Wichmann	. .	. —
Di Vico .	.	.	. 1848	Johnson .	. .	. —
Taylor	.	.	. —	Humboldt	. .	. —
Schumacher	.	.	. 1850	Bishop .	. .	. 1861

A.D.

1798—1804. *Humboldt* travels in America, and makes numerous observations.

1800. *Wollaston's* Circumpolar Catalogue. Bode's Maps and Catalogue. *Mudge* commences his great arc of the meridian, extending from the Isle of Wight to Clifton in Yorkshire. De Zach starts the *Monatliche Correspondentz*, which goes on for 12 years.

1801. Lalande's Catalogue. *Piazzi* discovers the minor planet Ceres. *Swanberg* begins to measure an arc in Lapland.

1802. Olbers discovers the planet Pallas. Lambton begins the measurement of an arc in India. Publication of Herschel's Catalogue of Nebulæ.

1803. Publication of Herschel's Discovery of Binary Stars.

1804. *Harding* discovers the planet Juno. Piazzi publishes the proper motion of 300 stars. De Zach's Solar Tables.

1805. Le Gendre enunciates the method of least squares. Commencement of researches in Stellar Parallax by several observers.

1806. *Méchain* and Delambre complete the French survey. Publication of Delambre's Solar Tables and Tables of Refraction; of Burg's Lunar Tables; of Carlini's

A.D.

Tables of Refraction; of Pond's Catalogue of North Polar Distances (altitude and azimuth). Herschel suspects the motion of the whole solar system towards the constellation Hercules. Publication of De Zach's Tables of Aberration and Nutation.

1807. Olbers discovers the planet Vesta. Extension of the French arc into Spain. Publication of *Cagnoli's* Catalogue, and Piazzi's Catalogue of 120 Stars.

1808. Researches of La Grange and Laplace on the Planetary Theory.

1809. *Troughton's* improvements in the graduation of instruments. *Ivory's* Theorems on the Figure of the Earth. Publication of Gauss's *Theoria Motûs*.

1810. Groombridge's Refraction Tables. Carlini's Solar Tables. *Lindenau's* Tables of Venus. *Bessel* appointed Director of the Observatory of Königsberg.

1811. Lindenau's Tables of Mars.

1812. Erection of Troughton's Mural Circle at Greenwich. *Burckhardt's* Lunar Tables.

1813. Lindenau's Tables of Mercury; and *Pond's* Catalogue of North Polar Distances (circle obs.).

1814. Publication of Piazzi's Catalogue of 7646 Stars. Foundation of the Königsberg Observatory. Commencement of the *Zeitschrift für Astronomie*, which goes on till 1818.

1815. Bessel's researches on Precession.

1816. Lindenau determines a new value of the Constant of Nutation. Poisson's researches on Planetary Perturbations.

1817. Delambre's Tables of Jupiter's Satellites. *Damoiseau's* researches on Halley's Comet.

1818. Publication of Bessel's *Fundamenta Astronomiæ*. De Zach starts the *Correspondance Astronomique*, which goes on till 1825.

1820. Foundation of the Royal Astronomical Society of London. *Reichenbach's* meridian circle erected at Königsberg. Publication of Brinckley's Tables of Refraction.

S

Commencement of the *Astronomische Nachrichten*, which valuable periodical is still in existence.

1821. Foundation of the Cape of Good Hope Observatory. Publication of *Bouvard's* Tables of Jupiter, Saturn, and Uranus. The practice of taking circle observations by reflection, introduced at the Greenwich Observatory. Researches of Poisson on the Precession of the Equinoxes.

1822. Foundation of the Paramatta Observatory, N. S. W. Publication of Harding's *Atlas Cœlestis*. *Argelander's* researches on the orbit of the comet of 1811.

1823. Foundation of the Cambridge Observatory. Researches by Ivory on Refraction. *Encke* suspects the existence of a resisting medium in space.

1824. Publication of *J. Herschel* and *South's* Catalogue of Double Stars. Encke discusses the observation of the transits of Venus in 1761 and 1767 for the determination of the solar parallax. Erection of the Dorpat Refractor.

1825. Commencement of the *Berlin Zones.* Jones's mural circle erected at Greenwich.

1826. Researches of Bessel on the oscillation of the pendulum. Discovery of *Biela's* Comet.

1827. Publication of the Royal Astronomical Society's Catalogue of Stars. The same Society commence the publication of their *Monthly Notices.*

1828. *Airy* discovers a long inequality in the motions of Venus and the Earth. *Kater's* vertical collimator. Publication of Damoiseau's Lunar Tables.

1829. Publication of Pond's Catalogue of 720 Stars. Researches of Poisson on the attraction of spheroids, and of *Pontécoulant* on the orbit of Halley's Comet.

1830. Sir J. Herschel's measures of. 1236 double stars. Publication of Bessel's *Tabulæ Regiomontanæ.*

1831. Sir J. Herschel's micrometrical measures of 364 double stars. Publication of Plana's *Theory of the Moon*, vol. i.

1832. Occurrence of a transit of Mercury. Sir J. Herschel's

A.D.

investigation of the orbits of binary stars. *Don Joaquim de Ferrer* determines the solar parallax from a discussion of the observation of the transit of Venus in 1769. Sir J. Herschel's Catalogue of 2017 Double Stars.

1833. Publication of Sir J. Herschel's Catalogue of Nebulæ in the Northern Hemisphere.[1] Airy obtains an important correction in the value of Jupiter's mass. Publication of the results of Lieut. *Foster's* pendulum experiments for determining the ellipticity of the Earth.

1834. Sir J. Herschel's researches on the satellites of Uranus. *Dawes's* micrometrical measures of 121 double stars. *Lubbock's* theory of the Moon.

1835. Sir *T. M. Brisbane's* Catalogue of 7385 Stars. Encke commences his researches on Planetary Perturbation. Encke obtains a correction of the value of the solar parallax as deduced from the transits of Venus in 1761 and 1769. Airy determines the time of the rotation of Jupiter. Sir J. Herschel's Catalogue of 286 Double Stars. *Johnson's* Catalogue of 606 Southern Stars. Airy appointed Astronomer Royal. Researches of *Rosenberger* and *Lehmann* on Halley's Comet.

1836. Publication of Baily's *Life of Flamsteed.* Publication of Damoiseau's Tables of Jupiter's Satellites.

1837. *Lamont's* researches on the satellites of Uranus. Researches by Pontécoulant on the lunar theory. *Henderson* determines the value of the Moon's equatorial parallax. Publication of W. Struve's *Mensuræ Micrometricæ* of 3112 Stars. Argelander's researches on the motion of the solar system in space. Hon. *J. Wrottesley's* Catalogue of 1318 Stars. Completion of the great Indian arc of the meridian.

1838. Lubbock's researches on the lunar theory, Part ii. Bessel determines the parallax of 61 Cygni. *Hansen's*

---

[1] See *Phil. Trans.*, vol. cxxiii.

A. D.

new method of investigating the lunar theory. *Robinson* determines the Constant of Nutation. Airy's Catalogue of 726 Stars. Lamont determines the mass of Uranus. Publication of La Caille's Catalogue of 9766 Southern Stars, by the British Association.

1839. *Le Verrier's* researches on the secular variations of the planets. Henderson determines the parallax of α Centauri. Foundation of the Imperial Observatory at Pulkova. Johnson appointed Director of the Radcliffe Observatory, Oxford. *Amici's* double-image micrometer.

1840. Foundation of the Cambridge (U. S.) Observatory. *Santini's* Catalogue of 1677 Stars. Airy's double image micrometer.

1841. Erection of Repsold's meridian circle at Königsberg. Researches of Hansen on the lunar theory.

1842. Foundation of the National Observatory, Washington (U. S.) *C. H. Peters* determines the Constant of Nutation; *Baily* determines the mean density of the Earth. Pearson's Catalogue of 520 Stars. Greenwich Catalogue of 1439 Stars.

1843. Hansen's new method of investigating the effects of planetary perturbation whatever be the eccentricity or inclination of the orbit. *Schwabe* detects a periodicity in the solar spots. W. Struve determines the Constant of Aberration. Adams commences his investigation on the orbit of Uranus which ultimately leads to the discovery of Neptune.

1844. *Sheepshanks* commences his researches to determine the length of the standard yard, which he continues till his death in 1855. Argelander concludes his northern hemisphere zone observations. *Taylor's* Catalogue of 11,015 Stars. Transmission of time by means of electric signals, commenced in the United States.

1845. Discovery of a new minor planet. In subsequent years many others are detected. Researches of

A.D.

Le Verrier on the theories of Mercury and Uranus.
Publication of the *British Association Catalogue of
8377 Stars*, and Smyth's *Cycle of Celestial Objects.*

1846. *Weisse's* reduction of Bessel's zone stars comprised
within 15° on each side of the equator. Airy's
measurement of the arc of parallel comprised between
Valencia and Greenwich. Discovery of the planet
Neptune. Publication of the results of the observations of the planets made at Greenwich between
1750 and 1850.

1847. Erection of the altazimuth at Greenwich. The British
Association publish the catalogue of 47,390 stars
contained in Lalande's *Histoire Céleste*. Hansen discovers 2 long inequalities in the Moon's mean motion.
Publication of Sir J. Herschel's *Results of Astronomical Observations made at the Cape of Good Hope* in
1833 and following years; of W. Struve's *Etudes
d'Astronomie Stellaire.* Researches of *Galloway* on
the motion of the solar system. *Lassell* discovers
the satellite of Neptune, and that satellite of Uranus
since called Ariel, whilst O. Struve discovers Umbriel.

1848. Researches of *Challis* for determining the orbit of a planet
or comet. *Jacob's* Poonah Catalogue of Double Stars.
Lassell in England, and *W. Bond* in America, discover
independently, the 8th satellite of Saturn, since called
Hyperion. Researches of *Wichmann* on the physical
libration of the Moon, and of C. H. Peters on stellar
parallax. Greenwich Catalogue of 2156 Stars.

1849. *Shortrede's* logarithms. Researches of *Powell* on irradiation. *Main* confirms the opinion of Bessel as to
the strictly elliptic form of Saturn.

1850. Publication of the Earl of *Rosse's* Observations on
Nebulæ. Main's catalogue of the proper motions
of 75 stars.

1851. Researches of C. H. Peters on the variability of the
proper motion of Sirius. Pendulum experiments of
Foucault for demonstrating the rotation of the Earth.

A. D.

Discovery of the dusky ring of Saturn. Erection of a new transit circle at Greenwich. Completion of the Russo-Scandinavian arc of the meridian. Dr. *Gould* starts the *Astronomical Journal*. *Oeltzen* commences the reduction of Argelander's zones, extending from 45° to 80° of north declination, which he finishes in the following year.

1852. Commencement of zone observations at the Cambridge (U. S.) Observatory. Publication of W. Struve's *Positiones Mediæ, &c.*, containing the mean position of 2874 stars whereof 2682 are double. Researches of *Villarceau* on the orbits of double stars. *Rümker's* Catalogue of 12,000 Stars. Researches of *Secchi* on the Earth's temperature. Observations with the reflex zenith tube commenced at Greenwich. Argelander's zone observations from 15° to 31° of south declination.

1853. Researches of Airy on ancient eclipses; of Adams on the secular inequality in the Moon's mean motion; of Hansen on the theory of the pendulum. Publication of the American Lunar Tables. Encke gives a new solution to the problem of Planetary Perturbation. Hansen's Solar Tables.

1854. The chronographic method of recording transits introduced at Greenwich. Researches of Lubbock on Refraction. Capt. Jacob's Catalogue of 1440 Stars. Airy's pendulum experiments in the Harton Colliery for determining the density of the Earth. Determination of the difference of the longitude of Greenwich and Paris by electric signals.

1855. Greenwich Catalogue of 1576 Stars. Bond's zone observations of 5500 small stars near the equator. Researches of Main on the value of the Constants of Aberration and Nutation, and on the rings of Saturn. Commencement of the publication of the *Annales* of the Paris Observatory, of the *American Nautical Almanac*, and of *Brunnow's Tables of Flora*.

1856. Researches of Challis on the problem of the 3 bodies; of Main on the diameter of the planets. Astronomical expedition to Teneriffe under *C. P. Smyth*.

1857. Researches of Airy on ancient eclipses. Publication of *Carrington's* Redhill Catalogue of Circumpolar Stars; of Hansen's Lunar Tables. *De La Rue*, Secchi, Bond, and others, obtain photographs of celestial objects. *Hoek's* investigation of the identity of the comets of 976 and 1556.

1858. De La Rue obtains a stereoscopic photograph of the Moon. Publication of Le Verrier's Solar Tables. Erection of a photoheliograph at the Kew Observatory. Occurrence of an annular eclipse which excited much interest in England. Completion of the calculations for determining the principal triangles of the trigonometrical survey of the British Isles, and deduction of final results relating to the figure, dimensions, and density of the Earth.

1859. Publication of Robinson's Places of 5345 Stars, observed at Armagh. Suspected discovery of a new planet revolving within the orbit of Mercury, and since named Vulcan. Numerous spots visible on the Sun during the summer months.

1860. Erection of a fine achromatic refractor at the Greenwich Observatory. Occurrence of a total eclipse of the Sun visible in Spain, to observe which a large party of astronomers sail from England in the "Himalaya," besides other parties from France, &c.

1861. Discovery of many new planets. Apparition of 2 comets visible to the naked eye, of which the 2nd, which appeared in June, had the longest tail on record — 105°.

BOOK X.

METEORIC ASTRONOMY.

—◆—

CHAPTER I.

Classification of the Subject.—Aërolites.—Summary of the Researches of Berzelius, Rammelsberg, and others.—Celebrated Aërolites.—Summary of Facts. — Catalogue of Meteoric Stones.—Arago's Table of Apparitions. — The Aërolite of 1492.—Of 1627.— Of 1795.—The Meteoric Shower of 1803.

THE phenomena, of which we are now about to speak, form a highly interesting and by no means unimportant branch of descriptive astronomy. We shall treat of the subject under three heads: —

1. Aërolites.
2. Fireballs.
3. Shooting Stars.

Of all cosmical meteors, those known as aërolites, meteorlites, or meteoric stones, are the rarest, but nevertheless not so rare as to prevent the most satisfactory evidence being given, that such occurrences have happened from time to time. It is to Chaldni that we owe much of our knowledge on this branch of the subject. Many of these meteoric stones, picked up in different parts of the world, have been subjected to chemical analysis at the hands of Berzelius, Rammelsberg, and others, whose deductions may be thus summed up: —

1. Meteoric stones are composed of elements all of which occur in terrestrial minerals.

2. Of the 65 elementary substances known, 19 have been found in meteoric stones.

3. The produce of a meteoric shower may be divided into meteoric iron and meteoric stone.

4. Meteoric iron is an alloy that has not been found among terrestrial minerals, and is composed of about 10 per cent. of nickel with small quantities of cobalt, manganese, magnesia, tin, copper, and carbon.

5. Meteoric stone is composed of minerals found abundantly in lavas and trap-rocks, and consequently of volcanic origin, a variable proportion of meteoric iron being usually admixed.

The circumstances attending the fall of aërolites differ considerably on different occasions. Not unfrequently the fall is attended by a loud detonation; but we must not therefore infer that every detonating meteor is indeed an aërolite, without the presence of positive proof to that effect. History records instances of considerable damage having been done to life and property by the explosion of these bodies: as for instance, from a Chinese catalogue, we learn that one which fell on January 14, 616 B.C., broke several chariots and killed 10 men. The chronicle of Frodoard informs us that in the year 944 A.D. globes of fire traversed the atmosphere, and burnt several houses. More recently, on the evening of November 13, 1835, a brilliant meteor was seen in the department of Aix (France). It traversed the country in a north-easterly direction, and burst near the castle of Lausière, setting fire to a barn and the stables, burning the corn and cattle in a few minutes. An aërolite was found near the place after the occurrence. Also on March 22, 1846, at 3 in the afternoon, a luminous sheaf, which traversed the air with great velocity and noise, fell on a barn in a village in the department of Haute Garonne, which instantly took fire, and was destroyed together with the adjoining stables and the beasts therein contained.[1] It is also related that the Emperor Jehangir had a sword forged from a mass of meteoric iron which fell at Jahlindu, in the Punjab, in 1620.[2]

---

[1] See Arago, *Ast. Pop.*, vol. iv. pp. 224–229, French Ed., where numerous other instances are given. In the English edition this and other important meteor catalogues are very improperly left out.

[2] *Phil. Trans.* vol. xciii. p. 202. 1803.

From the above and other similar observations, we learn three things.

1. That the fact is undoubtedly established, that from time to time masses of stone, of different sizes, and often of considerable weight, are seen passing through space, and are frequently precipitated upon the Earth's surface.

2. That these bodies rarely strike the Earth in a vertical or nearly vertical direction, but fall almost always in a direction very oblique to the plane of the horizon. This is ascertained by an inspection of the manner in which they penetrate the earth, which they often do to a considerable depth.

3. That they are endued with a very great velocity, similar, in fact, to the velocities which are found to characterise the planetary members of the solar system.

The ancients seem to have been well aware of the phenomena of which we are now treating, inasmuch as several things are mentioned by the classic writers as having fallen from heaven: we may refer to the Palladium of Troy, the image of Diana at Ephesus, the sacred shield of Numa, as examples. The ideas of the ancients, relative to the supposed celestial origin of these things, have often met with ridicule; but however fabulous the cases referred to may have been, still the moderns have been compelled, though reluctantly, to admit the fact of the actual transmission of stony substances from space, on to the surface of the Earth. The following catalogue of some of the more important recorded falls of meteoric stones, is taken chiefly from M. Izarn's work.[1]

Substance.	Period.	Place.
Shower of stones	About 650 B.C.	Rome.
Large stone	465 B.C.	River Negos, Thrace.
Three large stones	452	In Thrace.
Shower of stones	343	Rome.
Shower of iron	54	Lucania.
Shower of mercury	Date unknown	In Italy.
Mass of iron	,,	Abakauk, Siberia.
Large stone of 260 lbs.	1492 Nov. 7	Ensisheim, Upper Rhine.

[1] *Des Pierres Tombées du Ciel, ou Lithologie Astronomique.* Paris, 1803.

Substance.	Period.	Place.
About 1200 stones—one of 120 lbs., another of 60 lbs.	1510 . .	Padua, Italy.
Stone of 59 lbs. . .	1627 Nov. 27 .	Mont Vasier, Provence.
Sulphurous rain . .	1646 . .	Copenhagen.
Sulphurous rain . .	1658 . .	Duchy of Mansfield.
Shower of unknown matter.	1695 . .	Ireland.
Stone of 72 lbs. . .	1706 January .	Larissa, Macedonia.
Shower of fire . .	1717 Jan. 4 .	Quesnoy.
Shower of sand for 15 hours.	1719 April 6 .	In the Atlantic.
Shower of sulphur .	1721 October .	Brunswick.
Mass of stone . .	1750 . .	Niort, Normandy.
Shower of stones .	1753 July 3 .	Plaun, Bohemia.
Two stones weighing 20 lbs.	1753 September	Liponas, in Bresse.
Two stones of 200 and 300 lbs.	1762 . .	Near Verona.
A stone of 7½ lbs. .	1768 Sept. 13 .	Lucé, Le Maine.
A stone . . .	1768 . .	Aise, Artois.
A stone . . .	1768 . .	Le Cotentin.
Shower of stones . .	1789 July .	Barboutan, near Roquefort.
Extensive shower of stones.	1790 July 24 .	Near Agen.
About twelve stones .	1794 July 16 .	Sienna, Tuscany.
A stone of 56 lbs. . .	1795 Dec. 13 .	Wold Cottage, Yorkshire.
A stone of 10 lbs. . .	1796 Feb. 19 .	In Portugal.
A stone of 20 lbs. . .	1798 March 12	Sules, near Ville Franche.
A stone of about 20 lbs.	1798 March 17	Sâle, dep. of Rhone.
Shower of stones . .	1798 Dec. 19 .	Benares.
Mass of iron, 70 cubic feet.	1800 April 5 .	America.
Several stones, of from 10 to 17 lbs.	1803 April 26 .	Near L'Aigle, Normandy.
Shower of stones . .	1807 Dec. 14 .	Weston, Connecticut, U. S.
A stone of 1563 lbs. .	1810 . .	Santa Rosa, New Grenada.
A stone of 203 lbs. .	1821 June 15 .	Juvenas, Ardéche.
A large stone . .	1843 Sept. 16 .	Kleinwenden, Thuringia.

According to Arago, we find that the 206 falls of aërolites, of which we know the month of occurrence, were distributed

in the following manner throughout the 12 months of the
year : —

January	.	. 14	July	. .	. 23
February	.	. 10	August	. .	. 16
March .	.	. 22	September	.	. 17
April .	.	. 15 } 99	October	.	. 18 } 107
May .	.	. 20	November	.	. 20
June .	.	. 18	December	.	. 13

From an inspection of the above, it appears that the monthly
average from December to June (16) is less than the same
average from July to November, which is 19 per month, and
that, moreover, the months of March, May, July, and November exhibit maximum numbers: and we also deduce this
general fact — that the Earth, in its annual course round the
Sun, would seem to encounter a greater number of aërolites in
passing from aphelion to perihelion, or from the summer to
the winter solstice, than in going from perihelion to aphelion,
or from the winter to the summer solstice.

The circumstances connected with the occurrence which
stands 8th in the above list, are of more than average interest,
more especially from its having been long considered a poetical
romance of by-gone ages.   The following narrative was drawn
up at the time, by order of the Emperor Maximilian, and
deposited with the stone in the church at Ensisheim.   " In
the year of the Lord 1492, on Wednesday, which was Martin-
mas Eve, November 7, a singular miracle occurred ; for between 11 o'clock and noon, there was a loud clap of thunder,
and a prolonged confused noise, which was heard at a great
distance ; and a stone fell from the air, in the jurisdiction of
Ensisheim, which weighed 260 pounds, and the confused
noise was, moreover, much louder than here.   There a child
saw it strike on a field in the upper jurisdiction, towards the
Rhine and Jura, near the district of Giscano, which was sown
with wheat, and it did no harm, except that it made a hole
there ; and then they conveyed it from that spot, and many
pieces were broken from it which the landvogt forbade.
They therefore caused it to be placed in the church, with the

intention of suspending it as a miracle; and there came here
many people to see this stone. So there were remarkable
conversations about this stone; but the learned said they
knew not what it was; for it was beyond the ordinary course
of nature that such a large stone should smite the Earth, from
the height of the air, but that it was really a miracle of God;
for, before that time, never anything was heard like it, nor
seen, nor described. When they found that stone, it had
entered into the Earth to the depth of a man's stature, which
everybody explained to be the will of God that it should be
found; and the noise of it was heard at Lucerne, at Vitting,
and in many other places, so loud, that it was believed that
houses had been overturned: and as the King Maximilian
was here the Monday after S. Catherine's Day of the same
year, his Royal Excellency ordered the stone which had fallen
to be brought to the castle; and after having conversed a
long time about it with the noblemen, he said that the people
of Ensisheim should take it, and order it to be hung up in
the church, and not to allow anybody to take anything from
it. His Excellency, however, took two pieces of it, of which
he kept one, and sent the other to Duke Sigismund of Austria:
and they spoke a great deal about this stone, which they
suspended in the choir, where it still is; and a great many
people came to see it." This relic then remained in the
church for three centuries, when it was temporarily removed
during the turmoil of the French Revolution, to Colmar, but
it has since been restored. A fragment of it is in the British
Museum, and another piece may be seen at the Jardin des
Plantes, at Paris.

The fall of the aërolite of 1627 (No. 10), was witnessed by
the astronomer Gassendi: he states that when in the air it was
apparently surrounded by a halo of prismatic colours. This
being the only instance with which he was acquainted, Gassendi was led to attribute its origin to some one of the neighbouring mountains, from the summit of which a temporary
volcanic eruption had taken place.

The aërolite of December 13, 1795 (No. 28), is interesting
from the fact of its being one of the few instances recorded to

have taken place in this country. A loud explosion, followed
by a hissing noise, was heard through a considerable portion
of the surrounding district; a shock was also noticed, as if
produced by the falling to the Earth of some heavy body. A
ploughman saw the stone fall to the ground at a spot
not far distant from where he then was standing; it threw
up mould on every side, and after passing through the soil,
penetrated several inches deep into the solid chalk rock. It
fell on the afternoon of a mild but hazy day, during which
there was neither thunder nor lightning.[1]

One of the severest falls on record was that which happened in Normandy on April 26, 1803: (No. 34). It appears
that, at about 1 o'clock in the afternoon, a very brilliant fire-
ball was seen traversing the country with great velocity: and,
some moments afterwards, a violent explosion was heard,
which was prolonged for 5 or 6 minutes. The noise seemed
to proceed from a small cloud which remained motionless all
the time, but at a great elevation in the atmosphere; the
detonation was followed by the fall of an immense number
of mineral fragments, nearly 3000 being collected, the largest
weighing 17½lbs. The sky was serene, and the air calm, an
atmospheric condition that has frequently been noticed at the
descent of aërolites.[2]

---

[1] Howard, *Phil. Trans.*, vol. xcii. p. 174. 1802.
[2] A catalogue of 273 aërolites is given in Arago's *Ast. Pop.*, vol. iv.
pp. 184–203, Fr. Ed.

CHAPTER II.

*The Origin of Aërolites. — The Atmospheric Hypothesis. — The Volcanic
Hypothesis. — The Lunar Hypothesis. — The Planetary Hypothesis. —
The last named, probably the correct one.—Poisson's Theory.—Fireballs.
—Arago's Table of Apparitions. — Summary of Measurements.*

To account for the nature and origin of aërolites, the following hypotheses have been propounded : —

First.—It is supposed that the matter composing them has
been drawn up from the surface of the Earth in a state of
infinitely minute subdivisions, as vapour is drawn from liquids ;
that, being collected in clouds in the higher regions of the
atmosphere, it is there agglomerated and consolidated in
masses, and falls by its gravity to the surface of the Earth ;
being occasionally drawn from the vertical direction which
would be imparted to it by gravity by the effect of atmospheric currents, and thus occasionally striking the Earth
obliquely. We shall call this the *atmospheric hypothesis.*

Secondly.—It is supposed that meteoric stones are ejected
from volcanoes, with sufficient force to carry them to great
elevations in the atmosphere, in falling from which they acquire the velocity and force with which they strike the Earth.
The oblique direction with which they strike the ground is
explained by the supposition that they may be projected from
the volcanoes at corresponding obliquities, and that, by the
principles of projectiles, they must strike the Earth at nearly
the same inclination as that with which they have been ejected.
This we shall call the *volcanic hypothesis.*

Thirdly.—It has been suggested that the aërolites may be
bodies which have been ejected from lunar volcanoes, with
such a force that they may have departed from the Moon to a
distance so great as to come within such a distance of the

Earth, that the terrestrial attraction exerted upon them pre-
dominating over that of the Moon, they may have either
fallen down directly upon the Earth, or may have revolved
round it in a curvilinear orbit, with a motion constantly
retarded by the Earth's atmosphere, the consequence of which
would be that they would continually approach the Earth, and
at length fall upon its surface. We shall call this the *lunar
hypothesis*.

Fourthly.—It has been supposed that aërolites are planetary bodies; that they revolve in orbits round the Sun; that
these orbits intersect the annual path of the Earth; that when
the Earth passes through the point of intersection of its path
with their orbits, they either encounter it directly, and fall
upon its surface, or, entering its atmosphere, are rapidly retarded by the resistance of that fluid, and are then drawn to
the surface by the terrestrial attraction. This may be termed
the *planetary hypothesis*.

In a popular sketch like the present, it is of course impossible to adduce all the arguments that have been brought forward in support of and in opposition to the above theories;
we shall therefore pass on, merely remarking that the 4th on
the list seems fairly to represent the case, and is the explanation very generally adopted by men of science at the present
day. The periodicity before alluded to also seems to count-
enance this opinion. The luminous appearance which attends
their progress has been accounted for by supposing that their
rapid motion so condenses the atmospheric air lying in their
path, that it either itself becomes luminous, or acquires so
intense a heat as to render the stone incandescent. This surmise is supported by the well-known experiment of the fire
syringe. M. Poisson, the eminent French geometer, has
suggested that there exists above the atmosphere, a layer, as it
were, of electricity, and that the friction caused by the
passage of the aërolite is such as to decompose the electric
fluid, and thus produce a kind of spark, as occurs in the case
of an electric machine.[1]

---

[1] *Recherches sur la Probabilité des Jugements*, p. 6. Paris, 1837.

Fireballs appear to hold an intermediate position between aërolites and shooting stars. They appear suddenly, and after exhibiting a brilliant flame of light for a few seconds, as suddenly vanish. Their form is generally circular, or slightly oval, and of a perceptible magnitude. Not unfrequently they leave behind them a train of sparks, their own illuminating power being somewhat more feeble than that of the Moon. Sometimes they explode into fragments, which continue their course, or are precipitated, as we have already seen, upon the surface of the Earth in the form of aërolites.

If we classify the apparition of all the fireballs the dates of which are known, we find, according to Arago, that their number amounts to 813, distributed as follows: —

January	. . 55		July . . . 74		
February	. . 57		August . . 123		
March .	. . 48		September . . 64		
April .	. . 52	305	October . . 77	508	
May .	. . 50		November . . 90		
June .	. . 43		December . . 80		

Thus showing that the periodicity which prevails with the aërolites, also obtains with the fireballs, only in a much more marked manner.

Of the above 813 fireballs of which we possess any recorded account, 35 only, gave rise to aërolites the fall of which was actually witnessed. Small though this proportion undoubtedly is, yet we cannot but consider these 2 classes of phenomena to be intimately associated. It is, however, true, that cases have been known in which aërolites have fallen, which were not preceded by any luminous exhalation: an instance occurred on September 16, 1843, at the fall of the great aërolite of Kleinwenden.[1]

Many fireballs have been submitted to measurement as regards their size and distance, but, owing to the very sudden appearance, and in general the short visibility of these bodies, it seldom happens that the observer is able to attain to any

---

[1] *Compt. Rend.*, vol. xxv. p. 627.

great precision. The following results must, therefore, be received with caution : —

1. As to the height at the instant of apparition.

Greatest known.	Miles.	Least known.	Miles.
1844 October 27 . .	318·1	1846 March 21 . .	7·5
1718 March 19 . .	297·5	1852 April 2 . . .	10·0
1842 June 3 . . .	184·0	1754 August 15 . .	15·0

2. As to absolute diameter.

Greatest known.	Feet.	Least known.	Feet.
1841 August 18 . .	12,795	1852 April 2 . . .	105
1718 March 19 . .	8,399	1846 July 23 . . .	321
1837 January 4 . .	7,216	1850 July 6 . . .	705

3. As to velocity per second.

Greatest known.	Miles.	Least known.	Miles.
1850 July 6 . . .	47·22	1718 March 19 . .	1·67
1844 October 27 . .	44·74	1807 December 14. .	2·80
1842 June 3 . . .	44·74	1676 March 31 . .	3·11

It is desirable to remark, that the axial rotation of the Earth at any point situated on the terrestrial equator is 1524 feet per second, and that the Earth's orbital motion is 18·89 miles. We see, moreover, that the velocity of many of these fireballs is greater than that of any of the planets; it is also worthy of mention that the general direction of their motion is contrary to that of the Earth.[1]

---

[1] A catalogue of 854 fireballs is given in Arago's *Ast. Pop.*, vol. iv. pp. 230–279, Fr. Ed.

CHAPTER III.

Shooting Stars. — Have only recently attracted Attention. — To be seen in greater or less Numbers almost every Night.—Tabular Summary of the Results of the Observations of Coulvier-Gravier, Saigey, and Schmidt. — Early Notices of Meteoric Showers. — Shower of 1799. — Showers of 1831, 2, and 3. — The Meteors of 1833 divided into 3 Groups. — Table of Apparitions. — Singular Result. — Olmsted's Theory.—Herschel's Theory.

SHOOTING stars, although noticed in former times, have only within the last half century attracted any particular attention. This branch of the science may therefore be considered to be, comparatively, in its infancy. We must possess a long and carefully made series of observations before we are likely to be acquainted, with any degree of precision, with the physical nature of these objects. They were formerly considered to be merely atmospheric meteors, caused by the combustion of inflammable gases. This opinion has, however, now lost much, if not all, of its force, and they are now recognised as bodies which, although they become inflamed on coming in contact with the Earth's atmosphere, yet have their origin far beyond it.

It is now an established fact, that there is no night throughout the year on which shooting stars may be not seen; and that, on an average, from 5 to 7 may be noticed on a clear night every hour. These occasional meteors may be termed sporadic, in contradistinction to those swarms which appear at certain times of the year, and which are periodic. There is, moreover, an *horary* variation in their number, and the maximum occurs at 6 P.M., the mean at midnight, and minimum at 6 A.M., as shown by the following table [1] : —

---

[1] *Month. Not. R. A. S.,* vol. xvii. p. 47.

Hours P.M.	6–7.	7–8.	8–9.	9–10.	10–11.	11–12.
Mean no. of meteors .	3·3	3·5	3·7	4	4·5	5

Hours A.M.	12–1.	1–2.	2–3.	3–4.	4–5.	5–6.
Mean no. of meteors .	5·8	6·4	7·1	7·8	8	8·2

If we designate the numbers coming from the N. E. S. W. by those letters respectively, we find E. > 2 W.; N. = S. nearly, and that E. + W. = N. + S.

The following table contains the monthly mean of the hourly number of shooting stars as assigned by 3 eminent continental observers[1] : —

	MM. Coulvier-Gravier and Saigey.		M. Schmidt.	
January	... 3·6		... 3·4	
February	... 3·7		... ?	
March	... 2·7		... 4·9	
April	... 3·7	3·4	... 2·4	4·0
May	... 3·8		... 3·9	
June	... 3·2		... 5·3	
July	... 7·0		... 4·5	
August	... 8·5		... 5·3	
September	... 6·8		... 4·7	
October	... 9·1	8·0	... 4·5	4·7
November	... 9·5		... 5·3	
December	... 7·2		... 4·0	

Notwithstanding the discordances in the above results, both tables agree in showing that there are more shooting stars in the 2nd than in the 1st half of the year, a coincidence which we have already seen holds good both with aërolites and fire-balls. This has also been confirmed by the observations recorded in the Chinese annals.

---

[1] Quoted in Arago's *Pop. Ast.*, vol. ii. p. 505, Eng. Ed.

We now come to speak of the well-known and very beautiful showers of shooting stars seen at certain seasons in such
great abundance.   One of the earliest notices we find in
history of this phenomenon is by Theophanes the Byzantine
historian, who relates that, in November, 472 A.D., the sky at
Constantinople appeared to be on fire with flying meteors.
Condé, in his history of the dominion of the Arabs, speaking
of the year 902 A.D., states that in the month of October, on
the night of the death of King Ibrahim-Ben-Ahmed, an immense number of falling stars were seen to spread themselves
over the face of the sky like rain, and that the year in question
was thenceforth called the " Year of Stars."   In some Eastern
Annals of Cairo, it is related that, " In this year, in the month
Redjet [August, 1029] many stars passed, with a great noise,
and brilliant light ; " and in another passage it says, " In the
year 599, on Saturday night, in the last Moharrun [October
19, 1202], the stars appeared like waves upon the sky, towards the east and west; they, flew about like grasshop-
pers, and were dispersed from left to right; this lasted till
daybreak : the people were alarmed." It is also recorded that
a remarkable display took place in England and France on
April 4, 1095.   The stars seemed " falling like a shower of
rain from heaven upon the Earth," and an eye-witness, having
noticed where an aërolite fell, " cast water upon it, which was
raised in steam with a great noise of boiling."   In the Chronicle of Rheims, we read that the stars in heaven were
driven like dust before the wind, and Rastel says that, " By
the report of the common people in this kynge's time [William
II.,] divers great wonders were sene : and therefore the kynge
was told by divers of his familiars that God was not content
with his lyvyng; but he was so wilful and proud of mind,
that he regarded little their saying."
    In modern times, the earliest shower of falling stars of
which we have any detailed description is that of November
13, 1799, visible throughout nearly the whole of North and
South America.   It was seen in Greenland by the Moravian
missionaries.   Humboldt, then travelling with M. Bonpland,
in South America, says :—" Towards the morning of the 13th,

we witnessed a most extraordinary scene of shooting meteors. Thousands of bodies and falling stars succeeded each other during 4 hours. Their direction was very regular from north to south. From the beginning of the phenomehon there was not a space in the firmament equal in extent to 3 diameters of the Moon, which was not filled every instant with bodies or falling stars. All the meteors left luminous traces, or phosphorescent bands behind them, which lasted 7 or 8 seconds." Mr. Ellicott, an agent of the United States, at sea in the Gulf of Mexico, thus describes the scene : — " I was called up about 3 o'clock in the morning, to see the shooting stars, as they are called. The phenomenon was grand and awful. The whole heavens appeared as if illuminated with sky-rockets, which disappeared only by the light of the Sun after daybreak. The meteors, which at any one instant of time appeared as numerous as the stars, flew in all possible directions, except from the Earth, towards which they were all inclined more or less ; and some of them descended perpendicularly over the vessel we were in, so that I was in constant expectation of their falling on us." The same observer also states that his thermometer suddenly fell 24°, and the wind changed from S. to N. W., whence it blew with great violence for 3 days. Meteoric showers were also witnessed in North America, in the years 1814, 1818, and 1819.

Fine meteoric displays took place in 1831 and 1832, in both cases on November 13. Captain Hammond, of the ship " Restitution," then in the Red Sea, off Mocha, thus describes the latter: — " From 1 o'clock A.M. till after daylight, there was a very unusual phenomenon in the heavens. It appeared like meteors bursting in every direction. The sky at the time was clear, the stars and Moon bright, with streaks of light and thin white clouds interspersed in the sky. On landing in the morning, I inquired of the Arabs if they had noticed the above. They said they had been observing it most of the night. I asked if ever the like had appeared before. The oldest of them replied that it had not." This shower was seen from Arabia, westward to the Atlantic, and from the Mauritius to Switzerland.

By far the most splendid display of shooting meteors on record was that of November 13, 1833, and one which, from its recurring after so exact an interval of time, served to point out a periodicity in the phenomenon. It seems to have been visible over nearly the whole of the northern portion of the American continent, or, more exactly, from the Canadian lakes nearly to the equator. Over this immense area a sight of the most imposing grandeur seems to have been witnessed. The phenomenon commenced at about midnight, and was at its height at about 5 A.M. Several of the meteors were of peculiar form and considerable magnitude. One was especially remarked from its remaining for some time in the zenith over the falls of Niagara, emitting radiant streams of light. In many parts of the country the population were terror-struck by the beauty and magnificence of the spectacle before them. A planter of South Carolina thus narrates the effect of the phenomenon on the minds of the ignorant blacks : — " I was suddenly awakened by the most distressing cries that ever fell on my ears. Shrieks of horror and cries for mercy, I could hear from most of the negroes of the 3 plantations, amounting in all to about 6 or 8 hundred. While earnestly listening for the cause, I heard a faint voice near the door calling my name. I arose, and, taking my sword, stood at the door. At this moment, I heard the same voice still beseeching me to rise, and saying, ' O my God, the world is on fire ! ' I then opened the door, and it is difficult to say which excited me the most — the awfulness of the scene, or the distressed cries of the negroes. Upwards of 100 lay prostrate on the ground — some speechless, and some with the bitterest cries, but with their hands raised, imploring God to save the world and them. The scene was truly awful; for never did rain fall much thicker than the meteors fell towards the Earth; east, west, north, and south, it was the same." [1]

The meteors of which the above shower was composed, seem to have been seen of 3 different kinds : —

1. Phosphoric lines, apparently described by a point. These

---

[1] Quoted in Milner's *Gallery of Nature*, p. 140.

were the most abundant; they passed along the sky with immense velocity, as numerous as the flakes of a sharp snow storm.

2. Large fireballs, which darted forth at intervals across the sky, describing large arcs in a few seconds. Luminous trains marked their path, which remained in view for a number of minutes, and in some cases for half an hour or more. The trains were commonly white, but the various prismatic colours occasionally appeared, vividly and beautifully displayed. Some of these fireballs were of enormous size; indeed, one was seen larger than the Moon when full.

3. Luminosities of irregular form, which remained stationary for a considerable time. The one mentioned above as having been seen at the falls of Niagara was of this kind.[1]

For the last 25 years, the month of November has been distinguished by an unusual number of shooting stars; but none of the showers have ever equalled the one we have just described.

Subdividing the showers of shooting stars according to the month of the year, we obtain the following results : —

January	.	. 10 ⎫	July	.	. 14 ⎫
February	.	. 10 ⎪	August	.	. 56 ⎪
March .	.	. 12 ⎬ 55	September	.	. 19 ⎬ 163
April .	.	. 17 ⎪	October	.	. 13 ⎪
May	.	. 4 ⎪	November	.	. 29 ⎪
June	.	. 2 ⎭	December	.	. 17 ⎭

We thus find, and it is worthy of especial remark, that the coincidence we have already pointed out in the case of aëro-lites, fireballs and sporadic meteors also obtains with the showers of shooting stars — namely, that the Earth encounters a larger number of these bodies in passing from aphelion to perihelion, or from the summer to the winter solstice, than in passing from perihelion to aphelion, or from the winter to the summer solstice.[2]

---

[1] Quoted in Milner's *Gallery of Nature*, p. 141 (abridged).

[2] A catalogue of 221 meteoric showers is given in Arago's *Ast. Pop.*, vol. iv. pp. 292–314. Also a catalogue, extending from 538–1223 A.D., by Chasles, in *Compt. Rend.*, vol. i. pp. 499–509. 1841.

Professor Olmsted, of Yale College, U. S., has proposed the following theory, to account for the above phenomena:—That the meteors of November 13, 1833, emanated from a nebulous body, which was then pursuing its course, along with the Earth, around the Sun; that this body continues to revolve around the Sun in an elliptic orbit, but little inclined to the plane of the ecliptic, and having its aphelion near the orbit of the Earth; and, finally, that the body has a period of nearly 6 months, and that its perihelion is a little within the orbit of Mercury.[1]

The following summary, useful for amateur observers, is by Arago.

January.—It would seem from the recorded results, that we may look for a period of shooting stars somewhere about January 1—4.

February.—Modern observations do not indicate a period of shooting stars for February. The ancient showers of meteors, announced for this month by the chroniclers, seem to have failed for the last 8 or 9 centuries.

March.—Shooting stars have been perceived from time to time in this month.

April.—Apparitions of shooting stars are somewhat more numerous in this month than in the 3 preceding. We may look for them about April 4—11, and 17—25.

May.—Shooting stars are rare in May.

June.—Shooting stars are *very* rare in June.

July.—The apparitions of showers begin now to increase in number. We may expect them about July 26—29.

August.—Shooting stars are, as is well known, seen in great abundance in this month, particularly about August 9—11.

September.—Shooting stars are somewhat rare in September. We may, however, mention September 1±, and September 18—25 as possible periods.

---

[1] See Olmsted's theory, given in full in his *Mechanism of the Heavens*, pp. 329-341. Edin. Ed.

October. — Shooting stars occur about the middle of the month.

November. — Shooting stars, in past years, have appeared in remarkable numbers about November 11—13; but they are now less abundant than formerly.

December. — Showers of shooting stars may be looked for about December 5—15.

With reference to the periodicity Sir J. Herschel says: " It is impossible to attribute such a recurrence of identical dates of very remarkable phenomena to accident. Annual periodicity, irrespective of geographical position, refers us at once to the place occupied by the Earth in its annual orbit, and leads directly to the conclusion, that at that place it incurs a liability to *frequent* encounters or concurrences with a stream of meteors in their progress of circulation around the Sun. Let us test this idea, by pursuing it into some of its consequences. In the first place, then, supposing the Earth to plunge in its yearly circuit, into a uniform ring of innumerable small meteoric planets, of such breadth as would be traversed by it in one or two days; since, during this small time, the motions, whether of the Earth or of each individual meteor, may be taken as uniform and rectilinear, and those of all the latter (at the place and time) parallel, or very nearly so, it will follow that the relative motion of the meteors, referred to the Earth as at rest, will be also uniform, rectilinear, and *parallel*. Viewed, therefore, from the centre of the Earth (or from any point of the circumference, if we neglect the diurnal velocity, as very small compared with the annual), they will all appear to diverge from a common point, *fixed in relation to the celestial sphere*, as if emanating from a sidereal apex.

" Now this is precisely what happens. The meteors of the 12th—14th of November, or at least the vast majority of them, describe apparently arcs of great circles, passing through or near γ Leonis. No matter what the situation of that star, with respect to the horizon or to its east and west points, may be at the time of observation, the paths of the meteors all appear to diverge from that star. On the 9th—11th of

August, the geometrical fact is the same, the apex only differing; B Camelopardi, being for that epoch, the point of divergence. As we need not suppose the meteoric ring coincident in its plane with the ecliptic, and as for a *ring* of meteors we may substitute an elliptic annulus of any reasonable eccentricity, so that both the velocity and direction of each meteor may differ to any extent from the Earth's, there is nothing in the great and obvious difference *in latitude* of these apices at all militating against the conclusion.

"If the meteors be uniformly distributed in such a ring or elliptic annulus, the Earth's encounter with them in every revolution will be certain, if it occur once. But if the ring be broken—if it be a succession of groups revolving in an ellipse in a period *not* identical with that of the Earth, years may pass without a rencontre;· and when such happen, they may differ to any extent in their intensity of character, according as richer or poorer groups have been encountered.

"No other plausible explanation of these highly characteristic features (the annual periodicity and divergence from a common apex, *always alike for each respective epoch*) has been ever attempted, and, accordingly, the opinion is generally gaining ground among astronomers, that shooting stars belong to their department of science, and great interest is excited in their observation, and the further development of their laws." [1]

---

[1] *Outlines of Ast.*, p. 661.

APPENDICES.

APPENDIX I.

THE NOMENCLATURE OF THE MINOR PLANETS.

THIS is a subject on which we have a few words to say. In the early days of this branch of astronomical discovery, a sort of understanding was come to by astronomers that the names given to these bodies should as far as possible be those of ancient female divinities. So much for the theory which is unexceptionable. Now for the practice. This is bad in two ways:—
1. The original arrangement is constantly broken through; and 2, sufficient precautions are not taken to choose names, which cannot be mistaken (by reason of similarity of sound) for ones previously appropriated. With reference to the 1st, we have nothing particular to say against such names as *Parthenope*, *Massilia*, *Isis*, *&c.*, as they indicate where the discovery was made, but we most emphatically protest against the fawning servility which prompted such appellations as *Eugenia*, and *Maximiliana*, and *Angelina*. We have the highest opinion of the excellence of the Empress of France, nor do we doubt that King Maximilian of Bavaria has deserved well of his German subjects, but why should they be raised to the skies? In 1813 the Academy of Leipzic proposed to add the name of Napoleon I. to the constellations; astronomers, with much good sense, repudiated the idea. Has the race degenerated? The same remarks apply with equal force to *Angelina*. If Roman Catholics like to believe that Saint So-and-so performed, several centuries ago, such and such a miracle, let them; but it is going a little too far to ask all the world to immortalise some local celebrity by dedicating to him, her, or it, a planet.

T 3

In a practical point of view, the 2nd class of objections are the most important. In writing or speaking, mistakes are very apt to arise between, for instance, the following : —

Egeria and Hygeia.		Leto and Leda.	
Eugenia „ Egeria.		Pales „ Pallas.	
Hestia „ Vesta.		Thetis „ Metis.	
Isis „ Iris.		Thetis „ Themis.	
Lætitia „ Lutetia.		And perhaps others.	

We commend this subject to the consideration of planet-namers. If the present system goes on, the names of these objects will ere long become little better than senseless nuisances, which will one day have to be abandoned for the more business-like, but less pleasing arrangement of (1), (2), (3), &c.; just, in fact, as the symbols had to be given up some time since.

APPENDIX II.

A CATALOGUE OF ECLIPSES.

THE following Catalogue contains all the eclipses which occur during the second half of the 19th century, excepting solar eclipses hardly visible to any inhabited portion of the Earth, and lunar eclipses in which less than $\frac{1}{10}$ of the Moon's diameter is obscured. The time is approximately that of Greenwich, M. standing for morning, and A. for afternoon. Under the head of Locality the letter c points to the path followed by the central line; in cases where this passes very near the North or South Pole, it is not traced, but those places only are named where the eclipse will be visible (v.) The letters N.E. or S.E., following the name of a place, indicate the direction taken by the shadow after passing the parts in question.

Year.		Month and Day.	Hour.	Magni-tude.	Locality.
1851.	☾	Jan. 17	5 A.	0·46	W. China.
—	☉	Feb. 1	5 M.	...	C. Van Diemen's Land ; New Zealand.
—	☾	July 13	7½ M.	0·71	California.
—	☉	July 28	2½ A.	...	C. N.W. America ; Iceland ; Caspian Sea
1852.	☾	Jan. 7	6½ M.	1·33	Mexico.
—	☾	July 1	3 A.	1·46	Japan.
—	☉	Dec. 11	4 M.	..	C. Siberia ; Japan ; Mulgrave Island.
—	☾	Dec. 26	1 A.	0·66	New Caledonia.
1853.	☉	June 6	8 A.	...	Society Islands ; Gallegos ; Peru.
	☾	June 21	6 M.	0·19	C. Mississippi.
—	☉	Nov. 30	7¼ A.	...	Sandwich Islands ; Peru ; Rio Janeiro.
1854.	☾	May 12	4 A.	0·25	E. China.
—	☉	May 26	10 A.	...	C Ladrones ; N.W. America ; United States.
—	☾	Nov. 4	9½ A.	0·08	Russia.
—	☉	Nov. 20	10½ M.	...	C. Paraguay, S.E. ; V. S. Africa ; New Holland.

Year.		Month and Day.	Hour.	Magnitude.	Locality.
1855.	☽	May 2	4¼ M.	1·62	Canada.
—	☉	May 16	2¼ M.	...	V. N. Asia.
—	☽	Oct. 25	8 M.	1·56	New Albion.
1856.	☉	April 5	5 M.	...	V. New Holland ; C. New Zealand.
—	☽	April 20	9½ M.	0·79	Society Islands.
—	☉	Sept. 29	4 M.	...	V. N. Asia.
—	☽	Oct. 13	11¼ A.	0·96	France.
1857.	☉	Mar. 25	11 A.	...	C. New South Wales ; Pacific Ocean ; S. California.
—	☉	Sept. 18	6 M.	...	C. Greece ; India ; New Guinea.
1858.	☽	Feb. 27	10 A.	0·33	Poland.
—	☉	Mar. 15	0½ A.	...	C. Barbadoes ; Spain ; St. Petersburg.
—	☽	Aug. 24	2½ A.	0·46	New Guinea.
—	☉	Sept. 7	2¼ A.	...	C. Chili ; S.E. : V. S. Africa.
1859.	☽	Feb. 17	11 M.	1·62	Behring's Straits.
—	☉	Mar. 4	10 M.	...	V. Greenland.
—	☉	July 29	9½ A.	...	V. N. part of North America.
—	☉	Aug. 13	4½ A.	1·58	China.
1860.	☽	Feb. 7	2½ M.	0·77	Brazil.
—	☉	July 18	2 A.	...	C. New Mexico; Newfoundland ; Spain ; Upper Egypt.
—	☽	Aug. 1	5½ A.	0·40	Ava.
1861.	☉	Jan. 11	3½ M.	...	C. Isle of France ; New Holland, N.E.
—	☉	July 8	2 M.	...	C. Java; Caroline Islands; Society Is.
—	☽	Dec. 17	8½ M.	0·16	Nootka.
—	☉	Dec. 31	2½ A.	...	C. United States ; Cape Verd ; Sicily.
1862.	☽	June 12	6½ M.	1·21	Mexico.
—	☉	June 26	7 M.	...	V. Cape of Good Hope ; Van Diemen's Land.
—	☽	Dec. 6	8 M.	1·46	New Albion.
—	☉	Dec. 21	5½ M.	...	V. N. Asia.
1863.	☉	May 17	5 A.	...	V. North America and Europe.
—	☽	June 1	Midnt.	1·21	London.
—	☽	Nov. 25	9 M.	0·91	Pitcairn's Island.
1864.	☉	May 6	0½ M.	...	C. Borneo ; Sandwich Islands.
—	☉	Oct. 30	3½ A.	...	C. Gallipagos ; Rio Janeiro ; Cape of Good Hope.
1865.	☽	April 11	5 M.	0·13	Jamaica.
—	☉	April 25	3 A.	...	C. S. Pacific: Brazil; Cape of Good Hope.
—	☽	Oct. 1	11 A.	0·31	Italy.
—	☉	Oct. 19	5 A.	...	C. Slave Lake ; United States ; Cape Verd.
1866.	☉	Mar. 16	10 A.	...	V. N.E. Asia ; N.W. America.
—	☽	Mar. 31	5 M.	1 33	Jamaica.
—	☽	Sept 24	2½ A.	1·68	Van Diemen's Land.

Year.		Month and Day.	Hour.	Magnitude.	Locality.
1866.	⊙	Oct. 8	5 A.	...	N. of N. America ; N.W. Europe.
1867.	⊙	Mar. 6	10 M.	...	C. Cape Verd Islands; France; Tobolski.
—	☾	Mar. 20	9 M.	0·77	Pitcairn's Island.
—	⊙	Aug. 29	1 A.	...	C. Buenos Ayres, S.E.
—	☾	Sept. 14	1 M.	0·66	Canary Islands.
1868.	⊙	Feb. 23	2¼ A.	...	C. S. Pacific; Guiana ; N.E. Africa.
—	⊙	Aug. 18	5½ M.	...	C. Egypt; India; Caroline Islands.
1869.	☾	Jan. 28	1½ M.	0·46	Cape Verd Islands.
—	⊙	Feb. 11	Noon	...	V. S. Africa ; Madagascar.
—	☾	July 23	2 A.	0·56	New South Wales.
—	⊙	Aug. 7	10 A.	...	C. Manchu Tartary; New Albion; Mexico.
1870.	☾	Jan. 17	3 A.	1·25	Japan.
—	☾	July 12	11 A.	1·60	Italy.
—	⊙	Dec. 22	0½ A.	...	C. Mexico; Spain ; Black Sea.
1871.	☾	Jan. 6	9½ A.	0·66	Russia.
—	⊙	June 18	2½ M.	...	C. Java; New Guinea; Friendly Islands.
—	☾	July 2	1½ A.	0·33	Kamschatka.
—	⊙	Dec. 12	4½ M.	...	C. Persian Gulf; N. New Holland ; Mulgrave Island.
1872.	☾	May 22	11¼ A.	0·13	France.
—	⊙	June 6	3½ M.	...	C. Laccadives ; Pekin; Sandwich Islands.
—	⊙	Nov. 30	7½ A.	...	C. Friendly Islands ; Cape Horn, S.E.
1873.	☾	May 12	11¼ M.	1·46	Friendly Islands.
—	⊙	May 26	9½ M.	...	V. N. Atlantic; N. Europe; N. Asia.
—	☾	Nov. 4	4½ A.	1·50	China.
1874.	⊙	April 16	1½ A.	...	V. Cape of Good Hope.
—	☾	May 1	4½ A.	0·81	China.
1875.	⊙	April 6	7 M.	...	C. Kaffraria ; Maldives ; Philippine Is.
—	⊙	Sept. 29	1½ A.	...	C. United States; Sierra Leone; Mozambique.
1876.	☾	Mar. 10	6½ M.	0·29	Mexico.
—	⊙	Mar. 25	8 A.	...	C. Mulgrave Island; Nootka ; Greenland.
—	☾	Sept. 3	9½ A.	0·32	Russia.
—	⊙	Sept. 17	10 A.	...	C. New Guinea ; Cape Horn.
1877.	⊙	Feb. 27	7½ A.	1·62	E. Persia.
—	⊙	Mar. 15	3 M.	...	V. N. Asia.
—	⊙	Aug. 9	5 M.	...	V. N. Asia ; N. of N. America.
—	☾	Aug. 23	11½ A.	1·00	France.
1878.	☾	Feb. 17	11 M.	0·79	Behring's Straits.
—	⊙	July 29	9½ A.	...	C. Manchu Tartary ; Behring's Straits ; United States.
—	☾	Aug. 12	Midnt.	0·54	London.
1879.	⊙	Jan. 22	Noon	...	C. Peru; St. Helena ; Maldives.
—	⊙	July 19	9 M.	...	C. Guinea ; Abyssinia ; N.W. New Holland.

Year.		Month and Day	Hour.	Magnitude.	Locality.
1879.	☾	Dec. 28	4½ A.	0·14	China.
1880.	☉	Jan. 11	11 A.	...	C. Pellew Island; Scarborough Island; California.
—	☾	June 22	2 A.	1·06	New South Wales.
—	☉	July 7	1 A.	...	V. Cape of Good Hope.
—	☾	Dec. 16	4 A.	1·39	E. China.
—	☉	Dec. 31	2 A.	...	V. N. America; Europe.
1881.	☉	May 27	Midnt.	...	V. E. Asia; N.W. America.
—	☾	June 12	7 M.	1·31	New Mexico.
—	☾	Dec. 5	5½ A.	0·96	Ava.
1882.	☉	May 17	8 M.	...	C. Guinea; Persia; China.
—	☉	Nov. 10	Midnt.	...	C. Borneo; Norfolk Island; Easter Is.
1883.	☉	May 6	11⅓ A.	...	C. Philippine Island; Tonga Island; Pitcairn's Island.
—	☾	Oct. 16	7¼ M.	0·25	California.
—	☉	Oct. 30	Midnt.	...	C. N. Japan; Hawaii, S.E.
1884.	☉	Mar. 27	6 M.	...	V. N.E. Europe; N. Asia.
—	☾	April 10	Noon	1·25	New Zealand.
—	☾	Oct. 4	10½ A.	1·54	Greece.
—	☉	Oct. 19	1 M.	...	V. E. Asia; N. America.
1885.	☉	Mar. 16	6 A.	...	C. N. Pacific; Slave Lake; Baffin's Bay.
—	☾	Mar. 30	5 A.	0·83	W. China.
—	☉	Sept. 8	9 A.	...	C. Sidney; New Zealand, S.E.
—	☾	Sept. 24	8½ M.	0·75	Nootka.
1886.	☉	Mar. 5	10 A.	...	C. Torres Straits; Christmas Island; Gulf of Mexico.
—	☉	Aug. 29	1¼ A.	...	C. Honduras; Ascension Island; Kaffraria.
1887.	☾	Feb. 8	10½ M.	0·44	Sandwich Islands.
—	☉	Feb. 22	8 A.	...	V. New South Wales; C. S. America.
—	☾	Aug. 3	9 A.	0·42	Armenia.
—	☉	Aug. 19	6 M.	...	C. Norway; Lake Baikal; N. Pacific.
1888.	☾	Jan. 28	11½ A.	1·16	France.
—	☾	July 23	6 M.	1·00	Mississippi.
1889.	☉	Jan. 1	9 A.	...	C. Behring's Straits; Nootka; Hudson's Bay.
—	☾	Jan. 17	5⅓ M.	0·68	United States.
—	☉	June 28	9 M.	...	C. S. Africa; Madagascar, S.E.
—	☾	July 12	9 A.	0·46	Armenia.
—	☉	Dec. 22	1 A.	...	C. Carthagena; St. Helena; Abyssinia.
1890.	☉	June 17	10 M.	...	C. Cape Verd Islands; Smyrna; Pegu.
—	☉	Dec. 12	3 M.	...	C. Mauritius; New Zealand; Tahiti.
1891.	☾	May 23	7 A.	1·31	India.
—	☉	June 6	4½ A.	...	C. N.W. America; N. Pole; Russia.
—	☾	Nov. 16	0½ M.	1·44	Ireland.

Year.		Month and Day.	Hour.	Magnitude.	Locality.
1892.	⊙	April 26	10 A.	...	C. S. Pacific.
—	☾	May 11	11¼ A.	0·94	France.
—	⊙	Oct. 20	7 A.	...	V. N. America.
—	☾	Nov. 4	4½ A.	1·04	China.
1893.	⊙	April 16	3 A.	...	C. Easter Island ; Guiana ; N.E. Africa.
—	⊙	Oct. 9	9 A.	...	Sandwich Islands ; Peru.
1894.	☾	Mar. 21	2⅜ A.	0·25	New Guinea.
—	⊙	April 6	4½ M.	...	C. Egypt ; China ; Pacific.
—	☾	Sept. 15	4½ M.	0·21	Canada.
—	⊙	Sept. 29	5½ M.	...	C. Madagascar ; New South Wales ; New Zealand.
1895.	☾	Mar. 11	4 M.	1·56	Barbadoes.
—	⊙	Mar. 26	10 M.	...	V. Atlantic ; Europe ; N. Asia.
—	⊙	Aug. 20	0½ A.	...	V. N. Asia.
—	☾	Sept. 4	6 M.	1·54	Mississippi.
1896.	☾	Feb. 28	8 A.	0·83	E. Persia.
—	⊙	Aug. 9	4½ M.	...	C. Prussia ; E. Siberia ; Pacific.
—	☾	Aug. 23	7 M.	0·66	New Mexico.
1897.	⊙	Feb. 1	8 A.	...	C. New Caledonia ; Easter Is. ; Guiana.
—	⊙	July 29	4 A.	...	C. Gallipagos ; Barbadoes ; Guiana.
1898.	☾	Jan. 7	Midnt.	0·12	London.
—	⊙	Jan. 22	8 M.	...	C. Fezzan ; Socotra ; N. China.
—	☾	July 3	9½ A.	0·92	Russia.
—	⊙	July 18	7 A.	...	V. S. America.
—	☾	Dec. 27	Midnt.	1·33	London.
1899.	⊙	Jan. 11	11 A.	...	V. E. Asia ; N. America.
—	⊙	June 8	7 M.	...	V. N. Europe ; N. Asia.
—	☾	June 23	2⅜ A.	1·50	New Guinea.
--	☾	Dec. 17	1½ M.	0·96	Cape Verd Islands.
1900.	⊙	May 28	3 A.	...	C. Mexico ; Azores ; Egypt.
—	⊙	Nov. 22	8 M.	...	C. Benin ; Madagascar ; New South Wales.

APPENDIX III.

A CATALOGUE OF ALL THE COMETS WHOSE ORBITS HAVE HITHERTO
BEEN COMPUTED.

A NEW comet having been discovered, the first thing an astronomer
does, is to obtain 3 observations of it, whereby he may compute
the elements of the orbit. He then examines a catalogue of
comets, to see if he can identify the newly-found stranger with
any that have been before observed. The value of a complete
catalogue is therefore obvious, and as nothing of the kind has, as
far as we are aware, been published for some years, we have been
led to compile a new one.

In the preparation of the following, care has been taken that
only the most reliable orbits that were to be obtained should be
inserted, the general rule being to prefer the one which was
derived from the longest arc, other things being satisfactory.
Among the authorities consulted may be mentioned *Pingré,
Hussey, Olbers, Cooper, Hind, Arago,* and others.

From the Journals of the *Royal Astronomical Society of London,*
the *Academy of Sciences of Paris,* the *Astronomische Nachrichten,
&c.,* much valuable information has also been obtained.

PP denotes the time of perihelion passage expressed in Greenwich mean time, N.S., since 1582.

π denotes the longitude of the perihelion.

Ω denotes the longitude of the ascending node.

ι denotes the inclination of the orbit to the plane of the
ecliptic.

q denotes the perihelion distance expressed in semi-diameters of
the Earth's orbit.

ε denotes the eccentricity (of an elliptic orbit).

μ denotes the direction of motion, + direct, — retrograde.

The periods assigned in the column of " duration of visibility " are subject to much uncertainty, more especially in the case of the ancient comets.

No.	No.	Year.	PP.	π		Ω		ι		q
			d. h.	°	°	°	°	°		
I	I	370 B.C.	Winter	150—210		270—303		above 30		very sm.
2	2	136	April 29	230		220		20		1·01
3	3	68	July	300—330		150—180		70		0·80
4	4	11	Oct. 8 19	280		28		10 +		0·58
				°	′	°	′	°	′	
5	(4)	66 A.D.	Jan. 14 4	325	0	32	42	40	30	0·445
6	(4)	141	March 29 2	251	55	12	50	17	0	0·720
7		178	Sept. Beg.	290		190		18		0·5
8	(4)	218	April 6	....		....		....		....
9	5	240	Nov. 9 23	271	0	189	0	....		....
10	(4)	295	April 1 ±	....		....		44	0	0·372
11	(4)	451	July 3 12	....		....		....		....
12	6	539	Oct. 20 14	313	30	58 or 238		10	0	0·341
13	7	565 ii.	July 11 18	84	0	158	45	60	30	0·775
14	8	568 ii.	Aug. 29 7	318	35	294	15	4	8	0·907
15	9	574	April 7 6	143	39	128	17	46	31	0·963
16	(4)	760	June 11	....		....		....		....
17	10	770	June 6 14	357	7	90	59	61	49	0·642
18	11	837 i.	Feb. 28 23	289	3	206	33	10 or 12		0·580
19	12	961	Dec. 30 3	268	3	350	35	79	33	0·552
20	(4)	989 ii.	Sept. 11 23	264	0	84	0	17	0	0·568
21	(4)	1066	April 1 0	264	55	25	50	17	0	0·720
22	13	1092	Feb. 15 0	156	20	125	40	28	55	0·928
23	14	1097	Sept. 21 21	332	30	207	30	73	30	0·738
24	15	1231	Jan. 30 7	134	48	13	30	6	5	0·948

1. It is said to have separated into two parts.
3. It had a short, but brilliant tail.
4. An apparition of *Halley's Comet* (?), mentioned by Dion Cassius as having been suspended over Rome, previous to the death of Agrippa.
5. An apparition of *Halley's Comet* (?). It had a tail 8° long.
6. An apparition of *Halley's Comet*.
9. Elements somewhat doubtful. It had a tail 30° long.
11. Undoubtedly an apparition of *Halley's Comet*.
12. It had a tail 10 feet long ! !
13. A mean orbit. It had a tail 10° long.
14. Elements very reliable. On September 8 it had a tail 40° long.

ε	μ	Calculator.	Date of Discovery.	Discoverer.	Duration of Visibility.
1·0	—	Pingré	....	Greek obs.	(?.)
1·0	—	Peirce	....	Chinese obs.	5 weeks.
1·0	+	Peirce	68, July 23	Chinese obs.	5 weeks.
1·0	—	Hind	11, Aug. 26	Chinese obs.	9 weeks.
1·0	—	Hind	66, Jan. 31	Chinese obs.	7 weeks.
1·0	—	Hind	141, Mar. 27	Chinese obs.	4 weeks.
1·0	+	Hind			
1·0	...	Hind	218, April	....	6 weeks.
1·0	+	Burckhardt	240, Nov. 10	Chinese obs.	6 weeks.
1·0	...	Hind	295	....	7 weeks.
1·0	...	Laugier	451, May 17	Chinese obs.	13 weeks.
1·0	+	Burckhardt	539, Nov. 17	Chinese obs.	9 weeks.
1·0	—	Burckhardt	565, Aug. 4	Chinese obs.	15 weeks.
1·0	+	Laugier	568, Sept. 3	Chinese obs.	10 weeks.
1·0	+	Hind	574, May 2	Chinese obs.	13 weeks (?).
1·0	...	Laugier	760, May 16	Chinese obs.	8 weeks.
1·0	—	Laugier	770, May 26	Chinese obs.	10 weeks.
1·0	—	Pingré	837, Mar. 22	Chinese obs.	5 weeks.
1·0	—	Hind	962, Jan. 28	Chinese obs.	5 weeks.
1·0	—	Burckhardt	989, Aug. 5	Chinese obs.	5 weeks.
1·0	—	Hind	1066, April 2	Chinese obs.	6 weeks or +.
1·0	+	Hind	1092, Jan. 8	Chinese obs.	17 weeks.
1·0	+	Burckhardt	1097, Sept. 30	Chinese obs.	4 weeks.
1·0	+	Pingré	1231, Feb. 6	Chinese obs.	4 weeks.

15. Elements very uncertain.
16. An apparition of *Halley's Comet*.
17. It had a tail about 30° long.
18. Tolerably trustworthy. The maximum length of the tail was 80°, but it dwindled down to 3° in a fortnight.
20. Probably an apparition of *Halley's Comet*. Mentioned by several Saxon writers.
21. Possibly an apparition of *Halley's Comet*. This is the famous object which created such universal dread throughout Europe in 1066. In England it was looked upon as a presage of the success of the Norman invasion.
22. Elements satisfactory.
23. A tail 50° long was seen in China, and much bifurcated.

No.	No.	Year.	PP.			π		Ω		ι		q
				d.	h.	°	′	°	′	°	′	
25	16	1264	July	15	23	272	30	175	30	30	25	0·430
26	17	1299	March 31		7	3	20	107	8	68	57	0·318
27	18	1301 i.	September			180		60		80		0·333
28	(4)	— ii.	Oct.	23	23	312	0	138	0	13	0	0·640
29	19	1337 i.	June	15	1	2	20	93	1	40	28	0·828
30	20	1351	Nov.	25	23	69	0	Indeterminate.				1·0
31	21	1362 i.	March 11		4	219	0	249	0	21	0	0·456
32	22	1366	Oct.	13		66	0	212	0	6	0	0·958
33	(4)	1378	Nov.	8	18	299	31	47	17	17	56	0·583
34	23	1385	Oct.	16	6	101	47	268	31	52	15	0·774
35	24	1433	Nov.	5	4	262	1	110	9	77	14	0·329
36	(4)	1456	June	8	22	301	0	48	30	17	56	0·586
37	25	1457 ii.	Sept.	3	16	92	50	256	5	20	20	2·103
38	26	1468 ii.	Oct.	7	9	356	3	61	15	44	19	0·853
39	27	1472	Feb.	28	5	48	3	207	32	1	55	0·539
40	28	1490 {	Dec.	24	11	58	40	288	45	51	37	0·738
		{	Dec.	35	2	113	0	268		75		0·755
41	29	1506	Sept.	3	15	250	37	132	50	45	1	0·386
42	(4)	1531	Aug.	24	21	301	39	49	25	17	56	0·5670
43	30	1532 {	Oct.	19	14	135	44	119	8	42	27	0·6125
		{	Oct.	19	22	111	7	80	27	32	36	0·5091
44	31	1533 {	June	14	21	217	40	299	19	28	14	0·3269
		{	June	16	19	104	12	125	44	35	49	0·2028

25. One of the grandest comets on record. Its tail is said to have been 100° long.
26. Elements very doubtful.
27. Very uncertain.
28. Probably an apparition of *Halley's Comet.*
29. A fine comet. The elements assigned by Halley, Pingré, and Hind differ somewhat from those here given.
30. Very uncertain. No latitudes given.
31. Uncertain. The tail was 20 *feet* long, and the head was *the size of a wine-glass !*
32. Very uncertain.
33. An apparition of *Halley's Comet.*
34. Tolerably certain. The tail was 10° long.
36. An apparition of *Halley's Comet.* It had a splendid tail, 60° long. At one time the head was round, and the size of a bull's eye, and the tail like that of a peacock ! ! (*Chin. Obs.*)

ε	μ	Calculator.	Date of Discovery.	Discoverer.	Duration of Visibility.
1·0	+	Pingré	1264, July 14	Chinese & European	3 months.
1·0	−	Pingré	1299, Jan. 24	Chinese obs.	11 weeks.
1·0	+	Burckhardt	1301	(?)	(?).
1·0	−	Laugier	1301, Sept. 16	Chinese & European	6 weeks.
1·0	−	Laugier	1337, May	Chinese & European	3 or 4 months.
1·0	+	Burckhardt	1351, Nov. 24	Chinese obs.	1 week.
1·0	−	Burckhardt	1362, Mar. 5	Chinese obs.	5 weeks.
1·0	...	Peirce	1366, Aug. 26	Chinese obs.	Several days.
1·0	−	Laugier	1378, Sept. 26	Chinese obs.	6 weeks.
1·0	−	Hind	1385, Oct. 23	Chinese obs.	(?).
1·0	−	Hind	1433, Oct. 12	Chinese obs.	2 months.
1·0	−	Pingré	1456, May 29	European & Chinese	1 month.
1·0	+	Hind	1457, June	European obs.	3 months.
1·0	−	Laugier	1468, Sept.	European obs.	2 or 3 months.
1·0	−	Laugier	1471, Dec.	Regiomontanus	3 months.
1·0	+	Hind	} 1491, Jan.	Chinese obs.	(?).
1·0	−	Peirce			
1·0	−	Laugier	1506, July 31	Chinese obs.	2 weeks.
1·0	−	Halley	1531, Aug. 1 ±	P. Apian	5 weeks.
1·0	+	Méchain	} 1532, Sept. 22	P. Apian	16 weeks.
1·0	+	Halley			
1·0	+	Olbers	} 1533, June	P. Apian	2½ months.
1·0	−	Douwes			

37. Only approximate. It had a tail 15° long.
38. Uncertain. It had a tail 30° long.
39. A celebrated comet. When at its least distance from the Sun (3,300,000 miles) on January 21, it was quite visible in full daylight. It had a fine tail, which the Chinese say was *as long as a street!*
40. Uncertain.
41. Elements uncertain. It was *as large as a ball!* and had a tail from 3° to 5° long.
42. An apparition of *Halley's Comet.* It had a tail 7° long.
43. It had a tail several degrees long. Olbers has computed an orbit which agrees well with Halley's, but Méchain's is considered the best.
44. According to Olbers, both these orbits will satisfy the observations, and it is as yet impossible to decide between them. It had a tail 15° long.

No.	No.	Year.	PP.		π	Ω	ι	q
				d. h.	° ′	° ′	° ′	
45	(16)	1556	April	22 0	274 14	175 25	30 12	0·5049
46	32	1558	Aug.	10 12	329 49	332 36	73 29	0·5773
47	33	1577	Oct.	26 22	129 42	25 20	75 9	0·1775
48	34	1580	Nov.	28 13	109 11	19 7	64 51	0·5955
49	35	1582	May	6 16	245 23	231 7	61 27	0·2257
			May	7 8	281 26	214 42	59 29	0·0400
50	36	1585	Oct.	8 0	9 8	37 44	6 5	1·0948
51	37	1590	Feb.	8 0	217 57	165 37	29 29	0·5677
52	38	1593	July	18 13	176 19	164 15	87 58	0·0891
53	39	1596	July	25 5	270 54	330 20	51 58	0·5671
54	(4)	1607	Oct.	27 0	300 46	48 14	17 6	0·5841
55	40	1618 i.	Aug.	17 3	318 20	293 25	21 28	0·5129
56	41	— ii.	Nov.	8 8	3 5	75 44	37 11	0·3895
57	42	1652	Nov.	12 15	28 18	88 10	79 28	0·8475
58	(30?)	1661	Jan.	26 21	115 16	81 54	33 0	0·4427
59	43	1664	Dec.	4 11	130 41	81 14	21 18	1·0258
60	44	1665	April	24 5	71 54	228 2	76 5	0·1064
61	45	1668	Feb.	24 18	40 9	193 26	27 7	0·2511
			Feb.	28 19	277 2	357 17	35 58	0·0047
62	46	1672	March	1 8	46 59	297 30	83 22	0·6974
63	47	1677	May	6 0	137 37	236 49	79 3	0·2805
64	48	1678	Aug.	18 7	322 47	163 20	2 52	1·1453

45. A very fine comet, which was expected to return in 1860.
47. It had a tail 22° long. This comet formed the subject of the observations of Tycho Brahe for the detection of parallax.
48. Elements approximate. Observed also by Tycho Brahe.
49. Very uncertain. It had a faint tail 3° long, which resembled a piece of silk!!
50. This orbit was computed about 15 years ago, to see whether the comet of 1844 (ii.) was identical with this one.
51. It had a tail 7° long.
52. It had a tail 4⅛° long.
53. Discovered also by Tycho Brahe.
54. An apparition of *Halley's Comet*. It had a tail 7° long.
55. Somewhat uncertain. Seen at Lintz, August 27, and by Kepler, September 1.

ε	μ	Calculator.	Date of Discovery.	Discoverer.	Duration of Visibility.
1'0	+	Hind	1556, Feb. 28	P. Fabricius	10 weeks.
1'0	—	Olbers	1558, July 14	Landgrave Hesse	6 weeks.
1'0	—	Woldstedt	1577, Nov. 1	In Peru	12 weeks.
1'0	+	Pingré	1580, Oct. 2	Mœstlin	10 weeks.
1'0	—	Pingré	1582, May 12	Tycho Brahe	3 weeks.
1'0	—	Pingré			
1'0	+	C. A. Peters and Sawitsch	1585, Oct. 19	Tycho Brahe & Rothmann	4 weeks.
1'0	—	Hind	1590, Mar. 5	Tycho Brahe	3 weeks.
1'0	+	La Caille	1593, July 20	De Rissen	6 weeks.
1'0	—	Hind	1596, July 11	Mœstlin	5 weeks.
0'96708	—	Lehmann	1607, Sept. 11	Kepler	9 weeks.
1'0	+	Pingré	1618, Aug. 25	At Caschau	4 weeks.
1'0	+	Bessel	— Nov. 10	Harriott, etc.	3 weeks.
1'0	+	Halley	1652, Dec. 20	Hevelius	3 weeks.
1'0	+	Méchain	1661, Feb. 3	Hevelius	5 weeks.
1'0	—	Halley	1664, Nov. 17	In Spain	17 weeks.
1'0	—	Halley	1665, Mar. 27	At Aix	4 weeks.
1'0	+	Henderson ⎱	1668, Mar. 5	Gottignies, etc.	3 weeks.
1'0	—	Henderson ⎰			
1'0	+	Halley	1672, Mar. 2	Hevelius	7 weeks.
1'0	—	Halley	1677, April 27	Hevelius	12 days.
0'62697	+	Le Verrier	1678, Sept. 11	La Hire	4 weeks.

56. A splendid comet ; it had a tail, according to Longomontanus, 104° long, and of a reddish hue. Said to have been visible in the daytime.

57. Elements only approximate.

58. By some supposed to be identical with the comet of 1532 ; it was not reobserved, however, as was anticipated, about 1791.

59. It had a tail from 6° to 10° long.

60. It had a tail 25° long.

61. Seen chiefly in the southern hemisphere ; both orbits satisfy the observations, and it is impossible to say which is the correct one.

62. It had a tail about 1° long.

63. It had a tail about 6° long.

64. Elements only approximate.

No.	No.	Year.	PP.		π		Ω		ι		q
			d.	h.	°	′	°	′	°	′	
65	49	1680	Dec. 17	23	262	49	272	9	60	40	0·0062
66	(4)	1682	Sept. 14	19	301	55	51	11	17	44	0·5829
67	50	1683	July 12	17	86	31	173	18	83	47	0·5533
68	51	1684	June 8	10	238	52	268	15	65	48	0·9601
69	52	1686	Sept. 16	14	77	0	350	34	31	21	0·3250
70	53	1689	Nov. 29	4	269	41	90	25	59	4	0·1893
71	54	1695	Nov. 9	16	60	0	216	0	22	0	0·8435
72	55	1698	Oct. 18	16	270	51	267	44	11	46	0·6912
73	56	1699 i.	Jan. 13	8	212	31	321	45	69	20	0·7440
74	57	1701	Oct. 17	9	133	41	298	41	41	39	0·5926
75	58	1702 ii.	Mar. 13	14	138	46	188	59	4	24	0·6468
76	59	1706	Jan. 30	4	72	29	13	11	55	14	0·4258
77	60	1707	Dec. 11	23	79	54	52	46	88	46	0·8597
78	61	1718	Jan. 14	21	121	39	127	55	31	8	1·0254
79	62	1723	Sept. 27	15	42	52	14	14	50	0	0·9987
80	63	1729	June 13	6	320	31	310	38	77	5	4·0435
81	64	1737 i.	Jan. 30	8	325	55	226	22	18	20	0·2228
82	65	— ii.	June 8	7	262	36	123	53	39	14	0·8670
83	66	1739	June 17	10	102	38	207	25	55	42	0·6735
84	67	1742 i.	Feb. 8	4	217	35	185	38	66	59	0·7556
85	68	1743 i.	Jan. 8	4	93	19	86	54	1	53	0·8615
86	69	— ii.	Sept. 20	21	247	0	6	2	45	37	0·5229
87	70	1744	March 1	8	197	12	45	45	47	8	0·2220

65. A splendid comet, whose tail ultimately reached a length of from 70° to 90°. Halley conjectured that this was a return of the comet of 1106, 531 A.D. and 42 B.C., but this has since been shown to be unlikely. The orbit here given supposes a period of 8814 years; this, however, is subject to much uncertainty, inasmuch as the observations might possibly be satisfied by an 805 years' ellipse, or even by an hyperbolic orbit.

66. An apparition of *Halley's Comet*. It had a tail from 12° to 16° long.

67. It had a tail varying from 2° to 4°. An elliptic orbit; period assigned, 190 years.

69. Its nucleus was as bright as a 1st magnitude star, and it had a tail 18° long.

70. Observed very roughly in the East Indies. It had a tail 60° long. Pingré makes the $\Omega = 323° 45'$.

71. Observed still more imperfectly than the last in the southern hemisphere. It had a tail 18° long.

72. Uncertain.

ϵ	μ	Calculator.	Date of Discovery.	Discoverer.	Duration of Visibility.
0·99998	+	Encke	1680, Nov. 14	At Coburg	18 weeks.
0·96792	−	Rosenberger	1682, Aug. 15	Flamsteed	5 weeks.
0·98324	+	Clausen	1683, July 23	Flamsteed	6 weeks.
1·0	−	Halley	1684, July 1	Bianchini	2 weeks.
1·0	+	Halley	1686, Aug.	In India	1 month.
1·0	−	Vogel	1689, Dec. 10	Richaud	2 weeks.
1·0	+	Burckhardt	1695, Oct. 28	Jacob	3 weeks.
1·0	−	Halley	1698, Sept. 2	La Hire	4 weeks.
1·0	−	La Caille	1699, Feb. 17	Fontenay	2 weeks.
1·0	−	Burckhardt	1701, Oct. 28	Pallu	1 week.
1·0	+	Burckhardt	1702, April 20	Bianchini	2 weeks.
1·0	+	La Caille	1706, Mar. 18	D. Cassini	4 weeks.
1·0	+	La Caille	1707, Nov. 25	Manfredi	8 weeks.
1·0	−	Argelander	1718, Jan. 18	Kirch	3 weeks.
1·0	−	Spörer	1723, Oct. 12	At Bombay	9 weeks.
1·00503	+	Burckhardt	1729, July 31	Sarabat	25 weeks.
1·0	+	Bradley	1737, Feb. 6	In Jamaica	4 weeks.
1·0	+	Daussy	— Feb.	At Pekin	(?).
1·0	−	La Caille	1739, May 28	Zanotti	11 weeks.
1·0	−	La Caille	1742, Feb. 5	Cape of G. Hope	13 weeks.
0·72130	+	Clausen	1743, Feb. 10	Grischau	2 weeks.
1·0	−	D'Arrest	— Aug. 18	Klinkenberg	4 weeks.
1·0	+	Betts	— Dec. 9	Klinkenberg	4 months (?)

74. Observed also by Thomas at Pekin.
75. Very roughly observed; visible to the naked eye.
77. Discovered by D. Cassini, November 29.
79. Afterwards seen in Europe, with a faint tail 1° long.
80. Scarcely perceptible to the naked eye. The orbit is a hyperbolic one, and remarkable for its enormous perihelion distance, the greatest known.
82. Elements only approximate.
84. Visible to the naked eye, with a tail 6° or 8° long.
85. Very imperfectly observed. An elliptic orbit; period assigned, 5·436 years.
86. Very uncertain. Visible to the naked eye.
87. The finest comet of the 18th century. On February 15 it had a bifid tail, the eastern portion being 7° long, and the western 24°. Visible in a telescope in the daytime. Euler has calculated an elliptic orbit, to which he assigns a period of 122,683 years!! The statement of this comet having had six tails is believed to be a fabrication.

No.	No.	Year.	PP.		π		Ω		ι		q
			d.	h.	°	′	°	′	°	′	
88	(15?)	1746	Feb. 15	0	140	0	335	0	6	0	0·95
89	71	1747	March 3	7	277	2	147	18	79	6	2·1985
90	72	1748 i.	April 28	18	215	23	232	51	85	28	0·8404
91	73	— ii.	June 18	21	278	47	33	8	67	3	0·6253
92	74	1757	Oct. 21	7	122	58	214	12	12	50	0·3375
93	75	1758	June 11	3	267	38	230	50	68	19	0·2153
94	(4)	1759 i.	Mar. 12	13	303	10	53	50	17	36	0·5845
95	76	— ii.	Nov. 27	2	53	24	139	39	78	59	0·7985
96	77	— iii.	Dec. 16	21	138	24	79	50	4	51	0·9659
97	78	1762	May 28	8	104	2	348	33	85	38	1·0090
98	79	1763	Nov. 1	20	84	58	356	24	72	31	0·4982
99	80	1764	Feb. 12	13	15	14	120	4	52	53	0·5552
100	81	1766 i.	Feb. 17	8	143	15	244	10	40	50	0·5053
101	82	— ii.	April 26	23	251	13	74	11	8	1	0·3989
102	83	1769	Oct. 7	14	144	11	175	3	40	45	0·1227
103	84	1770 i.	Aug. 14	0	356	16	131	59	1	34	0·6743
104	85	— ii.	Nov. 22	5	208	22	108	42	31	25	0·5282
105	86	1771	April 19	5	104	3	27	51	11	15	0·9034
106	87	1772	Feb. 8	0	97	21	263	24	17	39	0·9118
107	88	1773	Sept. 5	14	75	10	121	5	61	14	1·1268
108	89	1774	Aug. 15	19	317	27	180	44	83	20	1·4328
109	90	1779	Jan. 4	2	87	14	25	4	32	30	0·7131
110	91	1780 i.	Sept. 30	22	246	35	123	41	54	23	0·0963

88. Elements uncertain, but they strongly resemble those of the comet of 1231. It passed very near the Earth.
89. Observed only during 1746.
90. Discovered by J. D. Maraldi, April 30. Visible to the naked eye, with a tail 2° long.
91. Very uncertain.
92. Elements tolerably reliable. It had a small tail.
94. The first *predicted* apparition of *Halley's Comet.* On May 5, its tail was 47° long.
95. Visible to the naked eye, with a tail 5° long.
96. This comet came near the Earth, and moved with great rapidity; it had a tail 4° long.
97. It had a small tail.
98. An elliptic orbit ; period assigned, 7334 years. Lexell makes it 1137 years.
99. Visible to the naked eye, with a tail 2½° long.

ϵ	μ	Calculator.	Date of Discovery.	Discoverer.	Duration of Visibility.
1·0	+	Hind	1746, Feb. 2	Kindermans	4 weeks.
1·0	−	La Caille	— Aug. 13	Chésaux	15 weeks.
1·0	−	Le Monnier	1748, April 26	At Pekin	9 weeks.
1·0	+	Bessel	— May 19	Klinkenberg	4 days.
1·0	+	Bradley	1757, Sept. 13	Bradley	5 weeks.
1·0	+	Pingré	1758, May 26	La Nux	5 months.
0·96768	+	Rosenberger	— Dec. 25	Palitzch	5 months.
1·0	+	La Caille	1760, Jan. 25	Messier	8 weeks.
1·0	−	La Caille	— Jan. 7	At Lisbon	14 weeks.
1·0	+	Burckhardt	1762, May 17	Klinkenberg	6 weeks.
0·99868	+	Burckhardt	1763, Sept. 28	Messier	8 weeks.
1·0	−	Pingré	1764, Jan. 3	Messier	6 weeks.
1·0	−	Pingré	1766, Mar. 8	Messier	9 weeks.
0·8640	+	Burckhardt	— April 1	Helfenzrieda	6 weeks.
0·9992	+	Bessel	1769, Aug. 8	Messier	16 weeks.
0·78683	+	Le Verrier	1770, June 14	Messier	15 weeks.
1·0	−	Pingré	1771, Jan. 10	La Nux	8 days.
1·00936	+	Encke	— April 1	Messier	15 weeks.
0·67692	+	Gauss	1772, Mar. 8	Montaigne	3 weeks.
1·0	+	Burckhardt	1773, Oct. 12	Messier	27 weeks.
1·02829	+	Burckhardt	1774, Aug. 11	Montaigne	11 weeks.
1·0	+	Zach	1779, Jan. 6	Bode	19 weeks.
0·99994	−	Clüver	1780, Oct. 26	Messier	5 weeks.

101. Discovered by Messier, April 8. An elliptic orbit; period assigned, 5·025 years. Visible to the naked eye, with a tail 3° or 4° long.

102. Visible to the naked eye, with a tail from 60° to 80° long. Bessel assigns 2090 years as the most likely period of revolution. He has shown that an error of 5″ either may increase the period to 2673 years, or diminish it to 1692 years.

103. The celebrated *Lexell's Comet*. The diameter of the head, July 1, was 2½°. It had also a small tail, and approached within 1,400,000 miles of the Earth.

104. It had a faint tail, 5° long.

105. The orbit of this comet is undoubtedly hyperbolic. It had a tail about 2° long.

106. The first recorded apparition of *Biela's Comet*.

107. Just perceptible to the naked eye.

109. Discovered by Messier, January 18.

110. An elliptic orbit; period assigned, 75,314 years.

No.	No.	Year.	PP.		π	Ω	ι	q
			d.	h.	° ′	° ′	° ′	
111	92	1780 ii.	Nov. 28	20	246 52	141 1	72 3	0·5152
112	93	1781 i.	July 7	4	239 11	83 0	81 43	0·7758
113	94	— ii.	Nov. 29	12	16 3	77 22	27 13	0·9610
114	95	1783	Nov. 19	13	49 31	55 12	47 43	1·4953
115	96	1784 i.	Jan. 21	4	80 44	56 49	51 9	0·7078
116	97	— ii.	March 10	0	137	35	84	0·637
117	98	1785 i.	Jan. 27	7	109 51	264 12	76 14	1·1434
118	99	— ii.	April 8	8	297 29	64 33	87 31	0·4273
119	100	1786 i.	Jan. 30	20	156 38	334 8	13 36	0·3348
120	101	— ii.	July 7	21	159 25	194 22	50 54	0·4101
121	102	1787	May 10	19	7 44	106 51	48 15	0·3489
122	103	1788 i.	Nov. 10	7	99 8	156 56	12 27	1·0630
123	104	— ii.	Nov. 20	7	22 49	352 24	64 30	0·7573
124	105	1790 i.	Jan. 15	5	60 14	176 11	31 54	0·7581
125	106	— ii.	Jan. 28	7	111 48	267 8	56 58	1·0632
126	107	— iii.	May 21	5	273 43	33 11	63 52	0·7979
127	108	1792 i.	Jan. 13	13	36 29	190 46	39 46	1·2930
128	109	— ii.	Dec. 27	6	135 59	283 15	49 1	0·9662
129	110	1793 i.	Nov. 4	20	228 42	108 29	60 21	0·4034
130	111	— ii.	Nov. 28	5	71 54	2 0	51 31	1·4951
131	(100)	1795	Dec. 21	10	156 41	334 39	13 42	0·3344
132	112	1796	April 2	19	192 44	17 2	64 54	1·5781
133	113	1797	July 9	2	49 27	329 15	50 40	0·5266
134	114	1798 i.	April 4	11	104 59	122 9	43 52	0·4847

111. Discovered by Olbers on the same day.

113. Visible to the naked eye, November 9, with a tail 3° long. It came very near the Earth.

114. An elliptic orbit ; period assigned, 5·613 years.

115. Visible to the naked eye, with a tail 2° long.

116. Not only are the elements uncertain, but it is doubtful whether the comet ever existed.

118. Visible to the naked eye, with a tail 8° long.

119. The first recorded apparition of *Encke's Comet*.

122. Visible to the naked eye, with a tail 2¼° long.

ε	μ	Calculator.	Date of Discovery.	Discoverer.	Duration of Visibility.
1·0	—	Olbers	1780, Oct. 18	Montaigne	3 days.
1·0	+	Méchain	1781, June 28	Méchain	3 weeks.
1·0	—	Méchain	— Oct. 9	Méchain	11 weeks.
0·6784	+	Burckhardt	1783, Nov. 19	Pigott	4 weeks.
1·0	—	Méchain	— Dec. 15	La Nux	23 weeks.
1·0	+	Burckhardt	1784, April 10	D'Angos	5 days.
1·0	+	Méchain	1785, Jan. 7	Messier	5 weeks.
1·0	—	Méchain	— Mar. 11	Méchain	5 weeks.
0·84836	+	Encke	1786, Jan. 17	Méchain	3 days.
1·0	+	Méchain	— Aug. 1	Miss Herschel	12 weeks.
1·0	—	Saron	1787, April 10	Méchain	7 weeks.
1·0	—	Méchain	1788, Nov. 25	Messier	5 weeks.
1·0	+	Méchain	— Dec. 21	Miss Herschel	4 weeks.
1·0	—	Saron	1790, Jan. 7	Miss Herschel	2 weeks.
1·0	+	Méchain	— Jan. 9	Méchain	3 weeks.
1·0	—	Méchain	— April 18	Miss Herschel	10 weeks.
1·0	—	Méchain	1791, Dec. 15	Miss Herschel	6 weeks.
1·0	—	Prosperin	1793, Jan. 8	Gregory	6 weeks.
1·0	—	Saron	— Sept. 27	Messier	15 weeks.
0·97342	+	D'Arrest	— Sept. 24	Perny	10 weeks.
0·84888	+	Encke	1795, Nov. 7	Miss Herschel	3 weeks.
1·0	—	Olbers	1796, Mar. 31	Olbers	2 weeks.
1·0	—	Olbers	1797, Aug. 14	Bouvard	3 weeks.
1·0	+	Burckhardt	1798, April 12	Messier	6 weeks.

124. Imperfectly observed on four occasions. Elements but approximate.
126. Visible to the naked eye, with a tail 4° long.
128. Discovered by Méchain and Piazzi, January 10. There was a trace of a tail to be seen.
130. Discovered by Miss Herschel, October 7. An elliptic orbit; period assigned, 422 years.
131. An apparition of *Encke's Comet*. It was just visible to the naked eye.
132. Very faint.
133. Discovered by Miss Herschel, and Lee on the same evening; by Rüdiger, August 15, and by Kecht, August 16.

U

No.	No.	Year.	PP.		π	Ω	ι	q
				d. h.	° ′	° ′	° ′	
135	115	1798 ii.	Dec.	31 13	34 27	249 30	42 26	0·7795
136	116	1799 i.	Sept.	7 5	3 39	99 32	50 56	0·8399
137	(56)	— ii.	Dec.	25 21	190 20	326 49	77 1	0·6258
138	117	1801	Aug.	8 13	183 49	44 28	21 20	0·2617
139	118	1802	Sept.	9 21	332 9	310 15	57 0	1·0941
140	119	1804	Feb.	13 15	148 53	176 49	56 44	1·0772
141	(100)	1805	Nov.	21 12	156 47	334 20	13 33	0·3404
142	(87)	1806 i.	Jan.	1 23	109 32	251 15	13 38	0·9068
143	120	— ii.	Dec.	28 22	97 2	322 19	35 2	1·0815
144	121	1807	Sept.	18 17	270 54	266 47	63 10	0·6461
145	122	1808 i.	May	12 22	69 12	322 58	45 43	0·3898
146	123	— ii.	July	12 4	252 38	24 11	39 18	0·6079
147	124	1810	Oct.	5 19	63 9	308 53	62 46	0·9691
148	125	1811 i.	Sept.	12 6	75 0	140 24	73 2	1·0354
149	126	— ii.	Nov.	10 23	47 27	93 1.	31 17	1·5821
150	127	1812	Sept.	15 7	92 18	253 1	73 57	0·7771
151	128	1813 i.	Mar.	4 12	69 56	60 48	21 13	0·6991
152	129	— ii.	May	19 10	197 43	42 40	81 2	1·2161
153	130	1815	April	25 23	149 1	83 28	44 29	1·2128
154	131	1816	March	1 8	267 35	323 14	43 5	0·0485
155	(87)	1818 i.	Feb.	7 9	95 7	254 4	20 2	0·7332

135. Discovered by Olbers, December 18. Elements only approximate.

136. Discovered by Olbers, August 26. At first faint, but afterwards visible to the naked eye, with a tail 10 long.

137. Probably a return of the comet of 1699. Visible to the naked eye, with a tail from 1° to 3° long.

138. Discovered at Paris, July 12. The observations were very rough.

139. Discovered by Méchain, August 28, and by Olbers, September 2.

140. Discovered by Bouvard, March 10, and by Olbers, March 12.

141. An apparition of *Encke's Comet.* Discovered by Pons, Huth, and Bouvard, October 20. Visible to the naked eye, with a tail 3° long.

142. An apparition of *Biela's Comet.* Discovered by Bouvard, November 16 and by Huth, November 22. Visible to the naked eye.

144. Discovered by Pons, September 20. It was visible to the naked eye, with a tail 5° long. An elliptic orbit; period assigned, 1714 years, which may, however, be extended to 2157 years, or reduced to 1403 years.

ϵ	μ	Calculator.	Date of Discovery.	Discoverer.	Duration of Visibility.
1·0	—	Burckhardt	1798 Dec. 6	Bouvard	1 week.
1·0	—	Burckhardt	1799, Aug. 7	Méchain	3 weeks.
1·0	—	Méchain	— Dec. 26	Méchain	10 days.
1·0	—	Burckhardt	1801, June 30	Reissig	3 weeks.
1·0	+	Olbers	1802, Aug. 26	Pons	6 weeks.
1·0	+	Bouvard	1804, Mar. 7	Pons	3 weeks.
0·84617	+	Encke	1805, Oct. 19	Thulis	3 weeks.
0·74578	+	Gambart	— Nov. 10	Pons	4 weeks.
1·0	—	Burckhardt	1806, Nov. 10	Pons	14 weeks.
0·99548	+	Bessel	1807, Sept. 9	Parisi	28 weeks.
1·0	—	Encke	1808, Mar. 25	Pons	1 week.
1·0	—	Bessel	— June 24	Pons	10 days.
1·0	+	Bessel	1810, Aug. 22	Pons	6 weeks.
0·99509	—	Argelander	1811, Mar. 26	Flaugergues	17 months.
0·98271	+	Nicolai	— Nov. 16	Pons	13 weeks.
0·95454	+	Encke	1812, July 20	Pons	10 weeks.
1·0	—	Nicollet	1813, Feb. 4	Pons	5 weeks.
1·0	—	Encke	— Mar. 28	Pons	6 weeks.
0·93121	+	Bessel	1815, Mar. 6	Olbers	25 weeks.
1·0	+	Burckhardt	1816, Jan. 22	Pons	11 days.
1·0	+	Pogson	1818, Feb. 23	Pons	4 days.

145. Discovered by Wisniewski, March 29.
146. Elements only approximate.
148. A very celebrated comet, conspicuously visible in the autumnal evenings of 1811. It had a tail 25° long, and 6° broad. The most reliable computations assign a periodic term of 3065 years, subject to an uncertainty of not more than 43 years. The orbit of this comet is liable to much planetary perturbation.
149. An elliptic orbit ; period assigned, 875 years. Visible to the naked eye.
150. An elliptic orbit ; period assigned, 70·68 years. Visible to the naked eye, with a tail 2° long.
152. Discovered also by Harding, April 3. Visible to the naked eye.
153. An elliptic orbit ; period assigned, 70·049 years. Bessel anticipates that planetary perturbation will bring it back to perihelion, 1887, February 9. It had a short tail.
154. Elements only approximate.
155. An apparition of *Biela's Comet*. The observations were few and indifferent.

No.	No.	Year.	PP.		π	Ω	ι	q
			d.	h.	° ′	° ′	° ′	
156	132	1818 ii.	Feb. 25	23	182 45	70 26	89 43	1·1977
157	133	— iii.	Dec. 4	22	101 55	89 59	63 5	0·8550
158	(100)	1819 i.	Jan. 27	6	156 59	334 33	13 36	0·3352
159	134	— ii.	June 27	17	287 5	273 42	80 44	0·3410
160	135	— iii.	July 18	21	274 40	113 10	10 42	0·7736
161	136	— iv.	Nov. 20	5	67 18	77 13	9 1	0·8925
162	137	1821	Mar. 21	12	239 29	48 40	73 3	0·0918
163	138	1822 i.	May 5	14	192 43	177 26	53 37	0·5044
164	(100)	— ii.	May 23	23	157 11	334·25	13 20	0·3459
165	139	— iii.	July 16	12	218 32	97 40	38 12	0·8367
166	140	— iv.	Oct. 23	18	271 40	92 44	52 39	1·1450
167	141	1823	Dec. 9	10	274 34	303 3	76 11	0·2265
168	142	1824 i.	July 11	12	260 16	234 19	54 34	0·5912
169	143	— ii.	Sept. 29	1	4 31	279 15	54 36	1·0501
170	(107)	1825 i.	May 30	13	273 55	20 6	56 41	0·8891
171	144	— ii.	Aug. 18	17	10 14	192 56	89 41	0·8834
172	(100)	— iii.	Sept. 16	6	157 14	334 27	13 21	0·3448
173	145	— iv.	Dec. 10	16	318 46	215 43	33 32	1·2408
174	(87)	1826 i.	Mar. 18	9	109 45	251 28	13 33	0·9025
175	146	— ii.	April 21	23	116 54	197 38	40 2	2·0111
176	147	— iii.	April 29	0	35 48	40 29	5 17	0·1881

157. Discovered by Bessel, December 22. It moved very rapidly. Rosenberger has computed a hyperbolic orbit.
158. An apparition of *Encke's Comet*, whose periodicity was now discovered.
159. A very brilliant comet, with a tail 7° long.
160. An elliptic orbit ; period assigned, 5·618 years. Considered by Clausen as a return of the comet of 1766 (ii).
161. Discovered by Pons, December 4. An elliptic orbit ; period assigned, 4·810 years. Clausen thought this comet might be identical with that of 1743 (i).
162. Discovered by Nicollet on the same day, and by Blainpain, January 25. Visible to the naked eye, with a tail 2½° long.
163. Discovered by Pons, May 14, and by Biela, May 17.
164. The first predicted apparition of *Encke's Comet*. Seen only in New South Wales.
165 Its apparent motion was very rapid.
166. Discovered by Gambart, July 16. An elliptic orbit ; period assigned, 5·444 years. Visible to the naked eye, with a tail 1½° long.

ϵ	μ	Calculator.	Date of Discovery.	Discoverer.	Duration of Visibility.
1·0	+	Encke	1817, Dec. 26	Pons	18 weeks.
1·0	−	Rosenberger	1818, Nov. 28	Pons	9 weeks.
0·84858	+	Encke	— Nov. 26	Pons	7 weeks.
1·0	+	Bouvard	1819, July 1	Tralles	16 weeks.
0·75519	†	Encke	— June 12	Pons	5 weeks.
0·68674	+	Encke	— Nov. 28	Blainpain	8 weeks.
1·0	−	Rosenberger	1821, Jan. 21	Pons	15 weeks.
1·0	−	Nicollet	1822, May 12	Gambart	7 weeks.
0·84446	+	Encke	— June 2	Rümker	3 weeks.
1·0	−	Heiligenstein	— May 31	Pons	2 weeks.
0·99630	−	Encke	— July 13	Pons	17 weeks.
1·0	−	Encke	1823, Dec. 1	In Switzerland	13 weeks.
1·0	−	Rümker	1824, July 15	Rümker	4 weeks.
1·00017	+	Encke	— July 23	Scheithauer	22 weeks.
1·0	−	Clausen	1825, May 19	Gambart	8 weeks.
1·0	+	Clausen	— Aug. 9	Pons	3 weeks.
0·84488	+	Encke	— July 13	Valz	8 weeks.
0·99536	−	Hansen	— July 15	Pons	12 months.
0·74657	+	Santini	1826, Feb. 27	Biela	8 weeks.
1·0	+	Clausen	1825, Nov. 6	Pons	22 weeks.
1·0	−	Clüver	1826, Mar. 29	Flaugergues	9 days.

167. Discovered by Pons, December 29, by Kohler, December 30, and by Santini, January 3. This comet had, in addition to the usual tail turned from the Sun, another turned *towards* it.

168. Seen only in the southern hemisphere.

169. Discovered by Pons, July 24, and afterwards by Gambart and Harding.

170. It had a tail 1¾° long. Elements resemble those of 1790 (iii).

171. Discovered by Harding, August 23. Orbit remarkable for its great inclination.

172. An apparition of *Encke's Comet*. Discovered by Plana, August 10, and by Pons, August 14.

173. Discovered by Biela, July 19. Very conspicuous early in October, with a bifid tail 15° long. An elliptic orbit ; period assigned, 4386 years.

174. An apparition of *Biela's Comet*, whose periodicity was now discovered. Found by Gambart, March 9.

176. Elements uncertain.

U 3

No.	No.	Year.	PP.		π		Ω		ι		q
				d. h.	o ′		o ′		o ′		
177	148	1826 iv.	Oct.	8 22	57 48		44 6		25 57		0·8528
178	149	— v.	Nov.	18 9	315 31		235 7		89 22		0·0268
179	150	1827 i.	Feb.	4 22	33 30		184 27		77 35		0·5065
180	151	— ii.	June	7 20	297 31		318 10		43 38		0·8081
181	152	— iii.	Sept.	11 6	250 57		149 39		54 4		0·1378
182	(100)	1829	Jan.	9 17	157 17		334 29		13 20		0·3455
183	153	1830 i.	April	9 7	212 11		206 21		21 16		0·9214
184	154	— ii.	Dec.	27 15	310 59		337 53		44 45		0·1258
185	(100)	1832 i.	May	3 23	157 21		334 32		13 22		0·3434
186	155	— ii.	Sept.	25 12	227 55		72 26		43 18		1·1836
187	(87)	— iii.	Nov.	26 2	110 0		248 15		13 13		0·8790
188	156	1833	Sept.	10 4	222 51		323 0		7 21		0·4584
189	157	1834	April	2 15	276 33		226 48		5 56		0·5150
190	158	1835 i.	Mar.	27 13	207 42		58 19		9 7		1·2041
191	(100)	— ii.	Aug.	26 8	157 23		334 34		13 21		0·3444
192	(4)	— iii.	Nov.	15 22	304 31		55 9		17 45		0·5865
193	(100)	1838	Dec.	19 0	157 27		334 36		13 21		0·3440
194	159	1840 i.	Jan.	4 11	192 11		119 57		53 5		0·6184
195	160	— ii.	Mar.	13 2	80 12		236 50		59 12		1·2204
196	(14)	— iii.	April	2 12	324 20		186 4		79 51		0·7420
197	161	— iv.	Nov.	13 15	22 31		248 56		57 57		1·4808
198	(100)	1842 i.	April	12 0	157 29		334 39		13 20		0·3450

177. The path of this comet crosses the ecliptic near the Earth's orbit.

178. Discovered by Clausen, October 26, and by Gambart, October 28. Visible to the naked eye, with a tail $\frac{1}{4}$° long.

180. Discovered also by Gambart.

181. At one time supposed to be a return of the comet of 1780 (i). An elliptic orbit; period assigned, 2611 years.

182. An apparition of *Encke's Comet*, afterwards visible to the naked eye.

183. Discovered in the southern hemisphere. Visible to the naked eye, with a tail 8° long.

184. Visible to the naked eye, with a tail $2\frac{1}{2}$° long.

185. An apparition of *Encke's Comet*. Discovered by Henderson, June 2. Only one observation was made in Europe.

186. Discovered by Harding, July 29.

187. The first predicted apparition of *Biela's Comet*.

ε	μ	Calculator.	Date of Discovery.	Discoverer.	Duration of Visibility.
1·0	+	Argelander	1826, Aug. 7	Pons	15 weeks.
1·0	−	Clüver	— Oct. 22	Pons	11 weeks.
1·0	−	Heiligenstein	— Dec. 26	Pons	5 weeks.
1·0	−	Heiligenstein	1827, June 20	Pons	4 weeks.
0·99273	−	Clüver	— Aug. 2	Pons	10 weeks.
0·84462	+	Encke	1828, Oct. 13	Struve	15 weeks.
0·99938	+	Hädenkamp and Mayer	1830, Mar. 16	D'Abbadie	22 weeks.
1·0	−	Wölfers	1831, Jan. 7	Herepath	9 weeks.
0·84541	+	Encke	1832, June 1	Mossotti	(?).
1·0	−	E. Bouvard	— July 19	Gambart	4 weeks.
0·75146	+	Santini	— Aug. 25	Dumouchel	18 weeks.
1·0	+	C. A. Peters	1833, Oct. 1	Dunlop	2 weeks.
1·0	+	Petersen	1834, Mar. 8	Gambart	6 weeks.
1·0	−	W. Bessel	1835, April 20	Boguslawski	5 weeks.
0·84503	+	Encke	— July 22	Kreil	9 weeks.
0·96739	−	Westphalen	— Aug. 6	Dumouchel	41 weeks.
0·84517	+	Encke	1838, Aug. 14	Boguslawski	16 weeks.
1·0	+	Lundahl	1839, Dec. 3	Galle	10 weeks.
0·99323	−	Loomis	1840, Jan. 25	Galle	9 weeks.
1·0	+	Petersen	— Mar. 6	Galle	3 weeks.
0·96985	+	Götze	— Oct. 27	Bremiker	16 weeks.
0·84479	+	Encke	1842, Feb. 8	Galle	15 weeks.

189. Discovered by Dunlop, March 16.
191. An apparition of *Encke's Comet.* Discovered by Boguslawski, July 30.
192. The second predicted return of *Halley's Comet.* It was visible to the naked eye during the whole of October, with a tail from 20° to 30° long.
193. An apparition of *Encke's Comet.* Discovered by Galle, September 16. Perceptible to the naked eye, November 7.
194. Perceptible to the naked eye, January 8.
195. An elliptic orbit; period assigned, 2423 years. Plantamour, however, makes it 13,864 years.
196. Probably a return of the comet of 1097. It had a tail 5° long.
197. An elliptic orbit; period assigned, 344 years, subject to an uncertainty of about 8 years. Possibly a return of the comet of 1490.
198. An apparition of *Encke's Comet.*

U 4

No.	No.	Year.	PP.		π	Ω	ι	q
				d. h.	° ′	° ′	° ′	
199	162	1842 ii.	Dec.	15 22	327 17	207 49	73 34	0·5044
200	163	1843 i.	Feb.	27 9	278 39	1 12	35 41	0·0055
201	164	— ii.	May	6 1	281 29	157 14	52 44	1·6163
202	165	— iii.	Oct.	17 3	49 34	209 29	11 22	1·6925
203	(48?)	1844 i.	Sept.	2 11	342 30	63 49	2 54	1·1864
204	166	— ii.	Oct.	17 8	180 24	31 39	48 36	0·8553
205	167	— iii.	Dec.	13 16	296 0	118 23	45 36	0·2512
206	168	1845 i.	Jan.	8 3	91 19	336 44	46 50	0·9051
207	169	— ii.	April	21 0	192 33	347 6	56 23	1·2546
208	(39)	— iii.	June	5 16	262 2	337 48	48 41	0·4016
209	(100)	— iv.	Aug.	9 15	157 44	334 19	13 7	0·3381
210	170	1846 i.	Jan.	22 2	89 6	111 8	47 26	1·4807
211	(87)	— ii.	Feb.	10 23	109 2	245 54	12 34	0·8564
212	171	— iii.	Feb.	25 7	116 28	102 37	30 57	0·6500
213	172	— iv.	March	5 12	90 27	77 33	85 6	0·6637
214	173	— v.	May	27 21	82 32	161 18	57 35	1·3762
215	174	— vi.	June	1 5	240 7	260 28	30 24	1·5287
216	175	— vii.	June	5 12	162 0	261 51	29 18	0·6334
217	176	—viii.	Oct.	29 17	98 35	4 41	49 41	0·8306
218	177	1847 i.	March	30 6	276 2	21 42	48 39	0·0425
219	178	— ii.	June	4 18	141 34	173 56	79 34	2·1161
220	179	— iii.	Aug.	9 8	21 17	76 43	32 38	1·4847

199. Small and faint.

200. One of the finest comets of the present century. It had a tail 60° long. The orbit is remarkable for its small perihelion distance. The period assigned is 376 years. This may be a return of the comet of 1668, but many others have also been supposed to be identical with it. (See Cooper's *Cometic Orbits*, pp. 162-9.)

202. Usually known as *Faye's Comet*. It had a very small tail. Period, 7·44 years.

203. Visible to the naked eye. An elliptic orbit ; period assigned, 5·469 years. It has not been observed since. Possibly identical with the comet of 1678.

204. Discovered by D'Arrest, July 9. Visible to the naked eye, November 10. Period, 102,050 years, subject to an uncertainty of 3090 years.

205. First seen in the southern hemisphere. It had a tail 10° long.

207. Discovered by Faye, March 6.

208. Discovered by Richter, June 6. A fine comet. Visible to the naked eye, with a tail 2½° long. A return of the comet of 1596. Period, 250 years.

ε	μ	Calculator.	Date of Discovery.	Discoverer.	Duration of Visibility.
1·0	—	Petersen	1842, Oct. 28	Laugier	4 weeks.
0·99989	—	Hubbard	1843, Feb. 28	Many observers	7 weeks.
1·00017	+	Götze	— May 2	Mauvais	21 weeks.
0·55596	+	Le Verrier	— Nov. 22	Faye	20 weeks.
0·61765	+	Brünnow	1844, Aug. 22	Di Vico	19 weeks.
0·99960	—	Plantamour	— July 7	Mauvais	35 weeks.
1·0	+	Hind	— Dec. 19	Wilmot	12 weeks.
1·0	+	Götze	— Dec. 28	D'Arrest	13 weeks.
1·0	+	Faye	1845, Feb. 25	Di Vico	9 weeks.
0·98987	—	D'Arrest	— June 2	Colla	4 weeks.
0·84743	-	Encke	— July 4	Walker	10 days.
0·99240	+	Jelinek	1846, Jan. 24	Walker	14 weeks.
0·75700	+	Plantamour	1845, Nov. 26	Walker	21 weeks.
0·79446	+	Hind	1846, Feb. 26	Brorsen	8 weeks.
0·96224	+	Peirce	— Feb. 20	Di Vico	10 weeks.
1·0	—	Argelander	— July 29	Di Vico	11 weeks.
0·72133	—	C. H. Peters	— June 26	C. H. Peters	4 weeks.
0·98836	—	Wichmann	— April 30	Brorsen	6 weeks.
1·0	+	Hind	— Sept. 23	Di Vico	3 weeks.
1·0	+	Pogson	1847, Feb. 6	Hind	11 weeks.
1·0	—	Von Littrow	— May 7	Colla	30 weeks.
1·0	—	Schweitzer	— Aug. 31	Schweitzer	13 weeks.

209. An apparition of *Encke's Comet.* Discovered by Di Vico, July 9, and by Coffin, July 10.
210. An elliptic orbit; period assigned, 2721 years.
211. An apparition of *Biela's Comet.* Discovered by Galle, November 28. It was at this return that the comet separated into two parts.
212. An elliptic orbit; period assigned, 5·58 years.
213. Discovered by G. P. Bond, February 26.
214. Discovered by Hind, 2 hours later.
215. Discovered by Di Vico, July 2. An elliptic orbit; period assigned, 12·8 years, subject to an uncertainty of 1 year.
216. Discovered by Wichmann, May 1. Visible to the naked eye, May 14. An elliptic orbit; period assigned, 400 years.
218. Visible in the daytime. It had a tail 1¾° long. The true elements are probably elliptical.

No.	No.	Year.	PP.		π		Ω		ι		q
				d. h.	o ′		o ′		o ′		
221	180	1847 iv.	Aug.	9 10	246	41	338	17	83	27	1·7671
222	181	— v.	Sept.	9 13	79	12	309	48	19	8	0·4879
223	182	— vi.	Nov.	14 9	274	14	190	50	71	53	0·3291
224	183	1848 i.	Sept.	8 1	310	34	211	32	84	24	0·3199
225	(100)	— ii.	Nov.	26 2	157	47	334	22	13	8	0·3370
226	184	1849 i.	Jan.	19 8	63	11	215	10	85	4	0·9599
227	185	— ii.	May	26 11	235	43	202	33	67	9	1·1593
228	186	— iii.	June	8 4	267	3	30	31	66	59	0·8946
229	187	1850 i.	July	23 12	273	24	92	53	68	12	1·0815
230	188	— ii.	Oct.	19 8	89	20	206	1	40	6	0·5647
231	(165)	1851 i.	April	3 11	49	42	209	30	11	21	1·6999
232	189	— ii.	July	9 0	324	10	149	19	14	14	1·1847
233	190	— iii.	Aug.	26 5	310	59	223	41	38	9	0·9843
234	191	— iv.	Sept.	30 19	338	46	44	29	74	0	0·1410
235	(100)	1852 i.	Mar.	14 18	157	51	334	23	13	7	0·3374
236	192	— ii.	April	19 13	280	0	317	8	48	52	0·9050
237	(87)	— iii.	Sept.	28 16	109	8	245	52	12	33	0·8606
238	193	— iv.	Oct.	12 15	43	12	346	13	40	58	1·2510
239	194	1853 i.	Feb.	24 6	153	21	69	49	20	19	1·0938
240	195	— ii.	May	10 8	201	12	41	12	57	53	0·9044
241	196	— iii.	Sept.	1 17	310	58	140	31	61	30	0·3067

221. A parabolic orbit best satisfies the observations.

222. Period assigned, 75 years.

223. Discovered by Di Vico, October 3, by Dawes, October 7, and by Madame Rümker, October 11.

225. An apparition of *Encke's Comet*. Discovered by Hind, September 13. Perceptible to the naked eye, October 6. On November 3 it had a tail more than 1° long.

226. A parabolic orbit satisfies the observation, but a period of 382,801 years has been assigned ! ! !

227. It had a small tail.

228. Discovered a few hours later by Bond, and by Graham, April 14. Period, 8375 years.

229. Visible to the naked eye, with a tail. Carrington has assigned a period of about 29.000 years.

230. Discovered by Brorsen, September 5, by Mauvais and Robertson, September 9, and by Clausen, September 14.

231. The first predicted apparition of *Faye's Comet*.

ϵ	μ	Calculator.	Date of Discovery.	Discoverer.	Duration of Visibility.
1·0	—	Von Littrow	1847, July 4	Mauvais	41 weeks.
0·97256	+	D'Arrest	— July 20	Brorsen	8 weeks.
1·0	—	D'Arrest	— Oct. 1	Miss Mitchell	13 weeks.
1·0	—	Sonntag and Quirling	1848, Aug. 7	Petersen	3 weeks.
0·84782	+	Encke	— Aug. 27	G. P. Bond	13 weeks.
1·0	+	Pogson	— Oct. 26	Petersen	20 weeks.
1·0	+	Goujon	1849, April 15	Goujon	24 weeks.
0·99783	+	D'Arrest	— April 11	Schweitzer	20 weeks.
1·0	+	Villarceau	1850, May 1	Petersen	17 weeks.
1·0	+	Reslhüber	— Aug. 29	G. P. Pond	9 weeks.
0·55501	+	Le Verrier	— Nov. 28	Challis	14 weeks.
0·70001	+	D'Arrest	1851, June 27	D'Arrest	14 weeks.
0·99685	+	Brorsen	— Aug. 1	Brorsen	8 weeks.
1·0	+	J. Breen	—Oct. 22	Brorsen	5 weeks.
0·84767	+	Encke	1852, Jan. 9	Hind	8 weeks.
1·0	—	Sonntag	— May 15	Chacornac	3 weeks.
0·75625	+	Santini	— Aug. 25	Secchi	5 weeks.
0·92475	+	Marth	— June 27	Westphal	24 weeks.
1·0	—	D'Arrest	1853, Mar. 6	Secchi	3 weeks.
1·0	—	Bruhns	— April 4	Schweitzer	10 weeks.
1·0	+	D'Arrest	— June 10	Klinkerfues	4 months.

232. Period, 6·441 years.

233. Discovered by Schweitzer, August 21. Period assigned, 5544 years.

234. It had a tail more than 1° long, and also a shorter one turned towards the Sun.

235. An apparition of *Encke's Comet*.

236. Discovered by Petersen, May 17, and by G. P. Bond, May 19. It was very small and faint.

237. An apparition of *Biela's Comet*. Theoretical elements.

238. Discovered also by C. H. Peters. Visible to the naked eye early in October. Period, 70 years.

239. Discovered by Schweitzer and C. W. Tuttle, March 8, and by Hartwig, March 10. Elements resemble those of the comet of 1664.

240. Visible to the naked eye in the beginning of May, with a tail 3° long.

241. Visible in the daytime, August 31 to September 4. In the south of Europe, a tail 15° long was seen.

No.	No.	Year.	PP.			π		Ω		ι		q
				d.	h.	°	′	°	′	°	′	
242	197	1853 iv.	Oct.	16	14	301	7	220	4	61	1	0·1725
243	198	1854 i.	Jan.	9	6	55	57	227	3	66	7	1·2002
244	199	— ii.	March	24	0	213	47	315	26	82	22	0·2770
245	(12)	— iii.	June	22	2	272	58	347	48	71	8	0·6475
246	200	— iv.	Oct.	27	9	94	20	324	34	40	59	0·8001
247	201	— v.	Dec.	16	1	165	52	238	19	14	10	1·3673
248	202	1855 i.	Feb.	5	17	226	33	189	40	51	12	1·2195
249	(21)	— ii.	May	30	5	237	36	260	15	23	7	0·5678
250	(100)	— iii.	July	1	5	157	53	334	26	13	8	0·3371
251	203	— iv.	Nov.	25	15	85	21	52	2	10	16	1·2248
252	204	1857 i.	March	21	8	74	49	313	12	87	57	0·7721
253	(171)	— ii.	March	29	5	115	48	101	53	29	45	0·6202
254	205	— iii.	July	17	23	249	37	23	40	58	59	0·3675
255	206	— iv.	Aug.	24	8	23	24	200	19	34	38	0·7427
256	207	— v.	Sept.	30	19	250	21	14	46	56	18	0·5651
257	208	— vi.	Nov.	19	1	44	15	139	18	37	50	1·1009
258	(189)	— vii.	Nov.	28	1	322	55	148	27	13	56	1·1696
259	(106)	1858 i.	Jan.	22	0	115	29	268	54	54	32	1·0274
260	(135)	— ii.	May	2	11	275	38	113	32	10	48	0·7689
261	209	— iii.	May	2	1	195	42	171	3	23	11	1·2090
262	210	— iv.	June	5	4	226	6	324	21	80	28	0·5462
263	(165)	— v.	Sept.	12	14	49	49	209	45	11	21	1·6999

242. Perceptible to the naked eye about the middle of the month. Elements resemble those of the comet of 1582.

243. Discovered by Klinkerfues, December 2.

244. First seen in the south of France, when very conspicuous, with a tail 4° long. Elements resemble those of the comet of 1799 (ii).

245. Discovered also by Van Arsdale. At the time of the PP. it was visible to the naked eye. The elements strongly resemble those of the comets of 961 and 1558.

246. Discovered also by several other observers. Probably a return of the comet of 1845 (i).

247. Discovered by Winnecke and Dien, January 15, 1855.

249. Discovered also by Dien and Klinkerfues. Probably a return of the comet of 1362. Period assigned, 493 years.

250. An apparition of *Encke's Comet*.

251. Discovered also by Van Arsdale.

ε	μ	Calculator.	Date of Discovery.	Discoverer.	Duration of Visibility.
1·0	—	Hoffmann	1853, Sept. 11	Bruhns	11 weeks.
1·0	—	Marth	— Nov. 25	Van Arsdale	12 weeks.
1·0	—	Hornstein	1854, Mar. 23	Many observers	6 weeks.
1·0	—	Klinkerfues	— June 4	Klinkerfues	10 weeks.
1·0	+	Bruhns	— Sept. 11	Klinkerfues	11 weeks.
1·0	+	Oudemans	— Dec. 24	Colla	16 weeks.
1·0	—	Winnecke	1855, April 11	Schweitzer	5 weeks.
0·99090	—	Donati	— June 3	Donati	2 weeks.
0·84779	+	Encke	— July 13	Maclear	5 weeks.
1·0	—	G. Rümker	— Nov. 12	Bruhns	7 weeks.
1·0	+	Pape	1857, Feb. 22	D'Arrest	9 weeks.
0·80160	+	Bruhns	— Mar. 18	Bruhns	11 weeks.
1·0	—	Pape	— June 22	Klinkerfues	3 weeks.
1·0	+	Villarceau	— July 25	C. H. Peters	5 weeks.
1·0	—	Bruhns	— Aug. 20	Klinkerfues	7 weeks.
1·0	—	Pape	— Nov. 10	Donati	5 weeks.
0·66014	+	Lind	— Dec. 10	Maclear	5 weeks.
0·82961	+	Bruhns	1858, Jan. 4	H. P. Tuttle	9 weeks.
0·75466	+	Winnecke	— Mar. 8	Winnecke	12 weeks.
1·0	+	Hall	— May 2	Tuttle	4 weeks.
1·0	—	Bruhns	— May 21	Bruhns	3 weeks.
0·55502	+	Bruhns	— Sept. 8	Bruhns	8 weeks.

252. Discovered also by Van Arsdale. Orbit decidedly parabolic.
253. An apparition of *Brorsen's Comet*, 1846 (ii).
255. Discovered by Dien, July 28, and by Habicht, July 30. Believed to revolve in an elliptic orbit, to which a period of 258 years has been assigned.
256. Faintly perceptible to the naked eye, September 20. It had a short tail. Elements resemble those of the comets of 1790 (iii) and 1825 (i). A period of 1618 years has been assigned by Villarceau.
257. Discovered a few hours later by Van Arsdale.
258. An apparition of *D'Arrest's Comet*. Period, 2331 days.
259. Discovered by Bruhns, January 11. Probably a return of the comet of 1790 (ii). Period assigned, 13·6 years.
260. An apparition of *Pons's Comet* of 1819 (iii). Period, 5·549 years.
262. Elements resemble those of the comet of 1799 (ii).
263. An apparition of *Faye's Comet*.

No.	No.	Year.	PP.	π	Ω	ι	q
			d. h.	° ′	° ′	° ′	
264	211	1858 vi.	Sept. 29 23	36 13	165 19	63 1	0·4822
265	212	— vii.	Oct. 12 19	4 13	159 45	21 16	1·4270
266	(100)	—viii.	Oct. 18 11	157 57	334 28	13 4	0·3407
*	(87)	1859 i.	May 24 0	109 33	245 43	12 23	0·8680
267	213	— ii.	May 29 5	75 9	357 7	84 9	0·2020
268	214	1860 i.	Feb. 16 17	173 45	324 3	79 35	1·1973
269	215	— ii.	March 5 17	50 16	8 56	48 13	1·3083
270	216	— iii.	June 16 2	161 32	84 40	79 18	0·2929
271	217	— iv.	Sept. 28 7	111 59	104 14	28 14	0·9537
272	218	1861 i.	June 3 4	243 3	29 51	79 55	0·9215
273	219	— ii.	June 11 15	249 14	278 58	85 32	0·8223

264. One of the finest comets of the present century. It became visible to the naked eye early in September, and was very conspicuously seen in Europe for about 6 weeks, when, owing to its rapid passage to the southern hemisphere, it became lost to view. It was seen at the Cape of Good Hope till March 4, 1859. During the first week in October it had a tail nearly 40° long. An elliptic orbit ; period assigned, 2101 years.

266. An apparition of *Encke's Comet*. It was very faint.

* Theoretical elements of *Biela's Comet*. It was unfavourably placed for observation, and escaped detection.

268. It does not appear that this comet was seen in Europe. Liais, who observed it in Brazil, states that it had a double nebulosity, and conjectures it to be identical with 1845 (ii), 1785 (i), and 1351.

270. Suddenly became visible towards the end of June. On the 22nd it had a tail 15° long. Liais has assigned a period of 1089 years.

ϵ	μ	Calculator.	Date of Discovery.	Discoverer.	Duration of Visibility.
0·99647	—	Bruhns	1858, June 2	Donati	7½ months.
1·0	—	Weiss	— Sept. 5	H. P. Tuttle	8 weeks.
0·84639	+	Powalky	— Aug. 7	Förster	10 weeks.
0·75349	+	Hubbard	...	. . . .	...
1·0	—	Hall	1859, April 2	Tempel	12 weeks.
1·0	+	Liais	1860, Feb. 26	Liais	2 weeks.
1·0	+	Seeling	— April 17	C. Rümker	7 weeks.
1·0	+	Moësta	— June 19	Several obs.	8 weeks.
1·0	—	Valz	— Oct. 23	Tempel	3 weeks.
0·99388	+	Pape	1861, April 4	Thatcher	8 weeks.
0·99390	+	Seeling	— May 13	Tebbutt	20 weeks.

271. Very faint, and only 4 obs. obtained.

272. Visible to the naked eye. It had a faint diffused tail. ¦Elliptic orbit ; period assigned, 1848 years.

273. First seen in the southern hemisphere. On July 2 it had a tail more than 100° long. It was then departing both from the Sun and from the Earth, and its brilliancy gradually diminished after that evening. On the whole, it may be regarded as having been one of the most magnificent comets on record. An elliptic orbit ; period assigned, 1568 years.

A SUMMARY OF THE PRECEDING CATALOGUE.[1]

FROM a careful inspection of the Catalogue just given we obtain certain deductions which will be tabulated in the present section.

It appears that 270 comet apparitions have been subjected to mathematical investigation, viz. : —

Known Elliptic Comets	18
Subsequent Returns	53
Elliptic Orbits not yet verified . . .	34
Parabolic Comets	159
Hyperbolic Comets	6
	270

Of known elliptic comets, we have the following, as the number of the apparitions of each : —

17	of Halley
16	of Encke
7	of Biela
3	of Faye

and 2 of each of the following :—

961: 1097: 1231: 1264: 1362 i: 1532: 1596: 1678: 1699 i: 1790 ii: 1790 iii: 1819 iii: 1946 iii: 1851 ii.

Elliptic orbits have been assigned *in the Catalogue,* to the following comets, but no 2nd returns have as yet taken place : —

1680: 1683: 1743 i: 1763: 1766 ii: 1769: 1770 i: 1780 i: 1783: 1793 ii: 1807: 1811 i: 1811 ii: 1812 i: 1815: 1819 iv: 1822 iv: 1825 iv: 1827 iii: 1830 i: 1840 ii: 1840 iv: 1843 i: 1844 ii: 1846 i: 1846 iv: 1846 vi: 1846 vii: 1847 v: 1849 iii: 1851 iii: 1852 iv: 1857 iv: 1858 vi.

---

[1] This summary was compiled in August 1860: it does not therefore include comet iv. 1860, *or any discovered subsequently.* The omission is quite immaterial as regards the general results.

Elliptic orbits have been assigned by some computers to the following comets; but the probability is not sufficiently great to warrant their being included in the preceding list : —

1585: 1744: 1773: 1826 ii: 1846 viii: 1847 i: 1847 iii: 1849 i: 1850 i: 1857 v: 1860 iii.

Hyperbolic orbits have been assigned by some computers to the following comets; but the probability is not sufficiently great to warrant their being definitely given as such : —

1723: 1773: 1779: 1818 iii: 1826 ii: 1840 i: 1843 i: 1844 iii: 1845 i: 1845 ii: 1849 iii.

The following have been supposed by some to be identical : —

1860 i	with	1845 ii, 1785 i, and 1351.
1858 iv	—	1799 ii.
1857 v	—	1825 i, and 1790 iii.
1854 iv	—	1558.
1854 ii	—	1799 ii.
1853 iv	—	1582 ii.
1853 i	—	1664.
1852 ii	—	1819 ii.
1844 i	—	1678.
1843 i	—	1668 and many others.
1840 iv	—	1490.
1827 iii	—	1780 i.
1819 iv	—	1743 i.
1819 iii	—	1766 ii.
1661	—	1532.

CLASSIFICATION OF THE DIRECTIONS OF HELIOCENTRIC MOTION.

I.

Of the 18 known elliptic comets, there are, whose motions are —

Direct	12 or 66·6 per cent.
Retrograde . . .	5 or 33·3 „
	18

II.

Of the 34 unverified elliptic comets, there are, whose motion
is —

    Direct . . . .   22 or 64·6 per cent.
    Retrograde . . .   12 or 35·4   „
                           —
                         34

III.

Of the 10 doubtful elliptic comets, there are, whose motion
is —

    Direct . . . .   8 or 80·0 per cent.
    Retrograde . . .   2 or 20·0   „
                         —
                         10

IV.

Of the 6 known hyperbolic comets, there are, whose motion
is —

    Direct . . . . ·   5 or 83·0 per cent.
    Retrograde . . .   1 or 17·0   „
                         —
                         6

V.

Of the 11 improbable hyperbolic comets, there are, whose motion is —

    Direct . . . .   8 or 72·7 per cent.
    Retrograde . . .   3 or 27·3   „
                         —
                         11

VI.

Then of the remaining 159 comets, probably parabolic (we here
include classes III. and V.), there are, whose motion is —

    Direct . . . .   73 or 47·1 per cent.
    Retrograde . . .   82 or 52·9   „
    Unknown . . .   4
                         —
                         159

Combining classes I. II. IV. and VI., we get —

> Direct .   .   .   .   111 or 22·1 per cent.
> Retrograde .   .   .   101 or 47·9   „
> Unknown .   .   .   4   ˌ
> ───
> 217

An examination of the preceding shows,—*That with comets revolving in elliptic orbits, there is a strong and decided tendency to direct motion; the same obtains with the hyperbolic orbits; with the parabolic orbits, there is a trifling preponderance the other way; and taking all the calculated comets together, the numbers are too nearly equal, to afford any indication of the existence of any general law governing the direction of motion.*

CLASSIFICATION OF INCLINATIONS.

Dividing the calculated comets into 4 classes, as before given, we shall find that the inclination of the orbits of every 100 are distributed as follows : —

Angle of Inclination.	Class I.	Class II.	Class IV.	Class VI.	Total.
o					
0—10	11·1	11·7	0·0	7·5	7·5
10—20	33·3	2·9	16·6	5·2	14·5
20—30	5·6	5·8	0·0	8·4	4·9
30—40	11·1	17·6	0·0	9·7	9·6
40—50	5·6	17·6	0·0	17·5	10·2
50—60	11·1	17·6	33·2	13·6	18·8
60—70	11·1	11·7	16·6	11·7	12·7
70—80	11·1	8·8	16·6	13·0	12·3
80—90	0·0	5·8	16·6	13·6	9·0
	100·0	99·5	99·2	100·5	99·5

An examination of the 1st column shows this fact : —
That there is a decided tendency in the periodic comets to revolve

in orbits but little inclined to the ecliptic, and therefore a low inclination is an eminently favourable indication of a periodic comet.

Combining the 4 classes we find:—*A decided disposition in the orbits to congregate in and around a plane inclined* 50° *to the ecliptic.*

CLASSIFICATION OF THE POSITIONS OF THE PERIHELIA AND NODAL POINTS.

Taking the longitudes of the perihelia and of the ascending nodes of 209 comets (deducting 8 not certainly determined), we shall find that in every 100 they are distributed as follows:—

	♊	♌
0—30	7·6	7·6
30—60	9·0	10·0
60—90	13·0	10·5
90—120	9·9	8·0
120—150	8·0	8·0
150—180	4·2	8·5
180—210	6·6	10·5
210—240	8·8	9·6
240—270	9·9	7·1
270—300	11·4	4·3
300—330	8·5	9·6
330—·360	2·8	5·7
	99·2	99·4

An examination of the 2nd column shows,—*that there is an evident tendency in the perihelia to crowd together in 2 opposite regions, between* 60°—120° *and* 240°—300°. A uniform distribution would give 16·6 perihelia to every arc of 60°: Now, in the 1st case, we have 22·9 or 38 per cent. above the mean; and in the 2nd, 21·3, or 29 per cent. above the mean. A further examination will show that the regions between 150°—180° and 330°—360° are correspondingly poor.

From the 3rd column it appears: — *That there is an evident,
though less marked tendency, in the nodes to come together in 2
regions (not, however, in this case exactly opposite) between* 30°—90°
and 180°—240°. A uniform distribution would give 16·6 nodes
to every arc of 60°: now, in the 1st case we have 20·5 or 23 per
cent. above the mean; and in the 2nd, 20·1 or 20 per cent. above
the mean. With the nodes the poor region seems to be between
270°—300.

CLASSIFICATION OF THE DISTANCES OF THE PERIHELIA.

Within a radius of 95,000,000 miles from the
  Sun . . . .   160 or 73·7 per cent.
Between 95,000,000—190,000,000 . .   52 or 24·0 „
  „   190,000,000—285,000,000 . .   4 or 1·7 „
  „   285,000,000—380,000,000 . .   0 or 0·0 „
  „   380,000,000—475,000,000 . .   1 or 0·4 „
Beyond 475,000,000 . . . . .   0 or 0·0 „
                                        ‾‾‾‾   ‾‾‾‾
                                        217   99·8

CLASSIFICATION OF THE PERIHELION PASSAGES ACCORDING TO THE
MONTHS OF THE YEAR.

Of the 268 known perihelion passages there occurred in: —

January 24
February 20
March 22
April 21
May 20
June 21
July 18
August 15
September 31
October 26
November 31
December 19

The monthly average is therefore 22·3. January, September,
October, and November only, are above the mean, all the others
below. The minimum is in August, which only exhibits 15·0 or
33 per cent. below the average; a circumstance doubtless due to
the long days and short nights, which more or less prevail during

the summer months. The quick rise in September is probably
due, less to the lengthening of the nights (and consequent increased opportunities for observation) than to the excellence of
that month for astronomical purposes. The advantages afforded by
the long winter nights are more or less neutralised by the frequent
inclement weather. Thus it happens that all the winter months
(January excepted) are below the average.

APPENDIX IV.

A CATALOGUE OF COMETS RECORDED, BUT NOT WITH SUFFICIENT PRECISION TO ENABLE THEIR ORBITS TO BE CALCULATED.

THE following Catalogue, it is almost needless to say, is founded upon that most valuable one of Pingré's, but this is the first time that a catalogue of the kind has been printed in such a full tabular form,—a great convenience for purposes of reference.

Our Catalogue is comprised in 29 octavo pages, whereas Pingré's extends to no less than 300 quarto pages : to his work, therefore, the reader who desires to know more of any of the following comets is referred ; its value is greatly enhanced by the copious references to his authorities which he gives : would that English authors, generally, would take a hint in that respect ! Pingré's *Cométographie* was compiled almost a century ago ; since his time, however, numerous and important accessions have been made to our stock of knowledge, more especially by E. Biot, who has translated several Chinese chronicles and catalogues of stars and comets, not previously properly understood, in some cases not understood at all. Hind's valuable Catalogues in the *Companion to the Almanac,* for 1859 and 1860, have also been consulted.

It is wholly impossible to give a list of the authorities ; the last column in the Catalogue will afford an idea of the number. Pingré gives references to *all* the works in which mention is made of any fact ; we, of course, have only been able to give one or two of the chief.

It may be well to add, that very great uncertainty hangs over the earlier comets, and to some extent over the later ones, more especially as regards the positions in which they are seen.

No.	Year.	Time of Visibility.	Duration of Visib. (Weeks.)	Path of the Comet.—Notes, &c.	Authority.	Reference.
1	B.C. 1770±	...	...	...	Varro	St. Augustine, *De Civit.* xxi. 8.
2	1194±	...	...	...	...	Fréret, *Acad. des Inscrip.* x. 357
3	975±	...	...	...	...	Pliny, *Hist. Nat.* ii. 23.
4	618 or 619	...	...	...	Sibyll. Orac.	Fréret, *Acad. des Inscrip.* x.
5	612	August	...	Amongst the seven stars in Ursa Major.	Chinese obs.	Gaubil.
6	533	Winter	...	Aquarius and the tail of Capricornus.	,,	Gaubil.
7	531	Winter	...	Aquarius.	,,	Gaubil; Mailla, *Hist. Gén.* ii. 193
8	524	Winter	...	Scorpio.	,,	Mailla, ii. 222.
9	481	End of the year	...	Scorpio.	,,	Pliny, ii. 22.
10	479	October	...	In shape like a horn.	European obs.	Pliny, ii. 59; Diog. Laër. *In Anax.*
11	465±	...	11	...	Chinese obs.	Couplet.
12	432	...	...	...	,,	Aristotle, *Meteor.* i. 6.
13	426 or 402	Winter	...	Arctic Pole.	European obs.	Mailla, ii. 267.
14	360	...	...	In the W.	Chinese obs.	Pliny, ii. 22.
15	345	...	...	In the N.W.	,,	...
16	344	...	...	Near the Equator. In Leo [?].	...	Diod. Sic. *Hist.* xvi. 11.
17	340	...	...	...	European obs.	Plut. *In Timol.*; Aristotle, i. 7.
18	304	...	...	...	Chinese obs.	Mailla, ii. 306.
19	302	...	...	...	,,	Mailla, ii. 306.
20	295	...	...	...	,,	Ma-tuoan-lin.

No.	Year.	Time of Visibility.	Duration of Visib.	Path of the Comet.—Notes, &c.	Authority.	Reference.
	B.C.					
21	239	June	...	In the N. thence Westward.	Chinese obs.	Ma-tuoan-lin.
22	237	April, May	...	Sagittarius.	„	Ma-tuoan-lin.
23	233	February	...	In the E.	„	Ma-tuoan-lin.
24	214 or 213	...	...	In the W.	„	Mailla, ii. 399.
25	203	August	...	Near Arcturus, "extended from E to W."	Chin. and Europ. obs.	Jul. Obseq. *Prodig. Suppl.*
26	202	...	...	...	European obs.	Jul. Obseq. *Suppl.*
27	171	...	...	...	Chinese obs.	Mailla, ii. 554.
28	168	...	...	...	European obs.	Jul. Obseq. *Suppl.*
29	166	...	...	...	„	Jul. Obseq. *Suppl.*
30	165	...	...	Described as being very brilliant.	„	Jul. Obseq. *Suppl.*
31	156	October	...	Aquarius, Equuleus and Pegasus.	Chinese obs.	Mailla, ii. 568.
32	154	January	...	...	...	Ma-tuoan-lin.
33	148 or 147	April or November	...	Head of Orion.	...	Mailla, ii. 584—8.
34	147	...	...	...	Chin. and Europ. obs.	{ Jul. Obseq.; Seneca, *Quæst. Nat.* vii. 15.
35	146	...	...	Possibly identical with the preceding.	„ „	
36	137 ii	June	...	Hercules.	Chinese obs.	See Pingré, *Comét.* i. 269.
37	— iii	Autumn	...	Possibly these two are identical with No. 2 in the Catalogue of Cal. Com.	„	See Pingré, *Comét.* i. 269.
38	136	...	...	...	European obs.	Jul. Obseq.
39	134	July—Sept.	10	P.P. took place about Aug. 15, Seen in N. and E.	Chinese obs.	Justin. *Hist.* xxxvii. 2.

No.	Year.	Time of Visibility.	Duration of Visib.	Path of the Comet.—Notes, &c.	Authority.	References.
			Weeks.			
40	B.C. 127	...	...	...	European obs.	Jul. Obseq. *Suppl.*
41	120 o?	Spring	...	In the E.	Chinese obs.	Mailla, iii. 46.
42	119	...	...	...	,,	Justin. xxxvii. 2.
43	118	May	...	Gemini.	,,	Mailla, iii. 61.
44	— ii.	May	...	Ursa Major.	,,	Mailla, iii. 61.
45	108	...	...	From Procyon towards Gemini.	,,	Ma-tuoan-lin.
46	102±	...	...	Near γ Boötis.	,,	Ma-tuoan-lin.
47	99	...	...	From W. to E.	European obs.	Jul. Obseq. Pliny, ii. 34.
48	93	...	...	...	,,	Jul. Obseq.
49	91	...	...	...	,,	Jul. Obseq.
50	86	Autumn	...	In Virgo [?].	Chinese obs.	Pliny, ii. 23.
51	84 or 83	Spring	...	In the N.W.	,,	Mailla, iii. 101.
52	75	...	...	...	European obs.	Pliny, ii. 35.
53	62	...	9 days	The tail is said to have extended from the horizon to the zenith.	Chin. and Europ. ob.	Dion Cass. *Hist. Roman.* xxxvii.
54	55	...	...	Passed northwards.	European obs.	Dion Cass. xxxix.
55	52	...	...	Passed westwards.	,,	Dion Cass. xl.
56	48 i.	...	...	Cassiopeia.	Europ. and Chin. ob.	Pliny, ii. 23; Lucan. *Phars.* i. 526; Mailla, iii. 155.
57	— ii.	September	...	Leo, Virgo. Possibly identical with the preceding. See Pingré.		

No.	Year	Time of Visibility.	Duration of Visib. Weeks.	Path of the Comet.—Notes, &c.	Authority.	Reference.
58	B.C. 43	May, June	...	Orion.	Chin. and Europ. obs.	Mailla, iii. 162.
59	42	...	...	...	European obs.	See Pingré, ii. 279.
60	41	...	...	...	?	See Pingré, ii. 279.
61	31	February	...	Pegasus.	Europ. and Chin. obs.	Dion Cass. l. 8; Mailla, iii. 178.
62	29	...	...	...	European obs.	Dion Cass. li.
63	4	Ver. Equinox	...	Head of Capricornus.	Chinese obs.	Mailla, iii. 214.
64	3 B.C.	April or May	...	Aquila. Possibly identical with the preceding.	„	Mailla, iii. 214.
65	10 A.D.	...	4	Aries. There are said to have been several comets this year.	European obs.	Dion Cass. lvi. 24.
66	14	Dec.—Jan.	3	...	Europ. and Chin. ob.	Dion Cass. lvi. 29; Mailla, iii. 240.
67	19	...	...	...	Chinese obs.	Couplet.
68	22	Dec.	1	Hydra.	„	Mailla, iii. 251.
69	39	March 13—April 30	7	From the Pleiades, through Aries towards Pegasus. It disappeared between γ Pegasi and α Andromedæ.	„	Mailla, iii. 326.
70	54	Autumn	...	First seen in the N.; moved to the zenith, and thence eastward.	Europ. and Chin. obs.	Dion Cass. lx. 35; Suetonius, *In Claud.*
71	55	Nov.—March 56	16	When first seen it was 2° long, and after moving to the S.W., disappeared 6° N.E. of γ Cancri.	Chinese obs.	Gaubil.

No.	Year.	Time of Visibility.	Duration of Visib.	Path of the Comet.—Notes, &c.	Authority.	Reference.
			Weeks.			
72	A.D. 60	Aug. 9—Dec. 22	18	First seen near α Persei with a tail 2° long; moved thence southward towards the feet of Virgo.	Europ. and Chin. ob.	Tacitus, *Annal.* xiv. 22; Mailla, iii. 352.
73	61	Sept. 27—Oct.	3 or+	First seen near ζ Boötis, with a tail pointing towards Corona Borealis.	„	Seneca, vii. 28.
74 i.	64 i.	May 3—July	11	First seen to the S. of η Virginis.	„	Gaubil.
75	— ii.	May—Oct.	25	From the N. by the W. to the Southern sky.	European obs.	Seneca, vii, 21, 29.
76	65	July 29—Sept.	8	From the mouth of Hydra towards α Leonis, α Persei (1) and β Leonis. The tail ("vapour") extended to ι Ursæ Majoris.	Chinese obs.	Ma-tuoan-lin.
77	69	...	...	Possibly the one referred to by Josephus, *Bell. Jud.* vi. 5.	European obs.	Dion Cass. lxv. 8.
78	70	Dec.—Jan. 71.	7	Near the head of Leo.	Chinese obs.	Gaubil.
79	71	March—April	8	From the Pleiades, apparently across Gemini and Cancer, into Leo.	„	Gaubil.
80	75	July		From Hydra, through the region S. of Coma Berenicis towards β Leonis.	„	Mailla, iii. 375.
81	76	Sept. 7—Oct.	6	From α Ophiuchi towards β Capricorni. Tail 3° long.	Europ. and Chin. ob.	Pliny, ii. 22.
82	77	Jan. 18—April	15	From β Arietis, through Andromeda, and possibly Cassiopeia towards Draco.	Chinese obs.	Ma-tuoan-lin.
83	79	Spring	...	...	European obs.	Dion Cass. lxvi. 17.

No.	Year.	Time of Visibility.	Duration of Visib.	Path of the Comet.—Notes, &c.	Authority.	Reference.
84	A.D. 84	June 4—July	Weeks. 6	From near Antares (α Scorpii) to Cassiopeia. Tail 3° long.	Chinese obs.	Hind.
85	101	Dec. 30	...	In Leo. Possibly not a comet at all, but only a new star.	,,	Hind.
86	104	June 10—July	3 or 4	First seen in the circumpolar regions; it then passed to the Pleiades.	,,	Hind.
87	107	Sept. 13	...	From Gemini southwards to Canis Major.	,,	Hind.
88	108	July 25	...	Ursa Major. Tail 2° long.	,,	Hind.
89	110	January	...	Seen near γ Eridani. The tail was 6° long, of a bluish tinge, pointing towards the N.E., in which direction the comet moved.	,,	Ma-tuoan-lin.
90	115	November	...	From Aquarius to the Pleiades.	,,	Hind.
91	117	January	...	From β Aquarius by α Equulei to Musca.	,,	Gaubil.
92	123	December	...	Seen near α Herculis and α Ophiuchi.	,,	Hind.
93	132	January 29	...	From Capricornus, through Sagittarius, Aquarius, Equuleus to Pegasus.	,,	Ma-tuoan-lin.
94	133	February 8	...	Seen to the S.W. of γ Eridani, the vapour was 50° and 2° broad.	,,	Hind.
95	149	October 19	3 days	Near the head of Hercules. Tail 5° long.	,,	Mailla, iii. 441.
96	161 i.	February 13	...	Near α Scorpii.	,,	Mailla, iii. 459.
97	— ii.	June 14	...	From α Pegasi to R.A. 14¾ h. (no Declination assignable). It remained stationary for some time, and when it reached the above R.A. it threw out a tail 5° long.	,,	Ma-tuoan-lin.

No.	Year.	Time of Visibility.	Duration of Visib.	Path of the Comet.—Notes, &c.	Authority.	Reference.
	A.D.		Weeks.			
98	180 i.	Aug. Sept.	3	From ι Ursæ Majoris towards Leo.	Chinese obs.	Ma-tuoan-lin.
99	— ii.	Winter	2 or 3 m.	From Canis Major to Hydra.	"	Mailla, iii. 506.
100	181 i.	March &c.	8	From Andromeda towards the Polar regions.	"	Ma-tuoan-lin.
101	— ii.	August	...	From λ Ursæ Majoris towards β Leonis.	"	Mailla, iii. 507.
102	188	March	...	...	"	Mailla, iii. 520.
103	190 +	...	...	...	European obs.	Herodian, *Hist.* 1.
104	192	Sept. or Oct.	...	In Virgo. A large comet.	Chinese obs.	Ma-tuoan-lin.
105	193	November	...	From α Virginis towards α Herculis and α Ophiuchi.	"	Ma-tuoan-lin.
106	200	November 6	1	Near δ Serpentis.	"	Mailla, iv. 35.
107	204	November	...	From Gemini to Leo.	Europ. and Chin. ob.	Dion Cass. lxxv. 16; Mailla, iv. 40.
108	206	February	...	In the circumpolar regions.	Chinese obs.	Mailla, iv. 43.
109	207	November 10	...	In Leo. Mailla assigns this comet to the previous year.	"	Mailla, iv. 45.
110	213	January	...	Near θ Geminorum.	"	Mailla, iv. 63.
111	222	November 4	1	Between β Virginis and σ Leonis. May have been a temporary star.	"	Gaubil.
112	225	December 9	...	Through Leo.	"	Ma-tuoan-lin.
113	232	December 4	...	First seen near σ Leonis.	"	Ma-tuoan-lin.
114	236 i.	November 30	...	Near σ Scorpii. Tail 3° long.	"	Ma-tuoan-lin.
115	— ii.	Dec. 15	...	In Hercules.	"	Ma-tuoan-lin.
116	238 i.	Sept. Oct.	6	First seen near ν Hydræ, passed thence to the E.; tail 3° long.	"	Ma-tuoan-lin.

No.	Year	Time of Visibility.	Duration of Visib.	Path of the Comet.—Notes, &c.	Authority.	Reference.
117	A.D. 238 ii.	Nov. 29,— Dec. 15.	Weeks. 2	From Pegasus to Ophiuchus.	Chinese obs.	Ma-tuoan-lin.
118	245	Sept. 18.—Oct.	3	Hydra all the time visible. Tail 2° long.	,,	Ma-tuoan-lin.
119	247	Jan. 16.—Mar.	8	First seen in Corvus. Tail 1° long.	,,	Ma-tuoan-lin.
120	248 i.	April	6	Seen in the Pleiades. The head was of a violet hue, and the tail extended 6° in a S.W. direction.	,,	Ma-tuoan-lin.
121	— ii.	August	...	From Crater to Corvus. It had a tail 2° long.	,,	Ma-tuoan-lin.
122	251	December 21— March 25	13	First seen near α and β Pegasi, and moved towards the W.	,,	Ma-tuoan-lin.
123	252	March 25	...	From Musca to Orion. The tail was 50° or 60° long, whitish, and stretched toward the S.	,,	Ma-tuoan-lin.
124	253	December— June 254	27	First seen in Corvus. The tail was 50° long, and proceeded towards the S.W. Probably the heliocentric motion was retrograde.	,,	Ma-tuoan-lin.
125	254	December	...	In December it is stated that a vapour very many tchang in length, emerged from near δ Sagittarii. Hind says that Pingré seems to doubt this being a comet, but we cannot find that Pingré says *anything.*	,,	Ma-tuoan-lin.
126	255	Jan. or Feb.	...	Seen near the horizon, in the N.W.	,,	Ma-tuoan-lin.
127	257	Nov. or Dec.	...	A white comet, seen near Spica Virginis.	,,	Ma-tuoan-lin.
128	259	Sept. or Oct.	1	From near β Leonis toward the S.W.	,,	Hind.

No.	Year.	Time of Visiblity	Duration of Visib.	Path of the Comet.—Notes, &c.	Authority.	Reference.
129	A.D. 262	Dec 2.—Jan.	Weeks. 6	Northward from χ Virginis. Tail said by Gaubil to have been 50° long. Ma-touan-lin says it was only seen 5 times.	Chinese obs.	Ma-tuoan-lin.
130	265	June	2	In Cassiopeia.	”	Ma-tuoan-lin.
131	268	February 18	...	First seen in Corvus. It advanced towards the N.W. and then towards the E. Probably the direction of the tail is here referred to. It was of a pale blue colour.	”	Ma tuoan-lin.
132	269	October	...	Seen within the circle of perpetual apparition.	”	Mailla, iv. 148.
133	275	January	...	Seen in Corvus.	”	Ma-tuoan-lin.
134	276	June 23—September	3 months	From Libra, towards Boötes, near Arcturus, into Leo, and thence to Ursa Major and Minor.	”	Ma-tuoan-lin.
135	277 i.	January	...	Seen first in the W.	”	Ma-tuoan-lin.
136	— ii.	June	...	First seen in Leo.	”	Ma-tuoan-lin.
137	278	May—Dec.	8 months	First seen in Gemini. A very large comet.	”	Ma-tuoan-lin.
138	279	April	4 or 5 months	From Hydra, through Leo, into the circumpolar regions.	”	Ma-tuoan-lin.
139	281 i.	September	...	First seen in Hydra.	”	Ma-tuoan-lin.
140	— ii.	December	...	First seen in Leo. It is possible that these two objects may be identical.	”	Ma-tuoan-lin.
141	283	April 22	1 day	Seen in the S.W.	”	Ma-tuoan-lin.
142	287	...	10 days	In Sagittarius. It had a tail 10 tchang long.	”	Ma-tuoan-lin.

No.	Year.	Time of Visibility.	Duration of Visib. Weeks.	Path of the Comet—Notes, &c.	Authority.	Reference.
143	A.D. 290	May	...	Seen in Tse-wei. It is doubtful whether this is a comet.	Chinese obs.	Ma-tuoan-lin.
144	301 i.	January	...	Seen in or near Capricornus, with a tail pointing to the W.	"	Ma-tuoan-lin.
145	— ii.	May	...	Seen near, with ω Capricorni or 110 Herculis.	"	Ma-tuoan-lin.
146	302	May	...	Visible in the morning.	"	Ma-tuoan-lin.
147	303	April	...	Seen in Ursa Major.	"	Ma-tuoan-lin.
148	304	May or June	...	Seen in the Hyades; that it was a comet is uncertain.	"	Ma-tuoan-lin.
149	305 i.	September	...	In the Pleiades according to Ma-tuoan-lin. De Mailla places it near the pole. It is not improbable that this and the following object may be identical.	"	Ma-tuoan-lin.
150	— ii.	November 22	...	In Ursa Major.	"	Ma-tuoan-lin.
151	329	August	3	First seen in the N.W.; passed thence to very near φ Sagittarii.	"	Ma-tuoan-lin.
152	336	February 16—May 30, 337	1¼ years	First seen near ζ Andromedæ, and afterwards in Aries. Of extraordinary magnitude.	Chin. and Europ. obs.	Mailla, iv. 349; Eutrop. Hist. Rom. x. 8.
153	340	March 25	1 ?	Seen near β Leonis.	Chinese obs.	Ma-tuoan-lin.
154	343	December 9	...	Its R. A. was about 13¾ h. No Decl. given. Of a white hue.	"	Ma-tuoan-lin.
155	350	January 7	...	In Virgo. It had a white tail, extending towards the W., and 10° long.	"	Ma-tuoan-lin.

No.	Year.	Time of Visibility.	Duration of Visib.	Path of the Comet.—Notes, &c.	Authority.	References.
	A.D.		Weeks.			
156	358	July 12	...	No positions given.	Chinese obs.	Ma-tuoan-lin.
157	363	Aug. or Sept.	...	From Virgo towards Hercules.	„	Mailla, iv. 413.
158	369	March—August	4 months	Seen in the circumpolar regions.	„	Ma-tuoan-lin.
159	373 i.	March 9		From Aquarius, through Libra, Virgo, Corvus, &c.	„	
160	— ii.	October 24	...	Seen near α Herculis and α Ophiuchi. Probably an apparition of *Halley's Comet*.	„	Ma-tuoan-lin.
161	374	January	...	Seen in Scorpio and Sagittarius. These positions will also accord with what *Halley's Comet* would have had at this return.	‚‚	Mailla, iv. 437.
162	375	November	...	...	European obs.	Ammian. Marcell. *Rer. Gest.* xxx. Marcellin. *Chronic.*, and others.
163	389	August	6	From the zodiacal region northwards towards Ursa Major and Minor. A brilliant comet with a long tail.	European obs.	
164	390	August 22—September 17	4	From near α and β Geminorun, through Leo to Ursa Major. It had a white tail 100° long. This comet, singularly enough, followed nearly the same path as the preceding; but it is certainly a different object.	Chin. and Europ. obs.	Marcellin.
165	392		...	No particulars are given.	Chinese obs.	Couplet.
166	393	April—October	6 months	First seen in Scorpio. It is not recorded that it had any motion during this long period of 6 months, so it may be a temporary star that is referred to.	„	

No.	Year.	Time of Visibility.	Duration of Visib.	Path of the Comet.—Notes, &c.	Authority.	Reference.
167	A.D. 395	August	Weeks.	In Aquarius.	Chinese obs.	Mailla, iv. 496.
168	400	March 19	...	From θ Andromeda, through Cassiopeia, Leo, and Virgo. It had a tail 30° long.	Chin. and Europ.obs.	Ma-tuoan-lin; Socrat. *Hist. Eccles.* vi. 6.
169	401	January	...	Seen in Cygnus.	Chinese obs.	Mailla, iv. 517.
170	402	November	2 months or +	From Cephus and Cassiopeia towards Ursa Major.	Chin. and Europ.obs.	Claudian, *De Bello Getico,* 228 et seq.
171	415	June	...	From Ophiuchus, through Hercules to Scorpio.	Chinese obs.	Pingré, ii. 598.
172	418 i.	June 24	...	Seen in Ursa Major.	,,	Ma-tuoan-lin.
173	— ii.	July 19—Oct.	4 months	First seen in Leo (?) and thence it passed by the tail of Ursa Major to the W. It had a tail, at first short, but which afterwards attained a length of 10 tchang or more.	Chin. and Europ.obs.	Marcellin ; & Mailla, iv. 590; Philostorg. *Epit. Hist. Eccles.* xii. 8.
174	419	Feb. 17	...	In Leo and Virgo.	Chinese obs.	Ma-tuoan-lin.
175	421	...	:	...	Chin. and Europ.obs.	Couplet.
176	422 i.	March 16	10 days	In Aquarius.	,,	*Chronicon Paschale ;* Gaubil.
177	— ii.	December 18	...	Near α and β Pegasi.	Chinese obs.	Ma-tuoan-lin.
178	423 i.	February 13	...	Near γ Pegasi, and α Andromedæ.	Chin. and Europ.obs.	Marcellin.
179	— ii.	December 14	...	Seen near α and β Libræ.	Chinese obs.	Ma-tuoan-lin.
180	432	...	...	From α Leonis to Arcturus (α Boötis), where it disappeared.	Chinese obs.	Ma-tuoan-lin..
181	436	June 21	...	In Scorpio near τ.		Gaubil.

No.	Year.	Time of Visibility.	Duration of Visib.	Path of the Comet.—Notes, &c.	Authority.	Reference.
			Weeks.			
182	A.D. 442	November 1	Several months	From Ursa Major, through Auriga, Taurus, and near π Ceti, to γ Eridani.	Chin. and Europ.obs.	Marcellin. Idatius. *Episcop. Chron.*
183	449	December 11	...	Seen near β Leonis.	Chinese obs.	Ma-tuoan-lin.
184	467	...	5 weeks	In shape like a trumpet.	European obs.	*Chron. Pasch.*
185	499	...	...	Details are wanting.	"	Zonaras, *Annal.* ii.56. Paris, 1686.
186	501	February 13	...	A grand comet seen near the horizon.	Chin. and Europ.obs.	Ma-tuoan-lin.
187	504	...	...	Described as very fine.	European obs.	Galfredus, *Britann.* viii. 4.
188	507	August 15	...	Seen in the N.E.	Chinese obs.	Gaubil.
189	519	Winter	...	A "fearful star" accompanied by a tail turned towards the W.	European obs.	Theophanes, *Chronographia*, p. 142. Paris, 1655. Malala, *Hist. Chron.* xvii.
190	520	October 7— December	8 weeks	Seen in the eastern heavens, bright like fire; towards the close of its apparition it was visible in the morning.	Chinese obs.	Gaubil.
191	524	...	4	Seen above the gate of Justin's palace.	European obs.	Cedrenus, *Compend. Hist.* p. 365. Paris, 1647.
192	530 or 531	September	3	Seen in Boötes and Ursa Major. A very fine comet, with a brilliant and lengthy tail. Probably an apparition of *Halley's Comet.*	Chin. and Europ.obs.	Theophanes, p.154; Malala, xviii.
193	533	March 1	...	There are no other particulars recorded, and it is doubtful whether it was a comet or not.	Chinese obs.	Ma-tuoan-lin.
194	534	...	...	From Leo, near ν and ξ Ursæ Majoris, through Pegasus to Andromeda.	"	Gaubil.

No.	Year.	Time of Visibility.	Duration of Visib. Weeks.	Path of the Comet.—Notes, &c.	Authority.	Reference.
195	A.D. 556	November	...	In the form of a lance.	European obs.	Malala, xviii.
196	560	October 9	...	It had a tail 4° long, directed towards the W.	...	Ma-tuoan-lin.
197	563	...	1 year	In the shape of a sword.	European obs.	Gregory Tours, *Hist. Franc.* iv.
198	565 i.	April	4	...	Chinese obs.	Ma-tuoan-lin.
199	568 i.	July 20—Aug.		In Gemini and Cancer. ...	"	Ma-tuoan-lin.
200	575	April 27		Seen near Acturus (α Boötis).	"	Ma-tuoan-lin.
201	581	January 20	...	Seen in the S.W.	Chinese obs.	Ma-tuoan-lin.
202	582	January	...	This seems to have been a very fine comet.	European obs.	Idatius, *Episcop. Chron.* vi. 14.
203	584	...	...	A column of fire suspended in the air.	"	*Chron. Turon.*
204	588	November 22	...	Near β Capricorni.	Chinese obs.	Ma-tuoan-lin.
205	591	...	4	...	European obs.	Boufin. *Rer. Hungar.* l. viii.
206	595	January 9	...	Seen in Aquarius, Pisces, Aries, &c.	Europ. and Chin. obs.	Simocat. *Hist.* vii.
207	602	...	...	No position or dates given.	European obs.	Theophan. p. 240.
208	605 i.	April—May	...	Date doubtful.	"	Paul Diac. *De Gest. Longobard.* iv. 33.
209	— ii.	Nov.—Dec.		Date doubtful.		Paul Diac. iv. 34.
210	607 i.	March 13	15	The Chinese account says that the comet passed from μ Geminorum, by κ Persei, Auriga, α Geminorum, β Leonis to α Herculis. This seems incredible.	Chinese obs.	Ma-tuoan-lin.
211	— ii. ?	April 4	...	It is not unlikely that this is identical with the preceding.	"	Gaubil.

No.	Year.	Time of Visibility.	Duration of Visib.	Path of the Comet.—Notes, &c.	Authority.	Reference.
	A.D.		Weeks.			
212	607 iii.?	October 21	...	From Virgo, through Leo to Gemini.	Chinese obs.	Ma-tuoan-lin.
213	608	...	...	From near α Aurigæ, by ο Ursæ Majoris into Scorpio, near β of which constellation it disappeared. Probably *Halley's Comet.*	"	Ma-tuoan-lin.
214	614	...	4	...	European obs.	Lubienietz. *Theat. Comet.*
215	615	July	...	First seen in the south-east of Wen-tchang, and it moved towards the circle of perpetual apparition. Its tail was 50° or 60° long, and had an undulatory motion.	Chinese obs.	Ma-tuoan-lin.
216	617 i.	July	1	Seen near β Leonis, and of a reddish yellow colour.	"	Ma-tuoan-lin.
217	— ii.	October		Seen near α and β Pegasi.	"	Ma-tuoan-lin.
218	622	...	...	Doubtful.	"	Lubienietz.
219	626	March		From the Pleiades towards γ Persei.	Chin. and Europ. obs.	Gaubil; *Chron. Pasch.*
220	632	June?	4	It was like a seam, and extended from N. to S.	European obs.	Cedrenus, p. 425.
221	633		...	In shape like a sword.	"	Weber, *Discurs. Curios.*
222	634	Sept. 22—Oct. 3	2	Mostly seen in Aquarius.	Chinese obs.	Gaubil.
223	639	April 30	...	Between the Hyades and Pleiades.	"	Gaubil.
224	641	July 22	4	In Coma Berenicis.	"	Mailla, vi. 93.
225	660	...	2	In Scorpio.	"	Lubienietz.
226	665	Sept. 27—29.	3 days	Seen near ζ Boötis. It had a tail 2 tchang long.	"	Ma-tuoan-lin.
227	667	May 24—June 12	3	In Auriga and Taurus, near the Pleiades.	"	Gaubil.

No.	Year	Time of Visibility.	Duration of Visib.	Path of the Comet.—Notes, &c.	Authority.	Reference.
			Weeks.			
228	A.D. 668	May	1	In Auriga.	Chinese obs.	Mailla, vi. 145.
229	673	...	10	Doubtful.	European obs.	*Vita S. Leodegar.*
230	674	...	...	Said to have been a great comet, though the observation is doubtful.	"	Alsted. *Thesaur. Chron.*
231	676 i.	Jan. 4	...	Near α Virginis. Tail 5° long.	Chinese obs.	Ma-tuoan-lin.
232	— ii.	September 4.—Nov. 1	8 or +	It moved from Gemini towards the N.E. It seems to have had a very brilliant tail, which the Chinese annals say was 3 tchang long.	Chin. and Europ. obs.	Gaubil ; Anastas ; *Hist. Eccles. in Dono.*; Paul. Diac. v. 31.
233	681	October 17—November 3	3	From α Herculis towards Altair (α Aquilæ). Tail was 90° long, according to one account; 5 tchang according to another.	Chinese obs.	Ma-tuoan-lin.
234	683	Apr. 20—May 15	4	In Auriga, and Taurus.	"	Ma-tuoan-lin.
235	684 i.	Sept.—Oct.	7	Seen in the W. in the evening, with a tail 10° long. This rough position will tolerably well describe that which *Halley's Comet* would have had, at its return in this year, and it is doubtless identical with it.	"	Ma-tuoan-lin.
236	— ii.	Nov. 11	...	Seen in the N. and resembled a half-moon.	European obs.	Ma-tuoan-lin.
237	685	January—February?	...	First seen near the Pleiades, and like the Moon partially covered by a cloud ; afterwards seen in the daytime, when its brilliancy was very great. Moved from W. to E. or from N. to S. (1) It is not unlikely this is identical with the last.	"	Calvisius, *Opus Chronol.*

No.	Year.	Time of Visibility.	Duration of Visib.	Path of the Comet.—Notes, &c.	Authority.	Reference.
	A.D.		Weeks.			
238	707	Nov. 16—Dec. 18	4	Seen in the W.	Chinese obs.	Ma-tuoan-lin.
239	708 i.	March 31	...	In the Pleiades.	"	Ma-tuoan-lin.
240	— ii.	September 21	...	In Tse-wei.	"	Ma-tuoan-lin.
241	711	...	11 days	...	Arabian obs.	Haly *in Centiloq.*
242	712	August	...	It emerged from the W. Its aspect was dull, and the tail turned toward the pole.	Chin. and Europ. obs.	Mailla, vi. 199.
243	716	...	...	The tail is also said to have been directed to the pole.	European obs.	Sabell. *Opera Omnia.* Ennead. viii. lib. vii.
244	729	January	2	Seen after sunset and before sunrise, with a tail pointing to the N.	"	Bede, *Hist. Eccles.* v.
245	730	Aug. 29—Sept.	1 day or +	From Auriga to near the Pleiades and Hyades.	Chinese obs.	Ma-tuoan-lin.
246	738	April 1	2 weeks	Seen within the circle of perpetual apparition, and specially in the "square" of Ursa Major.	"	Gaubil.
247	744	...	...	A great comet, observed in Syria.	European obs.	Theophan. p. 353.
248	762	...	...	Seen in the E. and in a shape like a beam.	"	Theophan. p. 365.
249	767	Jan. 22—Feb. 11	3	Seen near α and β Delphini.	Chinese obs.	Ma-tuoan-lin.
250	773	January 17	...	Seen near the belt of Orion, "its length traversed the heavens."	"	Ma-tuoan-lin.
251	813	August	...	...	European obs.	Theophan. p. 443.
252	815	April	...	Leo.	Chinese obs.	Ma-tuoan-lin.
253	817	February	...	Taurus.	Europ. and Chin. obs.	Ma-tuoan-lin.

No.	Year	Time of Visibility.	Duration of Visib.	Path of the Comet.—Notes, &c.	Authority.	Reference.
	A.D.		Weeks.			
254	821 i.	February 27		Crater, Leo.	Chinese obs.	Ma-tuoan-lin.
255	— ii.	July	...	In Taurus, near the Pleiades.	,,	Ma-tuoan-lin.
256	837 iii.	September	...	Aquarius, Equuleus, Pegasus.	Chin. and Europ.obs.	Ma-tuoan-lin ; Boeth.*Scot.Hist.*x.
257	838 i.	November	...	Corvus.	Chinese obs.	Ma-tuoan-lin.
258	— ii.	Nov. Dec.	...	Sagittarius, Scorpio.	,,	Ma-tuoan-lin.
259	839 i.	Feb. 7	...	Aquarius.	Chinese obs	Ma-tuoan-hn.
260	— ii.	Mar. 12—Apr. 13	4	Perseus.	,,	Ma-tuoan-lin.
261	840 i.	Mar. 20—Apr.	3	Pegasus, Andromeda.	,,	Ma-tuoan-lin.
262	— ii.	Dec. 3	...	In the E.	,,	Ma tuoan-lin.
263	841 i.	June, July	...	Sagittarius, Aquarius.	Europ. and Chin.obs.	Alber,Casin.*Chrom.*; Ma-tuoan-lin.
264	— ii.	Dec.—Feb. 842	...	Piscis Australis.	,,	Ma-tuoan-lin.
265	852	March	...	In Orion.	Chinese obs.	Ma-tuoan-lin.
266	855	August ?	3	...	European obs.	*Chrom. S. Mazent.*
267	857	Sept. 22	...	Scorpio.	Chinese obs.	Ma-tuoan-lin.
268	858	April	...	Its tail pointed to the E.	European obs.	Ptolom. *Hist. Eccles.* xvi. 9.
269	864	May, June	...	Aries.	Europ. and Chin.obs.	*Chrom. Floriac.*; Ma-tuoan-lin.
270	866	April	...	...	European obs.	Constant. Porph. *Incert.* iv. 126.
271	868	January	3	Aries, Musca.	Europ. and Chin.obs.	*Annal, Fuld.*; Ma-tuoan-lin.
272	869	September	...	Perseus, Caput Medusæ. It went to the N.E.	,,	Pontan.Hist.*Gelv.*; Ma-tuoan-lin.
273	873	...	4	In the N. and N.E.	European obs.	*Chron. Andegav.*
274	875	June	...	...	,,	*Annal. Fuld.*
275	877	March	2	In Libra.	,,	*Chron. Novalic.*

No.	Year	Time of Visibility.	Duration of Visib.	Path of the Comet.—Notes, &c.	Authority.	Reference.
	A.D.		Weeks.			
276	882	January	...	With a "prodigiously" long tail.	European obs.	*Annal. Fuld.*
277	885	...	...	Between Perseus and Gemini.	Chinese obs.	Ma-tuoan-lin.
278	886	June	...	Scorpio, Sagittarius, Ursa Major, Boötes	"	Ma-tuoan-lin.
279	891	April, May	...	Ursa Major, Boötes, &c.	Europ. and Chin.obs.	*Annal. Saxo.*; Ma-tuoan-lin.
280	892 i.	April—June	11	Scorpio	"	*Chron. Andegav.*; Ma-tuoan-lin.
281	— ii.	November	...	Sagittarius, Capricornus.	Chinese obs.	Ma-tuoan-lin.
282	— iii.	Dec. 28–30	3 d.	In the S.E.	"	Ma-tuoan-lin.
283	893	May	5	Ursa Major, Boötes.	"	Ma-tuoan-lin.
284	894	February	...	Gemini.	"	Ma-tuoan-lin.
285	894	December	6	In the E.	European obs.	Constant. Porph.
286	905	May, June	...	Gemini, Ursa Major, Leo, Virgo.	Europ. and Chin.obs.	*Chron. Florent.*; Ma-tuoan-lin.
287	912	March	2	Hydra. Possibly *Halley's Comet*, due about this year.	"	Sim. Log. *Annal.* p. 471; Ma-tuoan-lin.
288	923	October	...	Cancer.	Chinese obs.	Mailla, vii. 210.
289	928	December	3 d.	Capricornus.	"	Ma-tuoan-lin.
290	930	...	...	Cancer.	European obs.	Lubienietz.
291	936	September 21	...	Aquarius, Pegasus, Capricornus.	Chinese obs.	Ma-tuoan-lin.
292	939	July	1	...	European obs.	Luitprand, *Rer. Gest.* v. 1.
293	941	September	...	Hercules.	Chinese obs.	Ma-tuoan-lin.
294	942	October	3	Advanced from the E. to the meridian.	European obs.	*Chron. Andegav.*
295	943	November 5	...	Virgo.	Chin. and Europ. obs.	Ma-tuoan-lin.
296	945	...	...	...	European obs.	Frodoard. *Chron.*

No.	Year. A.D.	Time of Visibility.	Duration of Visib.	Path of the Comet.—Notes, &c.	Authority.	Reference.
297	956	March	...	Orion.	Chinese obs.	Ma-tuoan-lin.
298	959	Oct. 17—Nov. 1	...	...	European obs.	Constant. Porph. p. 289.
299	975	Aug.—Oct.	12	Hydra, Cancer, and Pegasus. Not unlikely identical with the comets of 1264 and 1556.	Europ. and Chin. obs.	Ma-tuoan-lin, &c.
300	981	Autumn	...	...	European obs.	*Chron. Engelh.*
301	985	...	...	...	"	Platin. *De Vit. Pontif.*
302	989 i.	February	2	Pegasus.	Europ. and Chin. obs.	*Annal. Saxo.*; Gaubil.
303	990	August	...	In the S.W.	"	Romuald.; Couplet.
304	995	August	...	...	European obs.	Florent. Vigorn. *Chron.*
305	998	February	...	In Pegasus.	Chinese obs.	Ma-tuoan-lin.
306	1000	December	9 d.	...	European obs.	Iperius. *Chron.* xxxiii.
307	1003 i.	February	...	...	"	Hepidan. *Annal.*
308	— ii.	December	4	Cancer, Gemini, Taurus, Orion.	Chinese obs.	Ma-tuoan-lin.
309	1005	Sept. Oct.	...	Seen in the north Polar region.	Europ. and Chin. obs.	Alpert. *Chron.*; Mailla, viii. 158.
310	1006	April	...	Scorpio.	Haly-ben-Rodoan	Cardan. *Astror.* ii. 9.
311	1012	...	3 m.	In the southern heavens.	European obs.	Hepidan.
312	1015	February	...	...	"	Protospat. *Chron.*
313	1017	...	4 m.	In Leo (?).	"	Sigebert, *Chron.*
314	1018	July, Aug.	5	Ursa Major, Leo, Hydra.	Europ. and Chin. obs.	Ditmar. *Chron.* viii. Ma-tuoan-lin.
315	1023	Autumn	...	Leo.	European obs.	Ademar. *Chron.*
316	1024	...	...	...	"	Cureus, *Annal. Siles.*
317	1033	March	...	In the N.E.	Europ. and Chin. obs.	*Fragm. Hist. Franc.*I.; Ma-tuoan-lin.

No.	Year.	Time of Visibility.	Duration of Visib.	Path of the Comet.—Notes, &c.	Authority.	Reference.
	A.D.		Weeks.			
318	1034	September	...	Hydra, Crater.	Europ. and Chin. obs.	Cedren. p. 737 ; Mailla, viii. 199.
319	1035 i.	September	12 d.	Hydra.	Chinese obs.	Ma-tuoan-lin.
320	— ii.	November	...	Pisces. Tail very faint.	,,	Ma-tuoan-lin.
321	1041	...	...	...	European obs.	Glycas, *Annal.* p. 316.
322	1044	October	4	Moved from E. to W.	,,	Glycas, p. 319.
323	1046	...	...	...	,,	Godell. *Chron.*
324	1049	March	16	Aquarius, Equuleus. Pingré remarks, "La route qu'on assigne à cette comète n'est pas naturelle."	Chinese obs.	Ma-tuoan-lin.
325	1056	September	...	Orion, Hydra. Tail 10° long.	,,	Ma-tuoan-lin.
326	1058	Easter week	...	...	European obs.	Hennenf. *Annal. Siles.*
327	1060	...	...	...	,,	Will. Malms. *Reg. Angl.*
328	1067	May	...	...	,,	*Chron. Andegav.* ii.
329	1071	...	4	...	,,	Cæsius, *Cat. Comet.*
330	1075	November	2	In Corvus, Taurus. Tail 7° long.	Chinese obs.	Mailla, viii. 285.
331	1080 i.	August	5	Coma Berenicis, &c. The account by M. is very confused.	,,	Ma-tuoan-lin.
332	— ii.	Aug. 27—Sept. 14	2	Hydra.	,,	Ma-tuoan-lin.
333	1096	October 7	...	In the S.	European obs.	*Annal. Saxo.*
334	1098	June	...	...	,,	Robert. *Hist. Hieros.* v.
335	1101	January	...	...	,,	*Synop. Chronol.*
336	1106	Feb. 4—March	8	Pisces, Andromeda, Aries, Taurus. A very fine comet.	Europ. and Chin. obs.	Matt. Paris, *Hist. Maj.*; Ma-tuoan-lin.

No.	Year.	Time of Visibility.	Duration of Visib.	Path of the Comet.—Notes, &c.	Authority.	Reference.
	A.D.		Weeks.			
337	1109	December	...	In the Milky Way. Tail pointing to the S.	European obs.	Hemingf. *Chron.* i. 33.
338	1110	May—July	1½ m.	Andromeda, Pisces, Aries, thence to the N.	Europ. and Chin. obs.	*Chron. Reg.*; Ma-tuoan-lin.
339	1113	May	...	A great comet.	English obs.	Matt. Paris; Matt. West.
340	1114	May, end of	...	...	„	M. Paris; *Annal. Waverl.*
341	1115	August	...	Aries.	Europ. and Chin. obs.	*Annal. Margan.*; Mailla, viii. 377.
342	1125	...	...	...	European obs.	Dubrav. *Hist. Bojem.* xi.
343	1126 i.	June, July	...	Hercules, Ursa Major.	Europ. and Chin. obs.	*Annal. Bosov.*; Ma-tuoan-lin.
344	— ii.	December	...	A very large comet.	Chinese obs.	Mailla, viii. 447.
345	1132 i.	January 5	...	...	„	Ma-tuoan-lin.
346	— ii.	October 2—27	3	Musca	Europ. and Chin. obs.	Florent. Vigorn.; Ma-tuoan-lin.
347	1138	August	...	...	Chinese obs.	Mailla, viii. 524.
348	1142	December	...	...	„	*Synop. Chronol.*
349	1145 i.	April—July	3 m.	Virgo, &c. The account of the path is much confused. However, there is no doubt but that this is a return of *Halley's Comet*, the PP. taking place on April 19.	Europ. and Chin. obs.	*Chron. Bossian.*; Ma-tuoan-lin.
350	— ii.	July	...	Orion, &c. (?)	Chinese obs.	Ma-tuoan-lin.
351	1146	...	...	Seen a long time in the W.	European obs.	*Annal. Hirsaug.*
352	1147 i.	January 11	...	Aquarius, Pegasus.	Chinese obs.	Ma-tuoan-lin.
353	— ii.	Feb. 17—Mar.	2	To the E. of σ Capricorni.	Europ. and Chin. obs.	*Hist. Episc. Virdun.* Ma-tuoan-lin.
354	1155	May 5	...	...	European obs.	*Chron. Admont.*
355	1156	July, August	...	In Gemini.	Chinese obs.	Ma-tuoan-lin.

No.	Year.	Time of Visibility.	Duration of Visib. Weeks.	Path of the Comet.—Notes, &c.	Authority.	Reference.
356	A.D. 1162	November	...	From the square of Pegasus, Aquarius, and Cetus. It had a tail 10° long.	Chinese obs.	Gaubil.
357	1165 l.	August	...	To the N.	European obs.	*Chron. Mailr.*
358	— ii.	August	...	To the S.	„	Boeth. xiii.
359	1181	July	22	Seen in Cassiopeia.	Europ. and Chin. obs.	*Chron. Mailr.*; Gaubil.
360	1198	November	2	...	English obs.	Coggesh. *Chron. Angl.*
364	1204	...	...	...	European obs.	Sicardi, *Chron.*
362	1208	...	2	" Regarded by the Jews as a sign of the approach of the Messiah."	„	Weichen. *Chron.*
363	1211	May	3	...	In Poland	M. Cromer, *Polon.* vii.
364	1214	March	...	...	European obs.	*Annal. Hirsaug.*
365	1217	Autumn	...	At first small; it then became a fine object, and afterwards diminished.	„	Ursperg. *Chron.*
366	1222	Aug.—Oct.	8	Boötes, Virgo, Libra.	Europ. and Chin. obs.	Roland, *Chrom.* ii. 3; Ma-tuoan-lin.
367	1223	July	...	Undoubtedly an apparition of *Halley's Comet.*	European obs.	*Chron. Franc.*
368	1230	Dec. 15—Mar. 30, 1231	15	Ophiuchus.	Chinese obs.	Ma-tuoan-lin.
369	1232	Oct. 18—Dec.	7	Virgo, Corvus. Its tail at one time was 40° long.	„	Ma-tuoan-lin.
370	1239	February	...	...	„	*Synop. Chronol.*
371	1240	Jan.—March	8	Pegasus, thence to the S.E. of α and β Cassiopeia.	Europ. and Chin. obs.	*Chron. Pad.* l.; Ma-tuoan-lin.
372	1250	December	...	...	European obs.	*Archiep. Trevir. Gesta*, No. 266.

No.	Year.	Time of Visability.	Duration of Visib.	Path of the Comet.—Notes, &c.	Authority.	Reference.
	A.D.		Weeks.			
373	1254	November	...	***	European obs.	Polyd. Virg. Hist. Angl. xvi.
374	1262	July, August	...	In the E. ***	,,	Crusius, Annal.Suev. p.iv.lib.ii.
375	1265	September	...	***	,,	Chron. Meltic.
376	1266	August	...	In Taurus.	European, Chinese, and Japan obs.	Nangis. Chron.
377	1269	Aug., Sept.	...	***	European obs.	Boeth. xiii.
378	1273	December	3	The Hyades, Ursa Major, Boötes to Arcturus.	Chinese obs.	Gaubil.
379	1274	***	...	***	European obs.	Guil. de Thoc. Vita S. Thom. x. 60.
380	1277	March	...	In the N.E. Tail 4° long.	Chinese obs.	Ma-tuoan-lin.
381	1285	April	...	Tail pointed to the W.	European obs.	Ptolom. Hist. xxiv. 17.
382	1293	November	4	In the 7 stars of Ursa Major. Tail 1° long	Chinese obs.	Ma-tuoan-lin.
383	1295	February	...	***	European obs.	Annal. Flandr. x.
384	1298	November ?	...	***	,,	Archiep. Trevir. Gesta.
385	1304	Feb.—May	10	Pegasus, Cygnus, to the circumpolar region.	Chin. and Europ. obs.	Ma-tuoan-lin.
386	1305	April	1	***	European obs.	Chron. Botham.
387	1313	April	...	Gemini.	Chin. and Europ. obs.	Ma-tuoan-lin; Mussat, Hist.i.xv.4.
388	1314	October	...	Virgo.	European obs.	Giov. Villan. Chron. ix. 54.
389	1315	Dec.—Feb. 1316	15	Leo, Virgo, Corvus, Pegasus (?).	Chin. and Europ. obs.	Mussat, ii. vii. 14; Ma-tuoan-lin.
390	1316	May	...	In the E.	European obs.	Chron. Rotom.
391	1334	August	...	With a brilliant tail 7½ feet long !	Chinese obs.	***
392	1337 ii.	July, August	2 m. ?	In Cancer.	European obs.	Giov. Villan. xi. 66.
393	1338	April	2	In Gemini, and from W. to E.	,,	Chron. Rotom.

No.	Year.	Time of Visibility.	Duration of Visib.	Path of the Comet.—Notes, &c.	Authority.	Reference.
394	A.D. 1340	Mar. 24—June 3	Weeks. 5	In Scorpio.	Europ. and Chin. obs.	Niceph. Gregor. *Hist. Byzan.* xi. 8 ; Ma-tuoan-lin.
395	1345	July	...	Ursa Major to Leo.	European obs.	N. Gregor. xv. 5.
396	1347	August	2 or +	Caput Medusæ.		*Chron. Nuremb.*
397	1356	Sept. 21—Nov.4	6	Hydra.	Chin. and Europ. obs.	Ma-tuoan-lin.
398	1360	March	1	Towards the E.	,,	Mailla, ix. 613. *Chron. Zwell.*
399	1362.ii.	June—August	6	Seen in the N. in the same R.A. as Capricornus, with a tail 1° long.	,,	Mailla, ix. 640.
400	1363	March	4	Towards the E.	,,	*Synop. Chronol.*
401	1368	Feb. 7—April	11	Taurus. It had a tail 8° long pointing towards Ursa Major.	,,	Ma-tuoan-lin; Walsingham, *Hist. Ang.*
402	1371	January	...	Towards the N. with a tail pointing in the opposite direction.	European obs.	Bonincontr. *Annal.*
403	1380	November	3 m.	...	Europ. and Jap. obs.	Kaempfer, *Hist. du Japon*, ii. 5; *Chron. Cite.*
404	1382. i.	March	...	...	Europ. and Chin. obs.	*Chron. Bothom.*
405	— ii.	August	...	Resembled a lance.	,,	Bonfin. Dec. ii. x.
406	— iii.	December	2	In the W.	European obs.	Walsingham.
407	1391	May	...	Ursa Major, small, faint, and with a faint tail. Ma-tuoan-lin says that two comets were visible this month.	,,	*Annal. Forolin.*
408	1402. i.	Feb.—April	2 m.	Aries, &c. A very large and brilliant comet, visible to the naked eye at mid-day.	,,	Pogg. *Hist. Florent.* iv. ; Ebendorff, *Chron. Austr.*

No.	Year.	Time of Visibility.	Duration of Visib. Weeks.	Path of the Comet.—Notes, &c.	Authority.	Reference.
409	A.D. 1402 ii.	June—Sept.	3 m.	Another very fine comet, also visible before sundown.	European obs.	Duc. Hist. Byzant. p. 341; Chron. Bossian.
410	1406	Some time in the first 6 m.	1 ?	Its tail was "on fire."	„	Chron. Bremen.
411	1407	Dec. 15	...	Very doubtful.	Chinese obs.	Ma-tuoan-lin.
412	1431	May 15 or 27	...	It had a tail about 5° long.	„	Ma-tuoan-lin.
413	1432	Feb. 2—Mar. 16	6	Its tail was about 10° long and "swept" the region near α Cygni. It disappeared on Feb. 12. On Feb. 29 another comet (doubtless the same, after its P.P.) became visible, and lasted 17 days.	„	Ma-tuoan-lin.
414	1436	Autumn	...	...	European obs.	Boeth. Hist. Scot. xvii.
415	1439 i.	Mar. 25—April	...	Hydra, Leo, Cancer. Tail 5° long.	Chinese obs.	Ma-tuoan-lin.
416	— ii.	July 12—Aug.	7	Tail 10° long. Perhaps identical with the preceding.	„	Ma-tuoan-lin.
417	1444	August 6—15	1	Leo, Virgo. The European accounts are very contradictory.	Europ. and Chin. obs.	Ma-tuoan-lin ; Cyp. Leovit. De Conjunct.
418	1449	Dec. 20—Jan. 1450.	3	Ophiuchus, Scorpio. It disappeared on Jan. 12 and reappeared on Jan. 19; for how long is not stated. In the interval it probably passed the P P.	„	Ma-tuoan-lin.
419	1452	March	...	In Taurus.	Chinese obs.	Ma-tuoan-lin.
420	1454	Summer	...	Seen after sunset, in form like unto a sword.	Phranza	Chron. de Reb. Constantinop. iii.

Y

No.	Year.	Time of Visibility.	Duration of Visib.	Path of the Comet.—Notes, &c.	Authority.	Reference.
	A.D.		Weeks.			
421	1457	Jan. 14—23.	1	In Taurus. It had a tail ¾° long.	Chinese obs.	Ma-tuoan-lin.
422	1458	June	...	In Taurus.	,,	Mailla, x. 236.
423	1460	August	...	With a very brilliant tail.	European obs.	Boeth. xviii.
424	1461	Aug. 5—Sept.	4 or +	Seen in the E. with a tail pointing to the S.W. It passed into Gemini.	Chinese obs	Ma-tuoan-lin.
425	1463	Winter	...	Virgo.	,	Gaubil.
426	1464	Spring	...	Leo. Possibly the same as the preceding.	,,	Gaubil.
427	1465	March	1 m. ±	In the N.W. Tail 30° long.	Chin. and Jap. obs.	Ma-tuoan-lin; Kaempf. ii. 5.
428	1467	October	...	Pisces, but little seen owing to rainy weather.	European obs.	Chron. S. Ægid.
429	1468	February	...	In Ursa Major.	Chinese obs.	Gaubil.
430	1471	Autumn	...	Virgo, Libra.	European obs.	Matt. Michov. Chron. Pol. iv. 6..
431	1476	Dec.—Jan. 1477	5	A "small" comet.	,,	Ripamont, Hist. Mediol.
432	1477	December	...	...	,,	Chron. Bossian.
433	1478	September	...	...	,,	Chron. Bossian.
434	1479	...	...	...	Arabian obs.	Lycosthenes, Chron. Prodig.
435	1491	January	...	...	European obs.	Chron. Bossian.
436	1500	May and July 10	...	Pisces, Pegasus, northwards to Draco.	Chin. and Europ. obs.	Ma-tuoan-lin.
437	1503	August	...	...	European obs.	Chron. Waldsass.
438	1512	March, April	...	...	,,	Chron. Magdeburg.
439	1513	Dec. Feb. 1514.	...	Passed from the end of Cancer to head of Virgo.	,,	Vicomerc. Commend. Aristot.
440	1516	January	1 ?	...	,,	Bizar. Genuen. Hist.
441	1518	April	...	...	,,	Cavitellius, Annal. Cremon.

No.	Year.	Time of Visibility.	Duration of Visib.	Path of the Comet.—Notes, &c.	Authority.	Reference.
	A.D.		Weeks.			
442	1520	February	...	...	Chinese obs.	Ma-tuoan-lin.
443	1521	April	...	Towards the Equator; it had a short tail.	Europ. and Chin. obs.	Vicomerc.
444	1522	...	...	Seen in the W.	European obs.	Mizald. *Cometographia*, ii. 11.
445	1523	July	...	Near α Ophiuchi.	Chinese obs.	Ma-tuoan-lin.
446	1530	November	...	...	European obs.	Mizald.
447	1538	January 17—30	2	In Pisces. Two positions are given by A. and G. (Pingré, i. 499.)	Apian and Gemma	Ap. *Astron. Cæsar.*; Gem. *De Divin. Charac.* i. 8.
448	1539	April 30—May	3	In Leo and Virgo. It had a tail 3° long. Two positions of this comet, also, are given by A. and G. (Pingré, i. 500.)	,,	Gem. i. 8.
449	1545	...	...	...	European obs.	Aretius, *Brevis Explicat.*
450	1554	June, July	4	From δ Ursæ Majoris towards θ in the same constellation.	Chinese obs.	Ma-tuoan-lin.
451	1557	Oct., Nov.	4	Ophiuchus, Sagittarius.	Chin. and Europ. obs.	Ma-tuoan-lin; Camerar. *Cometæ*.
452	1560	December	4	...	European obs.	Thuan. *Hist.* xxvii. 11.
453	1569	Nov. 2—28	4	Ophiuchus, Sagittarius, Capricornus. Its tail pointed towards the E.	Europ. and Chin. obs.	Kepler, *De Cometis*, 114; Ma-tuoan-lin.
454	1591	April	...	Pegasus, Aries.	Chinese obs.	Ma-tuoan-lin.
455	1618iii.	Nov. — Jan. 21 1619.	8	Centaurus.	Kepler; Longomon-tanus.	Kep. *Comet.* p. 58; Ma-tuoan-lin.
456	1625	Jan 26—Feb.	...	In Eridanus, Cetus.	Schickhardt	*Ast. Nach.* 31. April 1823.
457	1639	Oct. 27.	...	Canis Major. It had a small tail.	Placidus De Titis.	*Ast. Nach.* 171. Jan. 1830.
458	1647	Sept. 29—Oct.	1	Coma Berenicis, Boötes, Corona Borealis.	Hevelius	*Cometographia*, p. 463.

No.	Year.	Time of Velocity.	Duration of Visib.	Path of the Comet.—Notes, &c.	Authority.	Reference.
459	A.D. 1699 ii.	October 26	Weeks. 1 d.	Seen but once in Argo. Long. 122° 34', lat. S. 40° 38'. Passed to the S.	Kirch	*Miscell. Berolin.* v. 50.
460	1702 i.	Feb. 22—Mar. 1	1	Seen in Southern hemisphere.	Dutch Navig.	Struyck, *Vorvolg.*, p. 50.
461	1733	May 17	2	Seen at the Cape of Good Hope, &c. in the N.W.	Various Navig.	Struyck, p. 61.
462	1742 ii.	April 11	...	Seen at sea in the South Atlantic Ocean, in the S.E. with a tail 30° long.	Dutch Navig.	Struyck.
463	1748 iii.	April 24	...	Seen first at the Cape of Good Hope. This may be identical with Comet II. 1748, but it is impossible now to decide the point one way or another.	Dutch Navig. Kindermans.	Struyck, p. 100.
464	1750	January 21—25	4 d.	Near θ Pegasi.	Wargentin	*Tables Astron. de Berlin.* i. 35.
465	1846 ix.	October 18	1 day	Seen in Coma Berenicis, in the morning twilight and faint.	Hind	*Ast. Nach*, No. 582.
466	1849 iv.	Nov. 15 and 28	2	Seen on board ship in lat. 10° S., long. 30° W. Nucleus as light as Mars. Tail curved and turned towards the S.W.	J. M. Jenkins	*Month. Not. R.A.S.* x. 122 and 191.
467	1853	May 1	...	...	...	*Phil. Mag.* vii. 68.
468	1854	March 16	1 day	"A bright and nebulous object."	Brorsen	*Month. Not. R.A.S.* xiv. 178.
469	1856	August 7	1 day	In Virgo.	E. J. Lowe	*Month. Not. R.A.S.* xvii. 114.
470	1859	January	...	Seen in the N.W. sky.	At Panama	*Standard* Newspaper.
471	—	February—May	3 m.	Very faint even in a 14¾ inch refractor. In Feb. (no day given) its R.A. was 11h. 48m. Decl. +19° 49'. Its motion was very slow but seemingly in a northerly direction.	Slater	*Month. Not. R.A.S.* xix. 291.

APPENDIX V.

A CATALOGUE OF STARS, CLUSTERS, AND NEBULÆ WHICH CAN BE
OBSERVED WITH GREATER OR LESS FACILITY IN SMALL TELE-
SCOPES.

THINKING that a catalogue of objects for general observation by
amateurs might not be inappropriate, we have drawn up the
following list.

No.	Constellation.	R. A. 1860.	Decl. 1860.	Object.	Notes.	No. in Smyth's Cycle.
		h m s.	° ′			
1	22 H. Cassiopeiæ	0 19 17	+70 36	Double star and cluster	Mag. 8½ and 11	11
2	31 M. Andromedæ	0 35 9	+40 30	Elliptic nebula	Visible to the naked eye	24
3	32 M. Andromedæ		(In the	field,	with the preceding.	
4	78 H VIII. Cassiop.	0 35 23	+61 1	Cluster	... ...	25
5	251 P. O Piscium	0 52 13	+ 0 2	Double star	Mag. 8 and 9. Dist. 18″	35
6	ψ Piscium	0 58 11	+20 43	Double star	Mag. Both 5½. Dist. 30″	37
7	77 Piscium	0 58 34	+ 4 10	Double star	Mag. 7½ and 8. Dist. 32″	40
8	ζ Piscium	1 6 31	+ 6 50	Double star	Mag. 6 and 8. Dist. 23″	47
9	37 Ceti	1 7 19	− 8 41	Quadruple	... ...	48
10	α Ursæ Minoris	1 7 40	+88 34	Double star	Mag. 2⅜ and 9½. Dist. 19″. Polaris.	44
11	33 M. Trianguli	1 25 58	+29 58	Nebula	Large, faint and ill-defined	57
12	100 Piscium	1 27 25	+11 51	Double star	Mag. 7 and 8. Dist. 16″.	59
13	31 H VI. Cassiopeiæ	1 36 37	+60 32	Cluster	... ...	66
14	γ Arietis	1 45 50	+18 36	Double star	Mag. 4⅛ and 5. Dist. 8·8″	72
15	λ Arietis	1 50 8	+22 55	Double star	Mag. 5½ and 8. Dist. 37″	76
16	γ Andromedæ	1 55 18	+41 39	Double star	Mag. 3½ and 5¾. Dist. 11″. Triple in powerful telescopes	82
17	59 Andromedæ	2 2 28	+38 23	Double star	Mag 6 and 7½. Dist. 16″	87
18	66 Ceti	2 5 37	− 2 9	Double star	Mag. 7 and 8¾. Dist. 15″	90
19	33 H VI. Persei	2 9 20	+56 30	Cluster	Very fine	92
20	34 H VI. Persei		(In the	field,	with the preceding.)	

No.	Constellation.	R. A. 1860.	Decl. 1860.	Object.	Notes.	No. in Smyth's Cycle.
		h. m. s.	° ′			
21	o Ceti	2 12 16	− 3 37	Variable	... ..	94
22	19 ♅ V. Andromedæ	2 13 49	+41 41	Nebula	Elongated	95
23	227 H. Persei	2 23 18	+56 54	Cluster	Rather irregular.	101
24	30 Arietis	2 28 50	+24 2	Double star	Mag. 6 and 7. Dist. 38″	103
25	33 Arietis	2 32 29	+26 27	Double star	Mag. 6 and 9. Dist. 28·5″	105
26	34 M. Persei	2 32 2	+48 8	Double star and cluster	Very pretty	106
27	12 Persei	2 33 26	+39 36	Double star	Mag. 6 and 7¾. Dist. 23″	107
28	η Persei	2 40 30	+55 19	Double star	Mag. 5 and 8½. Dist. 28″	115
29	β Persei (Algol)	2 59 4	+40 24	Double and variable star	Mag. 2-4 and 11. Dist 55″	127
30	25 ♅ VI. Persei	3 5 2	+46 43	Cluster	... ...	128
31	η Tauri	3 39 10	+23 39	Double star and group.	Mag. 3 and 7. Dist. 115″. The *Pleiades*.	142
32	φ Tauri	4 11 44	+27 1	Double star	Mag. 6 and 8½. Dist. 56″	158
33	γ Tauri	4 11 49	+15 17	Star and group	The *Hyades*	159
34	χ Tauri	4 14 3	+25 18	Double star	Mag. 6 and 8. Dist. 19″	160
35	62 Tauri	4 15 33	+23 58	Double star	Mag. 7 and 8¼. Dist. 29″	161
36	τ Tauri	4 33 51	+22 41	Double star	Mag. 5 and 8. Dist. 62″	171
37	257 P. IV. Tauri	4 51 3	+14 20	Triple	Mag. 7, 8, and 10	177
38	23 Orionis	5 15 28	+ 3 24	Double star	Mag. 5 and 7. Dist. 32″	179
39	79 M. Leporis	5 18 39	−24 39	Nebula	A fine object 3′ in diameter	203
40	38 M. Aurigæ	5 20 21	+35 46	Cluster	... ...	204
41	δ Orionis	5 24 51	− 0 24	Double star	Mag. 2 and 7. Dist. 53″	211
42	1 M. Tauri	5 26 3	+21 55	Nebula	The " Crab."	212
43	θ Orionis	5 28	− 5 29	Multiple star and nebula	A splendid object	216
44	σ Orionis	5 31 43	− 2 41	Multiple	Mag. 4, 8, 7, &c.	222
45	35 M. Geminorum	6 0 33	+24 21	Cluster	... ...	236
46	15 Geminorum	6 19 25	+20 52	Double star	Mag. 6 and 8. Dist. 33″	247
47	20 Geminorum	6 24 7	+17 53	Double star	Mag. 8 and 8½. Dist. 20″	252
48	ν¹ Canis Majoris	6 30 15	−18 33	Double star	Mag. 6½ and 8. Dist. 17″	255
49	41 M. Canis Majoris	6 40 46	− 20 36	Cluster	" Superb "	265
50	27 ♅ VI. Monocerotis	6 44 32	+ 0 37	Cluster	... ...	267
51	34 ♅ VIII. „	7 7 55	− 10 1	Double star	Mag. 8 and 8½. Dist. 21″	279
52	19 Lyncis	7 11 24	+ 55 33	Triple	Mag. 7, 8, 8	281
53	12 ♅ VII. Canis Maj.	7 11 29	− 15 23	Cluster	... ...	284
54	α Geminorum	7 25 40	+ 32 12	Double star	Mag. 3 and 3½ 4·9″. Good test	292
55	159 P. VII. Camelopardi.	7 32 32	+ 65 29	Double star	Mag. both 8. Dist. 16″	297

No.	Constellation.	R.A. 1860.	Decl. 1860.	Object.	Notes.	No. in Smyth's Cycle.
		h. m. s.	° ′			
56	46 M. Argo Navis	7 35 25	− 14 30	Double star and cluster.	Mag. 8½ and 11. Dist. 15″. Diameter ⅓°.	302
57	64 ♅ IV. Argo Navis	7 35 39	− 17 53	Nebula	Brilliant planetary form; diameter 12″ or more.	303
58	93 M. Argo Navis	7 38 35	− 23 32	Cluster	... ...	307
59	2 Argo Navis	7 39 3	− 14 21	Double star	Mag. 7 and 7½. Dist. 17″	308
60	14 Canis Minoris	7 51 5	+ 2 36	Triple	Mag. 6, 8, and 9. Dist. 75″ and 115″.	310
61	44 M. Cancri	8 30 0	+ 29 15	Cluster	Præsepe	331
62	ι Cancri	8 38 13	+ 29 16	Double star	Mag. 5½ and 8. Dist. 30″; beautiful colours.	336
63	67 M. Cancri	8 43 32	+ 12 9	Cluster	... ...	339
64	159 P. X Hydræ	10 40 45	− 14 53	Double star	Mag. 8 and 9. Dist. 32″	386
65	97 M. Ursæ Majoris	11 6 34	+ 55 46	Nebula	A curious object 2′ 40″ in diameter.	402
66	12 Comæ Berenicis	12 15 27	+ 26 37	Double star	Mag. 5 and 8. Dist. 66″	444
67	δ Corvi	12 22 37	− 15 44	Double star	Mag. 3 and 8½. Dist. 24″	446
68	24 Comæ Berenicis	12 28 6	+ 19 9	Double star	Mag. 5½ and 7. Dist. 21″	451
69	196 P. XII. Virginis	12 44 6	− 9 35	Double star	Mag. 6½, 9½. Dist. 33″	458
70	94 M. Canum Venaticorum.	12 44 19	+ 41 53	Nebula	... ...	459
71	232 P. XII. Camelopardi.	12 48 13	+ 84 10	Double star	Mag. 6 and 6½. Dist. 22″	465
72	221 P. XII. Virginis	12 48 29	+ 12 15	Double star	Mag. 7½ and 9. Dist. 29″	463
73	12 Can. Venat.	12 49 29	+ 39 4	Double star	Mag. 2½ and 6½. Dist. 20″	466
74	25 P. XIII. Virginis	13 7 57	− 10 37	Double star	Mag. 7½ and 8½. Dist. 41″	475
75	ζ Ursæ Majoris	13 18 16	+ 55 39	Double star	Mag. 3 and 5. Dist. 14″. Alcor mag. 5 is distant 11½′	480
76	51 M. Can. Venat.	13 23 56	+ 47 54	Nebula	Rosse, " Spiral "	484
77	3 M. Can. Venat.	13 35 40	+ 29 4	Cluster	Very brilliant	492
78	220 P. XIII. Boötis	13 43 47	+ 21 58	Double star	Mag. 7½ and 8. Dist. 86″	497
79	τ Virginis	13 54 31	+ 2 13	Double star	Mag. 4½ and 8½. Dist. 78″	502
80	α² Libræ	14 42 35	− 15 27	Double star	Mag. 3 and 6. Dist. 229″	521
81	δ Boötis	15 9 51	+ 33 50	Double star	Mag. 3½ and 8½. Dist. 110″	537
82	5 M. Libræ	15 11 26	+ 2 45	Cluster	... ...	538
83	ι Draconis	15 21 49	+ 59 27	Double star	Mag. 3 and 9. Dist. 117″	545
84	β Scorpii	15 57 17	− 19 25	Double star	Mag. 2 and 5½. Dist. 13″	559
85	κ' Herculis	16 1 45	+ 17 25	Double star	Mag. 5¼ and 7. Dist. 31″	560
86	ν Scorpii	16 3 51	− 19 6	Double star	Mag. 4 and 7. Dist. 40″ B is also double under high powers.	561

No.	Constellation.	R. A. 1860.	Decl. 1860.	Object.	Notes.	No. in Smyth's Cycle.
		h. m. s.	°			
87	80 M. Scorpii	16 8 39	− 22 39	Cluster	Globular	564
88	σ Scorpii	16 12 40	− 25 15	Double star	Mag. 4 and 9½. Dist. 20″	568
89	23 Herculis	16 17 34	+ 32 40	Double star	Mag. 6 and 9. Dist. 36″	573
90	37 Herculis	16 32 42	+ 4 29	Double star	Mag. 6½, 7½. Dist. 69″	582
91	13 M. Herculis	16 36 40	+ 36 43	Cluster	Very fine	585
92	43 Herculis	16 39 6	+ 8 50	Double star	Mag. 5 and 9. Dist. 79″	588
93	12 M. Ophiuchi	16 39 68	− 1 42	Cluster	... ...	590
94	10 M. Ophiuchi	16 49 48	− 3 54	Cluster	Easily resolved	595
95	δ Herculis	17 9 17	+ 25 0	Double star	Mag. 4 and 8½. Dist. 25″	608
96	92 M. Herculis	17 12 51	+ 43 17	Cluster	Very easy	611
97	ν Serpentis	17 12 56	− 12 42	Double star	Mag. 4⅔ and 9. Dist. 50″	610
98	53 Ophiuchi	17 27 58	+ 9 41	Double star	Mag. 6 and 8. Dist. 41″	618
99	14 M. Ophiuchi	17 30 15	− 3 10	Cluster	... ...	621
100	61 Ophiuchi	17 37 32	+ 2 39	Double star	Mag. both 7½. Dist. 21″	623
101	μ Herculis	17 40 59	+ 27 48	Double star	Mag. 4 and 10. Dist. 30″	624
102	ψ Draconis	17 44 27	+ 72 13	Double star	Mag. 5½ and 6. Dist. 31′	625
103	23 M. Ophiuchi	17 48 44	− 18 59	Cluster	... ...	626
104	67 Ophiuchi	17 53 38	+ 2 56	Double star	Mag. 4 and 8. Dist. 55″	628
105	100 Herculis	18 2 10	+ 26 5	Double star	Mag. both 7. Dist. 14″	636
106	μ' Sagittarii	18 5 21	− 21 56	Multiple	Mag. 3½, 16, 9½, and 13 ; very difficult.	639
107	24 M. Clypei	18 8 48	− 18 27	Cluster	... ...	642
108	16 M. Clypei	18 10 52	− 13 50	Cluster	... ...	643
109	18 M. Clypei	18 11 45	− 17 11	Cluster	... ...	644
110	17 M. Clypei	18 12 31	− 16 15	Nebula	The " Horse Shoe "	645
111	40 Draconis	18 13 29	+ 79 59	Double star	5½ and 6. Dist. 20″	646
112	25 M. Sagittarii	18 23 24	− 19 10	Cluster	... ...	650
113	22 M. Sagittarii	18 27 38	− 24 1	Cluster	Very fine	654
114	26 M. Clypei	18 37 33	− 9 32	Cluster	... ...	658
115	ε Lyræ	18 39 41	+ 39 31	Multiple	5 and 6½, 5 and 5½, and three smaller ones, very difficult.	661
116	ζ Lyræ	18 39 55	+ 37 27	Double star	Mag. 5 and 5½. Dist. 44″	662
117	11 M. Antinoï	18 43 36	− 6 26	Cluster	Very pretty	664
118	β Lyræ	18 44 53	+ 33 12	Quadruple and A. variable.	Mag. 3, 8, 8½, and 9	666
119	57 M. Lyræ	18 48 21	+ 32 51	Annular neb.	... ...	669
120	θ' Serpentis	18 49 15	+ 4 1	Double star	Mag. 4⅔ and 5. Dist. 22″	670
121	ο Draconis	18 49 6	+ 59 13	Double star	Mag. 5 and 9. Dist. 30″	672
122	11 Aquilæ	18 52 39	+ 13 26	Double star	Mag. 7 and 10. Dist. 19″	673
123	15 Aquilæ	18 57 34	− 4 14	Double star	Mag. 6 and 7½. Dist. 35″	678

No.	Constellation.	R. A. 1860.	Decl. 1860.	Object.	Notes.	No. in Smyth's Cycle.
		h. m. s.	° ′			
124	η Lyræ	19 8 58	+38 54	Double star	Mag. 5 and 9. Dist. 28″	685
125	56 M. Lyræ	19 11 6	+29 56	Cluster	... ...	688
126	28 Aquilæ	19 13 8	+12 7	Double star	Mag. 6 and 10. Dist. 60″	690
127	β Cygni	19 25 4	+27 40	Double star	Mag. 3 and 7. Dist. 34″	700
128	241 P. XIX. Aquilæ	19 36 0	+ 8 3	Double star	Mag. 7½ and 9½. Dist. 27″	707
129	16 Cygni	19 38 6	+50 12	Double star	Mag. 6½ and 7. Dist. 37″	710
130	276 P. XIX. Cygni	19 40 31	+35 45	Double star	Mag. 8 and 8½. Dist. 15	712
131	278 P. XIX. Cygni	19 40 37	+34 40	Double star	Mag. 6 and 8. Dist. 39″	713
132	χ Cygni	19 41 6	+33 25	Double star	Mag. 5 and 9. Dist. 26″	715
133	57 Aquilæ	19 47 3	− 8 35	Double star	Mag. 6¼ and 7. Dist. 35″	723
134	320 P. XIX. Vulpec.	19 47 13	+19 58	Double star	Mag. both 7. Dist. 43″	724
135	27 M. Vulpeculæ	19 53 31	+22 20	Nebula	The "Dumb-bell"	729
136	θ Sagittæ	20 3 46	+20 30	Triple	Mag. 7, 9, and 8. Dist. 11″ and 70″.	734
137	o² Cygni	20 9 14	+46 19	Quadruple	Mag. 4, 16, 7½, and 5½. Dist. 15″, 106″, 338″.	739
138	σ Capricorni	20 11 19	−19 33	Double star	Mag. 5⅓ and 10. Dist. 54″	741
139	o² Capricorni	20 21 52	−19 3	Double star	Mag. 6 and 7. Dist. 22″	750
140	γ Delphini	20 40 10	+15 37	Double star	Mag. 4 and 7. Dist. 12″	762
141	15 M. Pegasi	21 23 11	+11 33	Cluster	... ...	785
142	2 M. Aquarii	21 26 11	− 1 27	Cluster	... ...	787
143	39 M. Cygni	21 27 11	+47 49	Cluster	... ...	788
144	β Cephei	21 26 47	+69 57	Double star	Mag. 3 and 8. Dist. 14″	789
145	3 Pegasi	21 30 45	+ 5 59	Double star	Mag. 6 and 8. Dist. 39″	790
146	30 M. Capricorni	21 32 25	−23 47	Cluster	... ...	791
147	248 P. XXI. Cephei	21 34 37	+56 51	Triple	Mag. 6, 8½, and 8¾. Dist. 12″ 20″.	792
148	65 P. XXII. Lacertæ	22 12 48	+37 4	Double star	Mag. 6½ and 9. Dist. 15″	808
149	δ Cephei	22 23 58	+57 42	Double and var.	Mag. 4½ and 7. Dist. 41″	825
150	ψ¹ Aquarii	23 8 32	− 9 51	Double star	Mag. 5⅓ and 9. Dist. 49″	833
151	94 Aquarii	23 11 33	−14 13	Double star	Mag. 6 and 8½. Dist. 14″	834
152	30 ⨳ V.. Cassiopeiæ	23 50 6	+55 56	Cluster	... ...	847

APPENDIX VI.

A CATALOGUE OF VARIABLE STARS.

THE following list has been compiled with some care; but variable star information is often of questionable authenticity, the accounts of different observers being quite as variable as the stars themselves, or even more so. This list will, it is hoped, be found to be the most complete ever published, special pains having been taken to make it so: as amateurs, having time and instruments at their command, may render good service by looking after these objects. The letter D appended to the name of a star signifies that its position is given for some other epoch than 1860: the symbol < signifies that the star's minimum magnitude fell *below* that given; but how much, is unknown.

No.	Star.	R.A, 1860.			Decl. 1860.		Period.	Change of Magnitude.		Discoverer.	
		h.	m.	s.	°	′	days.	from	to		
1	R Andromedæ, D	0	16	24	+37	46	...	6		Argelander	1860.
2	T Piscium	0	24	46	+13	46	242±	9·5	11	R. Luther	1855.
3	α Cassiopeiæ	0	32	35	+55	46	79·1	2	2·5	Birt	1831.
4	S Piscium	1	10	15	+ 8	11	...	9	13	Hind	1851.
5	R Piscium	1	23	25	+ 2	9	343	7	0	Hind	1850.
6	7 Arietis	1	48	2	+22	53	5 yrs.?	6	8	Piazzi	1798.
7	R Arietis	2	8	10	+24	24	186	8		......	
8	ο Ceti	2	12	17	− 3	37	331·3	2	12 <	D. Fabricius	1596.
9	ζ Persei	2	56	12	+38	18	30	...		? Schmidt.	
10	β Persei	3	0	24	+40	24	2·86	2·5	4	Montanari	1669.
11	λ Tauri	3	52	55	+12	5	3·95	4	4·5	Baxendell	1845.
12	R Tauri	4	20	38	+ 9	51	327	8	13·5<	Hind	1849.
13	S Tauri	4	21	32	+ 9	38	375	10		Oudemans	

No.	Star.	R.A. 1860.	Decl. 1860.	Period.	Change of Magnitude.		Discoverer.	
		h. m. s.	° ′	days.	from	to		
14	R Orionis	4 51 22	+ 7 55	237 ?	9	12·5	Hind	1848.
15	ι Aurigæ	4 51 56	+43 37	250±	3·5	4·5	Heis	1846.
16	R Leporis	4 53 14	−15 2	...	7		Schmidt	1855.
17	α Orionis	5 47 35	+ 7 22	196±	1	1·5	Sir J. Herschel	
								1836.
18	ζ Geminorum	6 55 48	+20 47	10·16	3·8	4·5	Schmidt	1847.
19	R Geminorum	6 58 56	+22 55	370	7·3	11	Hind	1848
20	R Canis Minoris	7 1 0	+10 14	1 year±	8	10	Argelander	1854.
21	S Canis Minoris	7 25 7	+ 8 37	340	7·5		Hind	1856.
22	S Geminorum	7 34 38	+23 47	294·07	9·2	13·5<	Hind	1848.
23	T Geminorum	7 40 54	+24 5	288 64	8·5	13·5<	Hind	1848.
24	U Geminorum	7 46 48	+22 22	103	9	14 <	Hind	1855.
25	R Cancri	8 8 57	+12 7	380	6	10 <	Schwerd	1829.
26	U Cancri	8 27 45	+19 23	...	9		......	
27	S Cancri	8 35 56	+19 32	9·48	8	10·5	Hind	1848.
28	S Hydræ	8 46 16	+ 3 36	256	8·5	13·5	Hind	1848.
29	T Cancri	8 48 40	+20 23	455±	8·5	12	Hind	1850.
30	T Hydræ	8 48 51	− 8 37	292	6·5	10·5	Hind	1851.
31	α Hydræ	9 20 42	− 8 3	55	2·5	3	Sir J. Herschel	
								1837.
32	ψ Leonis	9 36 6	+14 39	mny.yrs.	6	0	Montanari	1667.
33	R Leonis	9 40 2	+12 5	324	5	10	Koch	1784.
34	R Ursæ Majoris	10 34 41	+69 31	301·90	7	13	Pogson	1853.
35	η Argûs	10 39 38	−58 56	irreg.	1	4	Burchell	1827.
36	α Ursæ Majoris	10 55 3	+62 30	sme. yrs.	1·5	2	Lalande	1786.
37	R Comæ Berenicis	11 57 4	+19 34	1 year±	8		......	
38	δ Ursæ Majoris	12 8 28	+57 48	mny.yrs.	2	25	Unknown.	
39	T Ursæ Majoris	12 29 46	+60 17	270	...		Argelander.	
40	R Virginis	12 31 24	+ 7 46	146	6·5	11 <	Harding	1809.
41	S Ursæ Majoris	12 37 48	+61 52	226	7·5	12	Pogson	1853.
42	T Virginis	...	...	...	...		......	
43	U Virginis	12 44 0	+ 6 19	...	7·5		......	
44	21 Virginis	12 26 33	− 8 40	...	5·5		......	
45	V Virginis	13 20 36	− 2 28	8 mth.±	7		Goldschmidt	1857.
46	R (υ) Hydræ	13 22 4	−22 33	440	4	10 <	J. P. Maraldi	1704.
47	S Virginis	13 25 42	− 6 28	380·11	6	11	Hind	1852.
48	η Ursæ Majoris	13 42 1	+50 4	sme. yrs.	1·5	2	Lalande	1786.
49	* In Boötes	14 7 31	+19 43	...	9·7	14 <	Baxendell	1860.

No.	Star.	R.A. 1860.	Decl. 1860.	Period.	Change of Magnitude.		Discoverer.
		h. m. s.	° ′	days.	from	to	
50	S Boötis	14 18 1	+54 28	...	8	12	Argelander 1860.
51	R Boötis	14 31 1	+27 21	196	8		......
52	* In Libra	14 44 39	−11 45	...	8	9·5	Schumacher.
53	β Ursæ Minoris	14 51 9	+74 43	2 or 3 y.	2	25	W. Struve 1838.
54	* In Boötes	[A.N. 1281]	...	...	6		Hencke 1860.
55	S Serpentis	15 15 7	+14 49	335	8	10 <	Harding 1828.
56	R Coronæ	15 42 49	+28 35	323	6·2	12	Pigott 1795.
57	R Serpentis	15 44 15	+15 34	347	6·5	10 <	Harding 1826.
58	R Libræ	15 45 40	−15 49	722	10		Pogson 1855.
59	R Herculis	15 59 38	+18 45	310	8·5		......
60	R Scorpii	16 9 19	−22 35	...	9		Chacornac 1853.
61	S Scorpii	16 9 20	−22 33	1 year±	5	12	Chacornac 1855.
62	* In Hercules, D	16 17	+19 18	...	7	13	Hencke 1860.
63	30 Herculis	16 24 2	+42 11	...	5		......
64	* In Ophiuchus	16 25 43	−15 49	...	7	10	Pogson 1860.
65	S Ophiuchi	16 26 12	−16 52	232	9·3	13·5 <	Pogson 1854.
66	S Herculis	16 45 32	+15 11	305	7·5		......
67	Hind's Nova (1848)	16 51 39	−12 40	...	4·5	13·5 <	Hind 1848.
68	R Ophiuchi	16 59 44	−15 54	301	7·6	13 <	Pogson 1853.
69	α Herculis	17 8 15	+14 33	67	3·1	3·7	Sir W. Herschel 1795
70	T Herculis	18 3 48	+31 0	160	7·9		......
71	* In Serpens	18 21 59	+ 6 12	...	11	14 <	Baxendell 1860.
72	κ Coronæ Australis	18 23 48	−38 49	years	3	6	Halley 1676.
73	R Scuti Sobieskii	18 40 1	− 5 50	71·75	5	9	Pigott 1795.
74	β Lyræ	18 44 55	+33 12	12·91	3·5	4·5	Goodricke 1784.
75	13 Lyræ	18 51 4	+47 46	46	4·4	4·6	Baxendell.
76	R Aquilæ	18 59 38	+ 8 1	352	6·5		......
77	R Sagittarii	19 8 28	−19 33	467	8	12·8	Pogson 1858.
78	R Cygni	19 33 4	+49 53	416·72	8	14 <	Pogson 1852.
79	χ Cygni	19 41 7	+33 24	406	5	11 <	G. Kirch 1687.
80	η Aquilæ	19 45 9	+ 0 38	7·18	3·6	4·4	Pigott 1784.
81	η Cygni	19 51 3	+34 42	mny.yrs.	4·5	5·5	Sir J. Herschel 1842.
82	S Cygni, D	20 2 27	+57 34	...	9		Argelander 1860.
83	R Capricorni	20 3 28	−14 41	...	9·5	13·5 <	Hind 1848.
84	R Sagittæ	20 7 40	+16 18	72	10		......
85	34 Cygni	20 12 38	+37 36	18 yrs.±	3	6 <	Jansen 1600.

No.	Star.	R.A 1860.	Decl. 1860.	Period.	Change of Magnitude.		Discoverer.
		h. m. s.	° ′	days.	from	to	
86	24 Cephei (Hev.)	20 31 1	+88 42	...	5	11	Pogson 1856.
87	* In Delphinus, D	20 36 24	+16 34	...	8	11	Baxendell 1860.
88	* In Ursa Minor	20 38	+88 40	...	5	11	Pogson 1853.
89	U Capricorni	20 40 22	−15 18	420	10·5	13	Pogson 1858.
90	R Vulpeculæ	20 58 10	+23 16	130	8		......
91	T Capricorni	21 14 13	−15 45	274	9	14	Hind.
92	μ Cephei	21 39 43	+58 8	5 or 6 y.	3	6	Sir W. Herschel 1782.
93	S Pegasi	22 15 9	+ 7 19	...	8·5	13·5<	Hind 1848.
94	* In Aquarius	22 21 0	−10 42	...	8	0	Rümker.
95	δ Cephei	22 23 58	+57 41	5·36	3·7	4·8	Goodricke 1784.
96	β Pegasi	22 56 59	+27 19	31·5	2	2·5	Schmidt 1848.
97	R Pegasi	22 59 37	+ 9 46	378	8·5	13·5<	Hind 1848.
98	R Aquarii	23 37 15	−16 3	354	7	10 <	Harding 1810.
99	R Cassiopeiæ	23 51 18	+50 37	434·81	6	14 <	Pogson 1853.

APPENDIX VII.

THE following is a list of all the principal Catalogues of Stars which have ever appeared.

CATALOGUES OF ISOLATED STARS.

B. C.

128. HIPPARCHUS, containing 1022 stars, observed at Rhodes, incorporated by Ptolemy into his Μεγάλη Σύνταξις, or *The Almagest*, and by him reduced to the epoch of 137 A. D. Last edition by F. Baily, in *Memoirs*, R. A. S., vol. xiii. 1843.

A. D.

1437. ULUGH BEIGH, containing 1019 stars, observed at Samarcand. Last edition by Baily, in *Memoirs*, R. A. S., vol. xiii. 1843.

1602. TYCHO BRAHE, containing 777 stars, observed at Uraniburg, reduced to the year 1600. 2nd edition containing altogether 1005 stars, published by Kepler in 1627. Last edition by Baily, in *Memoirs*, R. A. S., vol. xiii. 1843.

1618. WILLIAM, LANDGRAVE OF HESSE, aided by Rothmann and Byrgius, containing 368 stars, reduced to the year 1593. Last edition by Flamsteed, in *Hist. Cœlest.* 1725.

1624. BARTSCH, JAMES, containing 136 southern stars.

1679. HALLEY, *Catalogus Stellarum Australium*, containing 341 southern stars, observed at St. Helena, reduced to 1677. Last edition by Baily, in *Memoirs*, R. A. S., vol. xiii. 1843.

1690. HEVELIUS, of Dantzic, *Prodromus Astronomiæ*, containing 1564 stars, reduced to 1660 (end of). Last edition by Baily, in *Memoirs*, R. A. S., vol. xiii. 1843.

A. D.

1725. FLAMSTEED, REV. JAMES, *Historia Cœlestis*, containing
2934 stars, observed at Greenwich, and reduced to 1690.
Last edition, considerably enlarged, by Baily in *Account
of Flamsteed*. 1835.

1725. SHARP, ABRAHAM, containing 265 southern stars, observed
at St. Helena; reduced to 1726. Published in Flamsteed's *Hist. Cœlest.* Ed. of 1725.

1757. LA CAILLE, containing 378 stars; reduced to January 1,
1750. Published in his *Fundamenta Astronomiæ.* Last
edition by Baily, in *Memoirs*, R. A. S., vol. v. 1833.

1763. LA CAILLE, containing 515 zodiacal stars; reduced to
1765. Edited by Bailly, of Paris.

1773. BRADLEY, REV. J., containing 389 stars, observed at Greenwich; reduced to 1760. Published in the *Nautical
Almanac.* 1773.

1775. MAYER, T., containing 998 stars, observed at Göttingen;
reduced to 1756. Edited by Lichtenberg, and published
at Göttingen in Mayer's *Opera Inedita.* Last edition by
Baily, in *Memoirs*, R. A. S., vol. iv. 1831.

1776. MASKELYNE, REV. N., containing 36 stars, observed at
Greenwich; reduced to 1770. Published in the *Green.
Obs.* 1776.

1792. DE ZACH, containing 381 stars; reduced to 1800. Published
at Gotha, in his *Tabulæ Motuum Solis.*

1800. WOLLASTON, REV. J., circumpolar stars, in *Fasciculus As-
tronomicus.*

1801. LALANDE, J. DE, containing 47,390 stars; reduced to
1800. Last edition by Baily, for the British Association.
1847.

1803. PIAZZI, containing 6748 stars, observed at Palermo.

1806. DE ZACH containing 1830 zodiacal stars, observed at See-
berg.

1814. PIAZZI, *Positiones Mediæ*, containing 7646 stars, observed
at Palermo; reduced to 1800.

1818. BRADLEY, REV. J., containing 3112 stars, observed at
Greenwich; reduced to January 1, 1750. Published by
Bessel in his *Fundamenta Astronomiæ.*

1824. FALLOWS, REV. F., containing 273 southern stars, observed
at the Cape of Good Hope; reduced to January 1, 1824.
Published in *Phil. Trans.* vol. cxiv.

A. D.

1826. ASTRONOMICAL SOCIETY OF LONDON, containing 2881 stars, compiled from various sources; reduced to January 1, 1830. Published in *Memoirs*, R. A. S., vol. ii. 1826.

1829. POND, J., containing 720 stars, observed at Greenwich; reduced to January 1, 1830. Published in *Green. Obs.* 1829.

1833. POND, J., containing 1112 stars, observed at Greenwich; reduced to January 1, 1830.

1835. BRISBANE, SIR THOMAS, containing 7385 southern stars, observed at Paramatta, N. S. W.

1835. JOHNSON, LIEUT., M. J., containing 606 southern stars, observed at St. Helena. Published by the H. E. I. Co.

1835. ARGELANDER, F., containing 560 stars, observed at Abo; reduced to January 1, 1830.

1836. WROTTESLEY, LORD, containing 1318 stars, observed at Blackheath; reduced to January 1, 1830. Published in *Memoirs*, R. A. S., vol. x. 1838.

1838. GROOMBRIDGE, S., containing 4243 circumpolar stars; reduced to 1810. Edited by Airy.

1838. AIRY, G. B., containing 727 stars, observed at Cambridge; reduced to January 1, 1830. Published in *Memoirs*, R. A. S., vol. xi. 1840.

1842. SANTINI, G., containing 1677 northern equatorial stars, observed at Padua; reduced to January 1, 1840. Published in *Memoirs*, R. A. S., vol. xii. 1842.

1842. AIRY, G. B., containing 1439 stars, observed at Greenwich; reduced to January 1, 1840. Published in *Green. Obs.* 1842.

1844. TAYLOR, containing 11,105 stars, observed at Madras.

1845. BRITISH ASSOCIATION, containing 8377 stars, compiled from various sources; reduced to January 1, 1850. Edited by Baily. [This is commonly considered to be the most useful ever published.]

1845. LA CAILLE, containing 9766 stars, observed at the Cape of Good Hope; reduced to January 1, 1750. The computations were carried on under the superintendence of Henderson, at the expense of H. M. Government.

1846. BESSEL (part i. equatorial regions), containing 31,895 stars, observed at Königsberg; reduced to the year 1825, by Weisse, at the expense of the Academy of Sciences at St. Petersburg.

A. D.

1848. AIRY, G. B., containing 2156 stars, observed at Greenwich; reduced, some to January 1, 1840, the remainder to January 1, 1845. Commonly called the 12-*Year Calendar*.

1849. FALLOWS, REV. F., containing 425 stars, observed at the Cape of Good Hope; reduced to January 1, 1830. Edited by Airy, and published in *Memoirs*, R. A. S., vol. xix. 1851.

1850. MAIN, REV. R., containing the proper motions of 877 stars, observed at Greenwich. Published in *Memoirs*, R. A. S., vol. xix. 1851.

1852. RÜMKER, containing 12,000 stars, observed at Hamburg.

1854. WROTTESLEY, LORD, containing 1009 stars, observed at Wrottesley; reduced to January 1, 1850. Published in *Memoirs*, R. A. S., vol. xxiii. 1854.

1855. AIRY, G. B. containing 1576, stars observed at Greenwich, reduced to 1850.

1856. COOPER, E. J., containing 60,066 stars near the ecliptic, observed at Markree in the year 1848–56. Printed at the expense of H. M. Government.

1857. CARRINGTON, R. C. containing 3735 circumpolar stars, observed at Redhill, reduced to January 1, 1855. Printed at the expense of H. M. Government.

1858. JACOB, CAPT., containing 317 stars, observed at Madras, for proper motions, reduced to January 1, 1855. Published in *Memoirs*, R. A. S. vol. xxviii. 1860.

1858. MAIN, REV. R., containing the proper motions of 270 stars, observed at Greenwich. Published in *Memoirs*, R.A.S., vol. xxxviii. 1860.

1859. ROBINSON, REV. T. R., containing 5345 stars, observed at Armagh, reduced to January 1, 1840. Printed at the expense of H. M. Government.

1860. JOHNSON, M. J., containing 6317 circumpolar stars observed at Oxford; reduced to January 1, 1845. Edited by the Rev. R. Main. [This may be regarded as one of the most valuable catalogues published for many years past.]

CATALOGUES OF DOUBLE STARS.

A. D.

1826. HERSCHEL, SIR J., 1st catalogue, containing 321 stars. *Memoirs*, R. A. S., ii. 475.

1829. HERSCHEL, SIR J., 2nd catalogue, containing 295 stars. *Memoirs*, R. A. S., iii. 47.

1829. HERSCHEL, SIR J., 3rd catalogue, containing 384 stars. *Memoirs*, R. A. S., iii. 201.

1831. HERSCHEL, SIR J., 4th catalogue, containing 1236 stars. *Memoirs*, R. A. S., iv. 331.

1833. HERSCHEL, SIR J., 5th catalogue, containing 2007 stars. *Memoirs*, R. A. S., vi. 1.

1836. HERSCHEL, SIR J., 6th catalogue, containing 286 stars. *Memoirs*, R. A. S., ix. 193.

1847. HERSCHEL, SIR J., 7th catalogue, containing 2013 stars. *Res. of Ast. Obs.* p. 171.

1837. STRUVE, W., *Mensuræ Micrometricæ*, containing 3112 stars, observed at Dorpat.

1845. STRUVE, W., containing 514 double and multiple stars, discovered at Pulkova with the great refractor.

CATALOGUES OF NEBULÆ (INCLUDING CLUSTERS).

1714. HALLEY, containing 6 nebulæ. Published in *Phil. Trans.* vol. xxix. p. 390.

1733. DERHAM, containing 16 nebulæ. Published in *Phil. Trans.* vol. xxxviii. p. 70.

1755. LA CAILLE, containing 42 nebulæ observed at the Cape of Good Hope. Published in *Mém. Acad. des Sciences.* 1755. p. 194.

1784 MESSIER, containing 103 nebulæ. Published in the *Connaissance des Temps.* 1784.

1786. HERSCHEL, SIR W., containing 1000 nebulæ observed at Slough. Published in *Phil. Trans.* vol. lxxvi. p. 471 *et seq.*

1789. HERSCHEL, SIR W., containing another 1000 nebulæ observed at Slough. Published in the *Phil. Trans.* vol. lxxix. p. 226 *et seq.*

1802. HERSCHEL, SIR W., containing 500 nebulæ observed at Slough. Published in *Phil. Trans.* vol. xcii. p. 503 *et seq.*

A. D.

1828 DUNLOP, containing 629 nebulæ observed at Paramatta, N. S. W. Published in *Phil. Trans.* vol. cxviii. p. 114.

1833. HERSCHEL, SIR J., containing 2306 nebulæ observed at Slough, whereof about 500 were new. Published in *Phil. Trans.* vol. cxxiii. p. 365 *et seq.*

1847. HERSCHEL, SIR J., containing 4015 nebulæ observed at the Cape of Good Hope. Published in *Results of Astronomical Observations, &c.*

1850. ROSSE, EARL OF. Engravings and notes on nebulæ observed at Birr Castle. Published in *Phil. Trans.* vol. cxl. p. 499 *et seq.*

MISCELLANEOUS CATALOGUES.

1844. SMYTH, ADMIRAL W. H., *Cycle of Celestial Objects*, containing 580 double stars, 20 binary systems, 80 triple and multiple stars, and 170 clusters and nebulæ, observed at Bedford; reduced to January 1, 1840. [A most interesting and instructive work.]

1859. WEBB, REV. T. W., *Celestial Objects for Common Telescopes.* [A popular handbook, on the plan of the preceding; well arranged, and most useful.]

APPENDIX VIII.

LIST OF OBSERVATORIES.

THE following list of observatories, as far as it goes, will, we hope, be found reliable, but it is probably far from being complete, information of this kind being somewhat difficult to obtain.* We only insert those observatories which are of an astronomical (as distinguished from those of a purely meteorological) character.

PUBLIC OBSERVATORIES — UNITED KINGDOM.

Place.	Kingdom.	Director.
Armagh.	Ireland.	Robinson, Rev. T. R.
Cambridge.	England.	Adams, J. C.
Chatham.	England.	
Dublin.	Ireland.	Hamilton, Sir W. R.
Durham.	England.	Chevallier, Rev. T.
Edinburgh.	Scotland.	Smyth, C. P.
Glasgow.	Scotland.	Grant, R.
Greenwich.	England.	Airy, G. B.
Liverpool.	England.	Hartnup, J.
Oxford (Radcliffe Obs.).	England.	Main, Rev. R.
Portsmouth.	England.	

PRIVATE OBSERVATORIES — UNITED KINGDOM.

Place.	County.	Proprietor.
Birr Castle.	King's County.	Rosse, Earl of.
Bradstones.	Lancashire.	Lassell, W.
Camden Hill.	Middlesex.	South, Sir J.
Clapton (Down's Road).	Middlesex.	Gorton, S.

---

* Additions or corrections, forwarded to the care of the Publisher, 50 Albemarle Street, will meet with due attention, and confer a favour.

Place.	County.	Proprietor.
Clifton.	Gloucestershire.	Burder, W. C.
Cranford.	Middlesex.	De La Rue, W.
Crumpsall Hall.	Lancashire.	Worthington, R.
Euston Road.	Middlesex.	Slater, T.
Forest Hill.	Kent.	Clarke, E.
Haddenham.	Buckinghamshire.	Dawes, W. R.
Hartwell House.	Buckinghamshire.	Lee, J.
Hartwell Rectory.	Buckinghamshire.	Lowndes, Rev. J.
Hardwick Rectory.	Herefordshire.	Webb, Rev. T. W.
Haverhill.	Suffolk.	Boreham, W. W.
Highbury.	Middlesex.	Burr, T. W.
Hove.	Sussex.	Howell, C.
Maresfield.	Sussex.	Noble, W.
Markree Castle.	Sligo.	Cooper, E. J.
Ochtertyre (Crieff).	Perthshire.	Murray, Sir W. K., Bart.
Ravensdale Park.	Louth.	Clermont, Lord.
Red Scar.	Lancashire.	Cross, Major.
South Villa (Regent's Park).	Middlesex.	Bishop, G.
Tarn Bank.	Cumberland.	Fletcher, I.
Tulse Hill.	Surrey.	Huggins, W.
Uckfield.	Sussex.	Prince, C. L.
Wrottesley Hall.	Staffordshire.	Wrottesley, Lord.

PUBLIC OBSERVATORIES — FOREIGN.

Place.	Country.	Director.
Albany (Dudley Obs.).	New York, N.U.S.	Gould, B. A.
Altona.	Denmark.	Peters, C. A. F.
Amherst (College).	Massachusetts, N.U.S.	
Ann-Arbor.	Michigan, N.U.S.	Brünnow, F.
Athens.	Greece.	Schmidt, J. F. J.
Bergen.	Norway.	Ästrand, J. J.
Berlin.	Prussia.	Encke, J. F.
Berne.	Switzerland.	Wolf, R.
Bologna.	States of the Church.	Respighi.
Bombay.	India.	Fergusson, E. F. T.
Bonn.	Rhenish Prussia.	Argelander, F.G.A.
Ben-zarea.	Algiers.	Bulard.
Breslau.	Silesia.	Galle, J. G.
Brussels.	Belgium.	Quetelet, A. J. L.
Buda.	Hungary.	

Place.	Country.	Director.
Cambridge (Harvard College).	Massachusetts, N.U.S.	Bond, G. P.
Cape of Good Hope.	Africa.	Maclear, Sir T.
Christiana.	Norway.	Hansteen, C.
Clinton (Hamilton College).	New York, N.U.S.	
Cincinnati.	Ohio, S.U.S.	
Coïmbra.	Portugal.	
Copenhagen.	Denmark.	D'Arrest, H.
Cracow.	Poland.	Weisse, W.
Dartmouth.	New Hampshire, N.U.S.	
Dessau.	Anhalt.	Schwabe, H.
Dorpat.	Russia.	Mädler, J. H.
Florence.	Tuscany.	Donati, G. B.
Geneva.	Switzerland.	Plantamour, E.
Georgetown (Jesuit College).	Maryland, N.U.S.	Curley, J.
Gotha.	Gotha.	Hansen, P. A.
Göttingen.	Hanover.	Klinkerfues, W.
Halle.	Prussian Saxony.	Rosenberger, O. A.
Hamburg.	Germany.	Rümker, K.
Haverford.	Pennsylvania, N.U.S.	
Helsingfors.	Finland.	Woldstedt.
Hudson (Western College).	New York.	
Kazan.	Russia.	Kowalsky.
Kieff.	Russia.	Fedorenko, J.
Königsberg.	Prussia.	Luther, E.
Kremsmünster.	Austria.	Reslhüber, A.
Leipzic.	Saxony.	Bruhns, C.
Leyden.	Holland.	Kaiser, F.
Lisbon.	Portugal.	Folque, F.
Madras.	India.	Pogson, N. R.
Madrid.	Spain.	Aguilar, A.
Mannheim.	Baden.	Schönfeld, E.
Marburg.	Hesse Cassel.	Gerling.
Marseilles.	France.	Valz, B.
Milan.	Lombardy.	Carlini, F.
Mitau.	Russia.	
Modena.	Modena.	Pierre.
Moscow.	Russia.	Schweitzer, G.
Munich.	Bavaria.	Lamont, J.
Naples.	Two Sicilies.	Capocci, E.

Place.	Country.	Director.
Newhaven (Yale College).	Connecticut, N.U.S.	Olmsted, D.
Nicolaieff.	Russia.	Knorre.
Padua.	Lombardy.	Santini, G.
Palermo.	Sicily.	Ragona, D.
Paris.	France.	Le Verrier, U. J. J.
Philadelphia (High School).	Pennsylvania, N.U.S.	
Philadelphia (Quakers).	Pennsylvania, N.U.S.	
Prague.	Bohemia.	Böhm, J.
Pulkova.	Russia.	Struve, F. G. W.
Quebec.	Canada East.	
Rome (College).	States of the Church.	Secchi, A.
Rome (University).	States of the Church.	Calandrelli, J.
St. Croix.	Antilles Islands.	Lang, Major A.
St. Fernando.	Spain (near Cadiz).	Marquez, F. De S. P.
Santiago.	Chili.	Moësta, C. G.
Shelbyville (Shelby College).	Kentucky, S.U.S.	
Speyer.	Bavaria.	Schwerd.
Stockholm.	Sweden.	Wrede, Baron
Sydney.	New South Wales.	Scott, W.
Tiflis.	Georgia.	Oblomicosky.
Toulouse.	France.	Petit, F.
Trevandrum.	India.	
Trieste.	Austria.	Scharb.
Tübingen.	Wirtemburg.	Zech.
Turin.	Sardinia.	Plana, Baron J.
Tuscaloosa.	Alabama, S.U.S.	
Upsala.	Sweden.	Svanberg.
Utrecht.	Holland.	Hoek, M.
Vienna.	Austria.	Von Littrow, K.
Warsaw.	Poland.	Baranowski.
Washington.	Columbia, N.U.S.	Gillis, J. M.
West Point.	New York, N.U.S.	
Williamstown (Williams College).	Massachusetts, N.U.S.	
Williamstown.	Victoria, N.S.W.	Ellery, R. L. J.
Wilna.	Russia.	Sabler, G.
Zurich.	Switzerland.	Wolf, R.

PRIVATE OBSERVATORIES.—FOREIGN.

Place.	Country.	Proprietor.
Bilk.	Rhenish Prussia.	Luther, R.
Buffalo.	New York, N.U.S.	Van Duzee.
Charleston.	South Carolina, S.U.S.	Gibbes, L.
Cloverden.	Massachusetts, N.U.S.	
Gotha.	Gotha.	Habicht.
Hobart Town.	Van Dieman's Land.	Abbott, F.
Newark.	New Jersey, N.U.S.	Van Arsdale.
New York.	New York, N.U.S.	Campbell.
New York.	New York, N.U.S.	Rutherford.
Olmutz.	Moravia.	Von Unkrechtsberg.
Senftenberg.		Von Senftenberg, Baron. (Observer, T. J. C. A. Brorsen.)
Sharon.	Pennsylvania, N.U.S.	

INDEX

TO

THE CHIEF REFERENCES TO SUBJECTS.

₊ *References in Book IX. are not included.*

In using this Index, the Table of Contents, at the commencement of the Volume, should also be consulted.

Z

INDEX

TO

THE CHIEF REFERENCES TO THE NAMES OF PERSONS.

* *Names occurring either in Book IX., or in any of the Appendices, are not noticed.*

THE END.

Printed in the United States
By Bookmasters